Molecular Manufacturing

This book explores the revolutionary potential of atomically precise manufacturing technologies to transform global healthcare, as well as practically every other sector across society. This forward-thinking volume examines how envisaged Factory@Home systems might enable the cost-effective domestic fabrication of (among myriad consumer products, including nutritious gourmet foods) advanced autonomous nanomedical devices that have the capacity to address most health challenges, including the disease of aging itself.

Molecular Manufacturing: The Future of Nanomedicine covers cutting-edge concepts including DNA-based manufacturing, quantum nanoscience applications, and the democratization of industrial capabilities through molecular manufacturing. The book presents compelling arguments for how these technologies might facilitate Global Health Care Equivalency – providing equitable access to advanced medical technologies regardless of wealth or geography.

Intended for researchers, healthcare professionals, policy makers, and technology enthusiasts who are interested in the convergence of nanotechnology, nanomedicine, and artificial intelligence, this book explores how emerging technologies will revolutionize healthcare delivery worldwide.

Molecular Manufacturing

The Future of Nanomedicine

Edited by

Frank J. Boehm

CRC Press
Taylor & Francis Group
Boca Raton London New York

CRC Press is an imprint of the
Taylor & Francis Group, an **informa** business

Designed cover image: Frank Boehm

First edition published 2026
by CRC Press
2385 NW Executive Center Drive, Suite 320, Boca Raton FL 33431

and by CRC Press
4 Park Square, Milton Park, Abingdon, Oxon, OX14 4RN

CRC Press is an imprint of Taylor & Francis Group, LLC

ISBN: 978-1-138-63200-4 (hbk)

ISBN: 978-1-041-20928-7 (pbk)
ISBN: 978-1-003-72489-6 (ebk)

DOI: 10.1201/9781003724896

Typeset in Times
by KnowledgeWorks Global Ltd.

I lovingly dedicate this book to the memory of my father, Josef Boehm, who ignited in me the flame of imagination by example of his unquenchable curiosity, quest for knowledge, and fascination with life and the universe; to my mother, Charlotte Boehm, whose amazing fortitude, generosity, and contagious enthusiasm for life has inspired all around her to never give up on their dreams and to always look up (we recently celebrated her 99th Birthday!); and to my sweeties, Liz Balfour and Jazmyn Balfour-Boehm, whose unconditional love and support has allowed me to realize my dreams. Further, to my amazing grandsons, Skye (SkyeGuy) Barney and Linden Barney (beautiful sons of Jazmyn Balfour-Boehm and Steve Barney), who are incarnations of pure love and joy, for whom I deeply hope and trust that the world they inherit will be one that is far more unified, loving, equitable, wise, and compassionate. I also dedicate this book to several colleagues whom I will always consider as "forces of nature," whose presence, passion, diligence, leadership, and kindness will be sorely missed – Veer Endra, Ioan Opris, Ned Seeman, Selwyn Super, and Jack Taunton.

Contents

About the Editor ix
List of Contributors xi
Preface xiii
Acknowledgments xv

Chapter 1 Quandary: Are Molecularly Manufactured Burgers Imbued with the Life Force? 1

Angelika Domschke and Frank J Boehm

Chapter 2 Factory@Home: Democratizing the Coming Industrial Revolution 13

Margaret Morris and Frank J. Boehm

Chapter 3 Quantum Nanoscience: Atomically Precise Manufacturing Meets Quantum Computation 43

Melanie Swan

Chapter 4 DNA-Based Manufacturing: Potential Strategies and Prospects 100

Frank J Boehm

Chapter 5 The Next Big Leap in Human Health: Bridging the Gap between Life and Healthspan 149

Bryan Sykes

Chapter 6 Subatomic Aggregates Are Us: Beyond Molecular Manufacturing 167

Angelika Domschke and Frank J Boehm

Index 207

About the Editor

Frank J Boehm comes from a diverse background that includes a 12-year stint as a full-time rock/pop musician in various bands that toured North America (1975–1987), and 17 years as a Senior Mechanical Designer involved in the design and development of dedicated robotic tooling for complex automated assembly systems (1987–2004). He has been further engaged as a Nanotechnology Consultant, Technology Scout, Product Designer, Research Associate, and Academic Writer/Editor for several international corporations and university laboratories (since 1996). Frank has been involved with nanotechnology and especially nanomedicine since 1996 (autodidactically trained), which inspired the development of numerous concepts and designs for advanced nanomedical diagnostic and therapeutic components, devices, and systems to potentially address myriad disease states. His aim is to develop and transform these concepts into real-world applications for global benefit. Frank serendipitously encountered the concept of nanotechnology on the Internet, and immediately became fascinated with its virtually limitless potential, particularly as it relates to the field of medicine. He passionately proceeded to evolve and textually articulate various advanced near-term and longer-term nanomedical concepts and designs. Concomitantly, he initiated correspondence with numerous nanotechnology and nanomedicine research scientists and thought leaders from across the globe and founded a startup company, NanoApps Medical, Inc. (2009), with the aim of investigating and developing advanced, innovative, and cost-effective nanomedical diagnostic and therapeutic devices and systems for the benefit of humanity. Upon recognizing the immense potential of nanomedicine to impart positive paradigm shifts across the medical domain (e.g., precisely targeted drug delivery, vascular/neurological/cellular plaque removal, eradication of cancers via hypothermic nanoparticles, completely non-invasive surgical procedures, enhancement of physiological systems, and extended longevity), Frank was deeply motivated to write more extensively on the topic. This culminated in the authoring/publication of his first book *Nanomedical Device and Systems Design: Challenges, Possibilities, Visions* (CRC Press, 2013). Subsequently, he authored/co-authored numerous book chapters and journal articles, including an in-depth paper entitled "Human Brain/Cloud Interface" (*Frontiers in Neuroscience*, 2019), and co-founded NanoApps Athletics, Inc. in the same year. Frank then set his focus on a far more expansive investigation into the possibilities of nanotechnology/nanomedicine coupled with AI/AGI/ASI and quantum computation via the simultaneous compilation of a set of three companion books (with multiple chapter contributors for each) that explore a trifecta of envisaged global-scale paradigm-shifting technology platforms *Global Health Care Equivalency in the Age of Nanotechnology, Nanomedicine and Artificial Intelligence* (CRC Press, 2025), *Molecular Manufacturing: The Future of Nanomedicine* (CRC Press, 2025), and *Nanomedical Brain/Cloud Interface: Explorations and Implications* (IOP Publishing, 2025).

List of Contributors

Frank J. Boehm
NanoApps Medical Inc.
NanoApps Athletics Inc.
NanoApps Consulting
Vancouver, British Columbia, Canada
Thunder Bay, Ontario, Canada

Angelika Domschke
Innovation in Biomedical Science and Healing Arts
Angelika Domschke Consulting, LLC
Duluth, GA, USA

Margaret Morris
Multidiscipline Researcher, Writer, Book Author (*Moon Base and Beyond*)
Geopolymerization and Diamond Microcircuitry Film (DMF) Expert
Detroit, MI, USA

Melanie Swan
UCL Centre for Blockchain Technologies
DIYgenomics
New York, NY, USA

Bryan Sykes
Public Health Advocate Research
Black Men's Health Alliance (BMH)
DiaBeatIt
South Africa

Preface

Since the publication of my first book (*Nanomedical Device and Systems Design: Challenges, Possibilities, Visions* – CRC Press, 2014), it became increasingly evident that synergies between concurrently rapid advances in nanotechnology and nanomedicine, in conjunction with the recent exponential progress in artificial intelligence (AI) would likely facilitate the capacity to actualize three envisaged global-scale paradigm-shifting technology platforms. Thus, I endeavored to explore these concepts through the generation of a series of three companion books in parallel. The rationale being that I believe these applications will be inextricably linked.

This second volume in this series (*Molecular Manufacturing: The Future of Nanomedicine* – CRC Press, 2025) explores what I believe will be the prerequisite technology platform and tipping point toward the emergence of mature iterations of what I envision as a single worldwide healthcare system, namely, Global Health Care Equivalency (GHCE). The first volume of this series (*Global Health Care Equivalency in the Age of Nanotechnology, Nanomedicine and Artificial Intelligence* – CRC Press, 2025) delves into the potential emergence of an equitable, cost-effective, non-monopolizable, globally distributed/decentralized, AI/QC-driven healthcare system that would operate on safe/secure nested quantum-encrypted blockchains. With the attainment of mature GHCE, which may be significantly expedited when advanced AI (and soon enough, the emergence of Artificial General Intelligence (AGI) and Artificial Superintelligence (ASI)) is coupled with quantum computation (QC) … "we can envisage a future world where any individual on the planet has access to the same advanced and cost effective nanomedical diagnostic and therapeutic technologies, no matter how wealthy or impoverished they are, no matter where they happen to reside, or under what conditions they live. Progress toward this goal will be incremental, with each successive wave of nanomedical technologies being more advanced than the previous wave" (NanoApps Medical, Inc., 2024).

In this regard, the advent of the capacity for hypothetical (for now) molecular manufacturing (MM) may enable the cost-effective domestic fabrication of advanced autonomous nanomedical devices (multiple components and systems of which I articulate in my first book), which may have the ability to address all that ails us, including the disease of aging.

Envisioned Factory@Home (F@H) MM appliances could be massively distributed and globally decentralized to facilitate GHCE, which would be but only one application (albeit a critically important one) of myriad possibilities. To quote from an essay I co-authored with Angelika Domschke https://forums.fqxi.org/d/2085-quandary-are-molecularly-manufactured-burgers-imbued-with-the-life-force-by-frank-josef-boehm-and-angelika-domschke.

> "The rapidly increasing sophistication and popularity of 3D printing that we are currently witnessing might be viewed as an early precursor of envisaged, exponentially more powerful and dynamic MM capabilities. These MM units might initially be manifest as microwave oven-sized 'factory-at-home' systems, which would utilize atomic and molecular feedstocks, rather than the primarily polymeric feedstocks

> that are employed in 3D printers today. The ultrafine positioning of specific molecular and atomic species might enable the cumulative assembly of practically anything we can imagine (e.g., healthy gourmet foods, running shoes, toothbrushes, physician-prescribed nanomedical diagnostic or therapeutic devices, photovoltaic cells, smartphones, computers, as well as components for building additional MM units) with atomic/molecular layer-by-layer precision. The parameters of what might be produced would be limited only by the human imagination, design inputs, and the dimensional constraints of a given MM unit. This will translate to enabling the capacity for the average person to produce practically any type of consumer product."

In terms of the progression of advances in F@H systems that may culminate with the advent of mature GHCE, initial feedstocks might consist of the full range of chemical constituents required to synthesize basic medicines. With further technological advances, standardized nanoparticles of various species might serve as the raw materials for enhancements by F@H systems, which are subsequently custom functionalized, replete with targeting molecules, to generate a dynamic range of highly precise diagnostic and therapeutic nanoparticles to address myriad conditions (from colds to cancers). More advanced iterations of F@H would rely on molecular and atomic feedstocks to enable the fabrication of passive and semi-autonomous nanodevices at molecular/atomic resolution that integrate nanoelectronics, sensing, communications, and computational capacities. Ultimately, the feedstocks of mature F@H system iterations would be comprised of pristine atomic species that enable the MM of autonomous atomic resolution nanomedical robots, endowed with further capabilities (e.g., propulsion, navigation, and a full armamentarium of cellular/organellar repair instrumentation). It is conceivable that mature MM may also enable the fabrication of specialized classes of autonomous self-installing/uninstalling "neuralnanorobots" with the capacity to interface the human brain with AI/AGI/ASI agents that reside in cloud/edge supercomputers or standalone wearable modules.

The third book in this series, *Nanomedical Brain/Cloud Interface: Explorations and Implications* (Institute of Physics Publishing (IOPP), 2025), explores how an envisaged nanomedically enabled brain/cloud interface might enable instantaneous access to the full range of human knowledge, significantly enhance our learning capacities, enable ultrahigh resolution fully sensorial immersive experiences, and empower us to keep pace/integrate with AI. Consequently, this technology platform may, in effect, precipitate even further enhancements and efficacies for MM and GHCE, which, from the author's perspective, would close the loop, as each of these three high technology platforms might synergistically feed into and cross-pollinate with the others; each having the capacity to reciprocate. Further, the broad tech-trusted acceptance and implementation of a mature B/CI may be nothing less than the next evolutionary step for humanity, one where we seamlessly merge with our technologies.

In due course, Recykl@Home systems would likely emerge that can disassemble myriad household wastes down to their atomic constituents, which are then recycled and repackaged as pristine atomic feedstocks to continuously resupply the F@H, which would complete the cycle. Extrapolating forward, the propagation of the potentially immense power of MM would likely catalyze a surge of positive transformations toward global abundance. Further, it may have the capacity to permeate almost every facet of human endeavor, while driving robust global-scale environmental remediation.

Acknowledgments

I am indeed deeply grateful to all the contributing authors and co-authors for their incredible efforts toward the creation and infusion of their amazing and inspiring intellects and life forces into this book, who are listed in the Contributors section. I would surely be remiss for not acknowledging K. Eric Drexler for presenting to the world his original vision of the boundless possibilities of nanotechnology, molecular manufacturing, and nanomedicine as articulated in his highly inspirational book *Engines of Creation*. Further, my true appreciation to Robert A. Freitas Jr. for further enlightening me as to the virtually limitless and exciting potential of nanomedicine with his acute insights and most excellent and groundbreaking *Nanomedicine* book series. Also, my sincere gratitude to Richard Satava (former DARPA Program Manager and Professor of Surgery at the University of Washington-Seattle) for initially inspiring me to articulate my nanomedical concepts on paper, while advising me to always endeavor to amply support their potential viability and promise. Finally, my special thanks and deep gratitude go to Amanda Scott (CEO NanoApps Medical, Inc., Creative Director Alias Studio) for her amazing and tireless efforts in support and promotion of my nanomedical concepts and publications as well as the latest progress in nanotechnology, nanomedicine, and AI through the creation and curation of our excellent NanoApps Medical, Inc. and NanoApps Athletics, Inc. sites, and social media. I also give many thanks and kudos to Heinz Hoenen for his exceptional contributions to our international social media presence.

I would like to express my deep appreciation to the following individuals for providing information, insights, and other forms of support in facilitating the realization of this project: Amara Angelica, Elizabeth Balfour, Jazmyn Balfour-Boehm, Steve Barney, Raj Bawa, Aicheng Chen, Angelika Domschke, Ted Duke, Claude Giroux, Urs Hafeli, David Harakal, Robert R. Hieronimus, Jonathan Kitzen, Adriana Marias, Kannan Mavila, Margaret Morris, Sabino Padilla, Michael Slaughter, Shannon Smith, Melanie Swan, Bryan Sykes, and Jack Taunton.

I sincerely apologize to any individuals whom I may have inadvertently overlooked.

My sincere thanks and gratitude go out to my publisher, Taylor & Francis Group/CRC Press, for having the confidence and trust in me to see this project through. Further, allowing me the complete freedom to explore some of the exciting possibilities of an envisaged future world of abundance through the prospective development of molecular manufacturing (MM). Among the endless possibilities/applications of MM, one of the most critical and beneficial for humanity and the planet may be the development of a single equitable, economical, non-monopolizable, tech-trusted, massively distributed and decentralized healthcare system (Global Health Care Equivalency (GHCE)). In particular, I wish to convey my appreciation to Michael Slaughter, Michele Smith, and Marc Gutierrez for their incredible patience and support throughout the writing process. In addition, I very much appreciate the efforts of Jennifer Petruzziello and the staff at Taylor & Francis Group/CRC Press, as well as Mayank Sharma, Saranya J. and the staff at KnowledgeWorks Global Ltd. in the production of this book.

Lastly, I will forever be grateful to my truly inspiring and wise father, Josef Boehm; my ever sweet and generous mother, Charlotte Boehm; my loving sister, Renata Swanson, and brother, John Boehm; Elizabeth Balfour, and our incredible daughter Jazmyn Balfour-Boehm and her husband Steve Barney, who blessed us with our beautiful and amazing grandsons Skye (SkyeGuy) Barney, and Linden Barney, for their unconditional love, encouragement, and unwavering support without which this book could not have been possible.

1 Quandary
Are Molecularly Manufactured Burgers Imbued with the Life Force?

Angelika Domschke and Frank J Boehm

(This essay was originally written for the Foundational Questions Institute's 2014 Essay Competition on "How Should Humanity Steer the Future?" https://forums.fqxi.org/d/2085-quandary-are-molecularly-manufactured-burgers-imbued-with-the-life-force-by-frank-josef-boehm-and-angelika-domschke Reproduced here with minor modifications.)

CONTENTS

1.1 Introduction – Humanity's Guiding Spirit 1
1.2 Current State of Humanity and Potential Trajectories 2
1.3 Superseding the Doom and Gloom 3
1.4 Nanotechnology and Molecular Manufacturing – Emergence of the Grand Equalizer? 4
1.5 Nanomedicine and Energy Fields – Radical Paradigm Shifts in Global Healthcare 6
1.6 Quandary – Are Molecularly Manufactured Burgers Imbued with the Life Force? 8
1.7 Integration of the 10 Billion Synapse World Mind 9
1.8 Conclusion – Humanity's Guiding Spirit 10
References 10

1.1 INTRODUCTION – HUMANITY'S GUIDING SPIRIT

It may be apt to commence with a quote from our book *Nanomedical Device and Systems Design: Challenges, Possibilities, Visions*:

> It would seem that the unfathomable wellsprings of cumulative human imagination, ingenuity, passion, and effort are not subject to any tether or constraint that cannot, over time, be unbound or circumvented.
>
> **Boehm, 2013**

DOI: 10.1201/9781003724896-1

Since its inception, humanity has been driven by an insatiable curiosity and quest for knowledge, while constantly striving to attain cherished ideals, which encompass innovation, advancement, adventure, freedom, honor, love, and hope. We invariably have a deep desire to challenge ourselves to be the best that we can possibly be, know no bounds, and to devise elaborate tools if necessary, to assist us in the attainment of our goals.

1.2 CURRENT STATE OF HUMANITY AND POTENTIAL TRAJECTORIES

We currently find ourselves in the midst of a complex, multifaceted convergence of momentous challenges (e.g., sociopolitical upheaval, a rapidly growing population, increasingly profound climactic effects due to anthropologically initiated global warming, coupled with strained water, food, and energy resources). These issues culminate in an overarching question: how will humanity continue to survive and flourish in the face of so many seemingly intractable forces that threaten to severely hobble, or perhaps even extinguish us as a species? To any objective observer with a historical overview of Earth and its inhabitants, it would seem blatantly evident that at this juncture, humanity is hurtling rapidly along an unsustainable path. One that will inevitably see the depletion of our (once thought) inexhaustible non-renewable fossil fuel resources, combined with ever higher loadings of CO_2 (and other greenhouse gases) that reside in the ambient atmosphere, with the consequence of further exacerbating global warming and its cascading negative environmental effects, and associated human tolls.

This situation, and its seemingly unavoidable dire consequences, was poignantly highlighted in a report from the U.N. Intergovernmental Panel on Climate Change (IPCC), which stated that, for us to stave off some of the more serious global warming threats that "... greenhouse-gas emissions will have to be cut by 40% to 70% compared with 2010 by mid-century, and to near zero by the end of this century." The report also points out that, "Only major institutional and technological change will give a better than even chance that global warming will not exceed this threshold." This "threshold" referred to is a median global temperature increase of "between 2 and 3.2 degrees above preindustrial levels by mid-century." Beyond this threshold, we can expect extensive and abrupt "high-impact changes—such as melting of the Greenland ice sheet ..." (Stocker et al., 2013; Naik, 2014). The authors propose that one of the critical initial prerequisites toward the establishment of a sustainable, clean, and flourishing future for humanity relates to our ability to evolve beyond our fossil-fuel–drenched energy mindset/paradigm toward the robustly expedited development of advanced solar technologies and other environmentally compatible renewable energy sources, via intense collaborative efforts, dedicated focus, and mass allocation of resources.

The Institution of Mechanical Engineers, UK, shared its thoughts on the fate of fossil fuels: "There are an estimated 1.3 trillion barrels of proven oil reserve left in the world's major fields, which at present rates of consumption will be sufficient to last 40 years. By 2040, production levels may be down to 15 million barrels per day – around 20%

of what we currently consume. It is likely by then that the world's population will be twice as large, and more of it industrialized (and therefore oil dependent)." When we contrast this (relatively) short-lived energy source with its diminishing returns (over time, despite technological advances, it will be less and less cost effective to extract oil from deeper and deeper recesses within the Earth) against the virtually limitless source of solar energy, the correct path appears abundantly obvious and clear. An *EcoWorld Magazine* article offers: "In full sun, you can safely assume about 100 watts of solar energy per square foot. If you assume 12 hours of sun per day, this equates to 438,000 watt-hours per square foot per year. Based on 27,878,400 square feet per square mile, sunlight bestows a whopping 12.2 trillion watt hours per square mile per year." In terms of the current (2025) power draw for individual homes, the World Energy Council estimates that the average American dwelling consumes 10,332 kWh per year or 28.4 kWh per day (Wilson, 2025).

The highest commercially available solar cell efficiency (for specialized applications) is presently 44.6% for a concentrator quadruple junction cell by the Fraunhofer Institute for Solar Energy Systems ISE, Freiburg, Germany (Helmers et al., 2024). Nanomaterials-based solar cells based on black silicon nanowires, which can be produced using standard processing technologies, have attained efficiencies of 25.2% (Liu et al., 2020). An InP nanowire-based solar cell developed in part at Lund University, Sweden, attained an efficiency of 15% (Otnes et al., 2018). Further, the Korea Institute of Energy Research developed flexible perovskite/$Cu(In,Ga)Se_2$ tandem solar cells that reached efficiencies of 23.64% (Jeong et al., 2025). The highest conversion efficacy to date was 26.5% for a bendable perovskite-silicon tandem solar cell created at Tokyo City University in 2025 (Shishido et al., 2025).

Nanotechnology is defined as the capacity for the controllable manipulation of matter at scales from 1 to 100 nm, where materials may exhibit unique and enhanced properties as compared to their bulk counterparts. With the advent of mature nanotechnology we are likely to see the development of molecular manufacturing (MM), through which virtually any consumer item, including foods, smartphones, high efficiency photovoltaic cells, advanced energy storage devices, etc. may be produced from atomic and/or molecular feedstocks on site, in microwave sized MM units (the personal nanofactory as envisaged by Drexler) (Drexler, 1986). Specialized MM systems will also make possible the fabrication of sophisticated autonomous nanomedical devices that have the capacity to address practically every human disease state, including the disease of aging (Phoenix & de Grey, 2007). Hence, the authors propose that once these disruptive technologies are manifest, they might serve as catalysts for dramatic positive paradigm shifts that touch practically every facet of human society, at least in terms of physical needs, and that nanotechnology may itself serve as a significant beneficial trajectory for humanity.

1.3 SUPERSEDING THE DOOM AND GLOOM

There is, of course, much more to consider than simply physical needs. Life demands motivated purpose. How will humanity extricate itself from being inundated by and mired within the negative sensationalist muck that permeates the daily news and myriad hypnotic doomsday scenarios? Has humanity always been so fixated on its own demise? Molly Downs, in an article for The Skyline View, explores the roots of this dilemma by initially quoting Dr. Michael Moynihan, Professor of Sociology at Skyline College, who offers his view as to how humanity has contemplated the Apocalypse for millennia: "It often represents a desperate attempt to reestablish order in people's lives and to cling to something that has meaning," Ms. Downs then reflects and concludes: "Why is America obsessed with the Apocalypse? Ironically, we may all have a deep desire to start over, to begin each day as if it were our first, not our last. Don't we always want that second chance and that opportunity to try again? Perhaps the obsession with the Apocalypse is not about endings at all. Perhaps it is, in fact, about beginnings." Yes! … new beginnings, an unwavering optimistic attitude, imagination and creativity, unceasingly seeking opportunities for advancement, enlightenment, harmony, health, synergy, adventure, fulfillment (happiness), liberating freedom … always love … and that eternally rising brilliantly incandescent core spark of life itself …. "hope." Are not these cherished ideals what all humans within their deepest selves aspire to, and hope for in the lives of others?

Can we, as a common species, ever dare "hope" to recreate/reshape our collective world view, global environment, our connection with amazing Mother Earth, and our own interrelations so as to manifest them toward an envisaged glorious, peaceful, and harmonious reality? In fact, well acknowledged scientists recommend that it would behoove humanity to strive to dwell in a state that encompasses an unwavering optimistic attitude, positive constructive imagination, and creativity. The emotional environment generated by these positive attitudes has well-documented beneficial effects on the biochemical processes of the human body (Lipton, 2007). When the body's own self-healing forces are engaged by positive expectations (e.g., placebo effect), they have strong potential to be as beneficial, or indeed, even more effective than conventional medical treatments (de la Fuente-Fernández & Stoessl, 2004).

We can anticipate that the biochemistries of belief and hope will be understood more deeply in the future, as ancient spiritual knowledge merges with Western science in broadening the focal point of awareness. Continually enhanced, highly sensitive measurement techniques are developed to enlarge our view into the molecular, nanometric, and quantum domains as they unite to provide a clearer understanding of our biology. In the future, it is likely that we will be better equipped to elucidate the underlying quantum fabric of our consciousness as well as the interconnectedness between the human mind/spirit/essence and the material world (Popp & Beloussov, 2013). We surmise that the immense power of cumulative optimistic, positive, and creative attributes will have the effect of not only creating an increasingly healthy population but will also spread and be amplified due to the connective nature of consciousness (Hunt, 1996).

1.4 NANOTECHNOLOGY AND MOLECULAR MANUFACTURING – EMERGENCE OF THE GRAND EQUALIZER?

The still nascent discipline of nanotechnology comprises a rapidly evolving and fundamentally disruptive and enabling set of capabilities that have extremely strong prospects for imparting extensive and primarily positive impacts on virtually every sector and aspect of human societal life on a global scale. It involves capacities for the ultrafine controllable manipulation of matter at the atomic and molecular levels toward the fabrication of unique materials, components, devices, and systems with novel attributes with dramatically improved performance. As espoused by Eric Drexler in his 1986 book, *Engines of Creation* (Drexler, 1986) and in his latest book, *Radical Abundance* (Drexler, 2013), myriad and significant facets of human society are likely to undergo dramatic paradigm shifts.

The prerequisite and prime driver of these possibly dramatic changes will be ushered in with the advent of mature MM, which is integrated with advanced artificial intelligence (AI). The rapidly increasing sophistication and popularity of 3D printing that we are currently witnessing might be viewed as an early precursor of envisaged, exponentially more powerful and dynamic MM capabilities (Phoenix, 2003; Phoenix, 2005). These MM units might initially be manifest as microwave oven-sized "factory-at-home" (F@H) systems, which would utilize atomic and molecular feedstocks, rather than the primarily polymeric feedstocks that are employed in 3D printers today.

The ultrafine positioning of specific molecular and atomic species might enable the cumulative assembly of practically anything we can imagine (e.g., healthy gourmet foods, running shoes, toothbrushes, physician prescribed nanomedical diagnostic or therapeutic devices, photovoltaic cells, smartphones, computers, as well as components for building additional MM units) with atomic/molecular layer by layer precision. The parameters of what might be produced would be limited only by the human imagination, design inputs, and the dimensional constraints of a given MM unit. This will translate to enabling the capacity for the average person to produce practically any type of consumer product.

Issues to be considered here may pertain to who will control the production, supply, distribution, and cost of MM "inks" (atomic/molecular feedstock cartridges). It is conceivable that these supplies might also be produced at home/on site by the user via an addition appliance (e.g., dedicated atomic/molecular disassemblers and highly sophisticated recycling and atomic/molecular sortation units, for the sole purpose of generating MM feedstock cartridges). An additional consideration will relate to who will administer and regulate the data that is required to direct what is produced by the MM units (perhaps for the most part, MM software data will be open source, save for highly encrypted files for nanomedical devices that are custom prescribed to align with individual patient requirements, and other items related to national and global security).

When MM evolves to the stage where it becomes available to practically any individual on the planet, it will undoubtedly translate to dramatic paradigm shifts as relates to conventional manufacturing facilities that generate commercial goods and foods, which will impact retail sectors, transportation infrastructures, energy production and distribution grids, banking and monetary systems, and underlying social constructs. The arrival of MM will likely enable the domestic on-site fabrication of ultrahigh efficiency solar cells, which will translate to a new paradigm of decentralized energy self-sufficiency.

When coupled with considerable anticipated improvements in other renewable energies, such as wind, wave, geothermal, energy storage, ubiquitous piezo energy harvesting (e.g., via integrated zinc oxide nanowire arrays (Wang & Song, 2006; Ou et al., 2016; He et al., 2023), and factors such as enhancements in the electronic efficiencies of domestic, commercial and industrial heating, cooling, appliances, etc., humanity's reliance on fossil fuels is likely to undergo a dramatic reduction. Hence, for the most part, every dwelling, which in the future might conceivably itself be fabricated on site via multiple specialized MM systems, worldwide would be off grid. This considerable paradigm shift, though somewhat unthinkable at present, could invariably lead to progressively diminished requirements for centralized energy grids, thus they might come to be considered as icons of an earlier era and relegated to obsolescence.

A critical question to explore in regard to MM will relate to how we can fail-safe MM units to prevent their malicious use. This possibility might be addressed via inbuilt programming that would intuitively establish parameters on exactly what can be made; with a default dwell mode status being engaged should these parameters be exceeded. That said, if humanity is empowered with the ability to fabricate and possess practically any material object that they desire, would this translate to the absence of discontent on a global scale, when humans eventually; ideally, arrive at a place where there are no longer any marginalized victims? If all of humanity could be endowed with optimally healthy bodies and selectively lengthy lifespans, would we finally be fulfilled and truly happy?

These are important questions. In essence, for the human species and all of its current constructs, nanotechnology may indeed be raised to attain the status of the "grand equalizer." How will we relate these tenets to the impoverished, marginalized, and suffering in the world, where currently, at least 80% of humanity lives on less than $10 a day? If we posit that nanotech might serve as the grand equalizer for humanity as far as the provision of all manner of consumer goods, and that nanomedicine may have the capacity for making each of us robustly healthy, the logical first step will be to assist the majority of humanity, to significantly raise their standards of living, with the aim of attaining a human equilibrium; an even playing field. This will also serve to exponentially augment humanity's collective intellectual resources, which may be applied to the resolution of serious challenges beyond those associated with climate change (e.g., development of human undersea habitats to facilitate housing for increasing populations, advance space travel and the colonization of other planets, addressing multiple issues brought on by the capacity for dramatically increased longevity).

1.5 NANOMEDICINE AND ENERGY FIELDS – RADICAL PARADIGM SHIFTS IN GLOBAL HEALTHCARE

There is a potent and rapidly increasing drive toward the development of smaller (less invasive), smarter, more efficacious, and cost effective medical diagnostics and therapeutics. These trends are evolving in conjunction with the emergence of what may currently be considered as alternative diagnostics and therapeutics, such as the manipulation of energy fields associated with the human body, and indeed permeate and intimately connect us with the universe. The nascent discipline of nanomedicine holds great potential for the development of advanced medical procedures that may, in the not-too-distant future, negate the requirement of invasive surgeries, eradicate practically every disease state and condition brought on by pathogens, toxins, or microorganisms, and finally conquer the disease of aging. Aubrey de Grey (President and CSO of Longevity Escape Velocity (LEV) Foundation) proposes that through various rejuvenative strategies, the first person to have the option of living to 1000 is likely alive today (Zolman, 2018; Palmer, 2022; Longevity by Design, 2025).

These approaches might include the eradication of lipofuscin from the lysosomes ("garbage disposal" organelles) of human cells. Lipofuscin accumulates over time within cells to the point that it can begin to negatively impact the proper functionality of non-dividing cells (e.g., brain, liver, kidney, heart muscle, adrenals, eyes, nerve cells), and is associated with a range of age-related conditions. Thus, if lipofuscin can be efficaciously extricated from these cells, it may (hypothetically) have the effect of revitalizing them, with potentially significant beneficial health effects (Boehm, 2013).

Though it may sound somewhat naïve or utopianist, the authors envisage that advanced nanomedicine might serve as a positive and benevolent facilitator toward the eventual attainment of health care homeostasis/equilibrium globally; enabling access to beneficial advanced medical technologies for those in remote regions and the developing world, while significantly reducing medical expenditures in the developed world. These considerable savings might be reallocated, for instance, to further advancing and expediting clean and renewable energy harvesting and generating technologies and water purification/seawater desalination technologies for the large-scale provision of clean potable water, which will be increasingly precious and in high demand.

A prudent developmental strategy for nanomedicine might involve the establishment of stringent standardized protocols to ensure the safe development of nanomedical components, devices, and systems, validation of their robust efficacy, and the assurance of appropriate post-use trajectories (e.g., capacity for complete recyclability or biodegradability, benign environmental fates). This would require that nanomedical components and devices be designed to be comprised of non-toxic nanometric elements from their inception. A critical element in all of this would be the establishment of genuine transparency and interactivity, initiated and conveyed by those who are engaged in the development of these nanotechnologies and nanomedicines, as relates to the general public. After all, we will all be in this together.

On another front, we can expect the development of ever faster and more sensitive measurement techniques to assist us in deciphering the inherent kinetics and connectivities of nanoscale and quantum domains (likely significantly expedited by surging advances in ANI [Artificial Narrow Intelligence], and soon enough, AGI [Artificial General Intelligence] and ASI [Artificial Superintelligence] [OECD, 2023; Wheeler, 2025]) and how they relate to the macroworld. Indeed, various aspects of diverse research fields have already been elucidated in minute detail. Examples may be found in particle physics: Higgs Boson (Cho, 2012; Salam et al., 2022; Pokorski, 2024), which will help us to understand, among other of our deep queries; the origin of mass, and the nature of dark matter and dark energy (Womersley, 2005).

In terms of structural biology, a milestone has been recently achieved toward deciphering the extremely complex protein units of the yeast mitochondrial large ribosomal subunit, at the atomic level, utilizing electron microscope detectors with unprecedented speed and sensitivity (Amunts et al., 2014). In the realm of biophysics, we are learning to decode energetic forces and pervasive, albeit subtle, energy fields at subatomic levels. At this juncture, interdisciplinary cross-pollination has not yet been optimized, given the complexity of each area in conjunction with the different personalities involved, who may establish a certain breadth of focus as relates to their particular area of research. However, as cross communication between these fields continues to increase, a cumulative effect can be expected that may abruptly and exponentially enhance integrative knowledge between these disciplines.

Concomitant with our increased knowledge of how to create at the nanoscale will arise an exponentially deeper understanding of our own highly complex physiological processes. Hence, more symbiotic and holistic approaches to medicine will also accelerate toward continually less invasive procedures. An excellent illustration of a symbiotic approach that harnesses the body's own functions utilized nanotechnology fused with cell biology in the visionary work of Spatz, who employed gold nanoparticles to "train" human immune cells outside the body to boost the immune system when the now specifically trained immune cells were introduced back into the human host (Platzman et al., 2013).

1.6 QUANDARY – ARE MOLECULARLY MANUFACTURED BURGERS IMBUED WITH THE LIFE FORCE?

The potential plethora of MM foods may hold great promise in the provision of optimal sustenance for immense human populations yet to come, as the precise control of food constituents will be possible. The UN has estimated that by 2050 the global population will reach 9.8 billion and then 10.3 billion by 2084 (Ritchie & Rodés-Guirao, 2024). With our growing molecular scale understanding of food ingredients and food-related biochemistries, specific beneficial biological molecular level food compounding will be standardized. Thus MM is likely to replace high volumes of low-quality processed food items, which may contain large amounts of unhealthy "unidentified edible objects" which are under increasingly intense scrutiny by a rapidly expanding health conscious population. MM may also offer an alternative to meat products that are derived from animals that are kept under unacceptable conditions on factory farms (Robbins, 2012), and/or have been fed or treated with ingredients that might compromise human health. The analytical aspect of MM foods may also spur the development of accompanying technologies that would enable the consumer to custom-generate unique food items via health optimizing or enhanced molecular "recipes" that precisely align with their unique physiological requirements.

The prospect, however, of replacing a "natural" grass-fed free range bison burger with an MM "artificial" meat item may elicit, for many people (not only gourmets), a wincing feeling that … "something is just not right." Might this reaction be a reminiscence of the life force that is innate to all living things? Would we be missing a most critical aspect of foods? We have proven to get by without the traditional honoring of the spirit of the buffalo (integration or transference of their strength and wisdom), which was celebrated by early Native Americans, in our model bison burger.

Even if molecular food fabrication is successful in replicating the identical compositions and chemistries of essential food ingredients; amino acids, proteins, fats, etc., atom by atom, and succeeds in fusing them into a digestible food item, is the energetic imprint of the life force something that is essential and quantifiable for our health? Does it play a vital role in sustaining our own life force, as an aspect yet to be understood? Might this aspect be considered as being transferable via constituent "live" atomic/molecular feedstocks, by virtue of integrated "living" electrons and underlying quantum energies? Measurements of the energy fields in plants and humans are of growing interest as we begin to understand the subtle yet ubiquitous interconnectedness between these energetic fields within and between organisms, and acknowledge their importance in sustaining health. Advances in this domain will likely enhance our knowledge as relates to the importance of the energetic profiles of food and biochemistry, which may translate to not only physically and chemically, but also energetically precise biomimetic MM.

1.7 INTEGRATION OF THE 10 BILLION SYNAPSE WORLD MIND

Humans have an innately deep-seated need for communication, which is a prime driver behind the incessant globalization that we are witnessing today, facilitated in no small measure by mobile phones and the internet. There are currently ~4.7 billion smartphones in use, which is set to expand to ~5.8 billion by 2028 (Backlinko & Semrush, 2025), while the present number of mobile phones is more than double the human population at ~18.2 billion (Laricchia, 2023).

These technologies have enabled us to communicate with others in the most far-flung regions on Earth. We propose that nanotechnology will take this immense human dialogue to unprecedented levels. Nanotechnology-enabled MM, in conjunction with advanced AI, will likely enable the transition from our use of extraneous devices to those that are integrated directly with our brains via imperceptible nanomedical implants. In essence, these embedded "nanonetworks" may serve as wireless interfacial conduits to an omnipresent Internet (Zax, 2011; Mozzini-Alister, 2024), which may also have (optional) direct links to our optic nerves and other senses. This would give us the capacity, if we so desire, to experience fully immersive stories, movies, or indeed, if consent and access is granted by other parties, segments of the lives of any individual on the planet, via non-intrusive "transparent shadowing," where one would "plug in" to another person's experiences (Martins et al., 2019). This would take human collaboration and empathy to unprecedented levels, as one might literally walk in someone else's shoes and directly "sample" their experiences. The proviso here would be that the thoughts/emotions of the "host" would remain completely private and inaccessible.

1.8 CONCLUSION – HUMANITY'S GUIDING SPIRIT

It seems appropriate to conclude with a final quote from our book as well:

> In the larger frame, we live day-to-day on our miraculous and beautiful blue world (at times oblivious to the overwhelming odds for us not existing at all) in the midst of the deep and inconceivably vast expanse of darkness speckled with trillions of stars. Where do our boundaries lie? It may be suggested that any constraint imposed upon the human imagination and the extent to which we may explore and advance within ourselves and into the without, are entirely self-imposed.
>
> **Boehm, 2013**

The title of our essay pertains to a central and critical question, with the proviso that nanotechnology and AI-driven MM will indeed come to fruition, of whether humanity will ultimately reject or embrace these advanced technologies in the interests of cumulative survival and the creation of a vibrant, healthy, creative, and fulfilled humanity. When we consider that an extensive variety of conventional food items, including non-animal derived "meats" (the "blueprints" of which would be meticulously derived from "real" foods and optimized for nutritional value and discerning tastes) might be readily available to virtually every individual on the planet via MM, this in itself is likely to convey dramatic positive impacts toward a sustainable and flourishing human future. However, will they be accepted by humanity?

One need not look further than the controversies spun around genetically modified (GM) (Prakash, 2014; Cabelkova et al., 2024) foods to see the understandable and natural cautiousness with which we approach any new technology and its implications, which we do not yet fully understand. A report released by the Pew Research Centre based on a national phone survey of 1,001 adults in the US, revealed that only 20% of respondents "… would be willing to eat meat that was grown in a lab" (Smith, 2014; Rabie, 2019; Karalis et al., 2020; Fan et al., 2021). This is indeed indicative of the substantial gap that will have to be traversed toward the acceptance of MM foods, which may coalesce as a pivotal in defining our collective fate. In short, despite all of our shortcomings, the authors firmly believe that humanity possesses the capacities to efficaciously identify, rationally assess and resolve the most recalcitrant of challenges, and subsequently, to implement prudent and positive modifications toward the continuation and expansion of our prosperity; perhaps in the future, to the edges of our galaxy, or indeed the periphery of the universe.

REFERENCES

Amunts A, Brown A, Bai XC, Llácer JL, Hussain T, Emsley P, et al. Structure of the yeast mitochondrial large ribosomal subunit. *Science*. 2014 Mar 28;343(6178):1485–1489.

Backlinko and Semrush. Smartphone Usage Statistics (as of January 30, 2025) Backlinko. https://backlinko.com/smartphone-usage-statistics (accessed April 06, 2025).

Boehm FJ. *Nanomedical Device and Systems Design: Challenges, Possibilities, Visions*. CRC Press (Taylor & Francis Group), Boca Raton, FL. November 18, 2013.

Cabelkova I, Sanova P, Hlavacek M, Broz D, Smutka L, Prochazka P. The moderating role of perceived health risks on the acceptance of genetically modified food. *Front Public Health*. 2024 Jan 25;11:1275287.

Cho A. Particle physics. Higgs boson makes its debut after decades-long search. *Science*. 2012 Jul 13;337(6091):141–143.

de la Fuente-Fernández R, Stoessl AJ. The biochemical bases of the placebo effect. *Sci Eng Ethics*. 2004 Jan;10(1):143–150.

Drexler KE. *Engines of Creation: The Coming Era of Nanotechnology*. Anchor Books. 1986. https://web.archive.org/web/20090105170239/http://e-drexler.com/d/06/00/EOC/EOC_Table_of_Contents.html (accessed April 05, 2025).

Drexler KE. *Radical Abundance: How a Revolution in Nanotechnology Will Change Civilization*. PublicAffairs. 2013.

Fan Z, Mu Y, Sonstegard T, Zhai X, Li K, Hackett PB, Zhu Z. Social acceptance for commercialization of genetically modified food animals. *Natl Sci Rev*. 2021 Apr 21;8(8):nwab067.

He S, Xie D, Wang B, Zhu M, Hu S. Photocatalytic fuel cell based on integrated silicon nanowire arrays/zinc oxide heterojunction anode for simultaneous wastewater treatment and electricity production. *J Colloid Interface Sci*. 2023 Nov 15;650(Pt B):1993–2002.

Helmers H, Höhn O, Lackner D, Schygulla P, Klitzke M, Schön J, et al. Advancing solar energy conversion efficiency to 47.6% and exploring the spectral versatility of III-V photonic power converters. *Proc. SPIE 12881, Physics, Simulation, and Photonic Engineering of Photovoltaic Devices XIII*. 1288103. March 08, 2024.

Hunt V. *Infinite Mind*. Malibu Publishing Co, Malibu, California. 1996.

Jeong I, Lee TL, Tran HV, Hwang I, Hwang J, Lee A, et al. Flexible and lightweight perovskite/Cu(In,Ga)Se_2 tandem solar cells. *Joule*. 2025;9(3):101794.

Karalis DT, Karalis T, Karalis S, Kleisiari AS. Genetically modified products, perspectives and challenges. *Cureus*. 2020 Mar 18;12(3):e7306.

Laricchia F. Number of mobile devices worldwide 2020–2025. Statista. March 10, 2023. https://www.statista.com/statistics/245501/multiple-mobile-device-ownership-worldwide/#statisticContainer (accessed April 06, 2025).

Lipton BH. *The Biology of Belief: Unleashing the Power of Consciousness*. Matter, & Miracles, Hay House, Inc, Vista, California. 2007.

Liu Y, Gao W, Ran C, Dong H, Sun N, Ran X, et al. All-inorganic Sn-based perovskite solar cells: Status, challenges, and perspectives. *ChemSusChem*. 2020 Dec 17;13(24):6477–6497.

Longevity by Design. The Path to a 1,000-Year Lifespan with Dr. Aubrey de Grey. January 8, 2025. https://blog.insidetracker.com/longevity-by-design-podcast-aubrey-de-grey (accessed March 05, 2025).

Martins NRB, Angelica A, Chakravarthy K, Svidinenko Y, Boehm FJ, Opris I, Lebedev MA, Swan M, Garan SA, Rosenfeld JV, Hogg T, Freitas RA Jr. Human brain/cloud interface. *Front. Neurosci*. 2019;S13:112.

Mozzini-Alister C. COVID-19, social media dependency and desire for omnipresence: Ways forward. *Adv Exp Med Biol*. 2024;1458:145–155.

Naik G. U.N. Climate Change Report Says Worst Scenarios Can Still Be Avoided. *The Wall Street Journal*, April 13, 2014. https://www.wsj.com/articles/SB10001424052702303663604579498721229249160?msockid=1a57e40117d16b281101f4e216726ae2 (accessed April 05, 2025).

Otnes G, Barrigón E, Sundvall C, Svensson KE, Heurlin M, Siefer G et al. Understanding InP nanowire array solar cell performance by nanoprobe-enabled single nanowire measurements. *Nano Lett*. 2018 May 9;18(5):3038–3046.

Ou C, Sanchez-Jimenez PE, Datta A, Boughey FL, Whiter RA, Sahonta SL, Kar-Narayan S. Template-assisted hydrothermal growth of aligned zinc oxide nanowires for piezoelectric energy harvesting applications. *ACS Appl Mater Interfaces*. 2016 Jun 8;8(22):13678–13683.

Palmer RD. Three tiers to biological escape velocity: The quest to outwit aging. *Aging Med (Milton)*. 2022 Dec 12;5(4):281–286.

Phoenix C. Design of a primitive nanofactory. *J Evol Technol.* October 2003;13, https://www.jetpress.org/volume13/Nanofactory.htm
Phoenix C. Molecular Manufacturing: What, Why and How. Center for Responsible Nanotechnology. May 17, 2005.
Phoenix C, de Grey AD. A model of aging as accumulated damage matches observed mortality patterns and predicts the life-extending effects of prospective interventions. *Age (Dordr).* 2007 Dec;29(4):133–189.
Platzman I, Janiesch JW, Spatz JP. Synthesis of nanostructured and biofunctionalized water-in-oil droplets as tools for homing T cells. *J Am Chem Soc.* 2013 Mar 6;135(9):3339–3342.
Pokorski S. After the Higgs boson discovery: A turning point in particle physics. *Philos Trans A Math Phys Eng Sci.* 2024 Feb 5;382(2266):20230090.
Popp FA, Beloussov L. *Integrative Biophysics: Biophotonics.* Springer Science & Business Media, Secaucus, New Jersey. 2013.
Prakash CS. A look at the recent news from around the world on genetically modified food and crops. *GM Crops Food.* 2014 Jan–Mar;5(1):1–3.
Rabie P. The truth about lab-grown meat. *Scienceline.* January 16, 2019 https://scienceline.org/2019/01/the-truth-about-lab-grown-meat/ (accessed May 07, 2925).
Ritchie H, Rodés-Guirao L "Peak global population and other key findings from the 2024 UN World Population Prospects." Published online at OurWorldinData.org. 2024. https://ourworldindata.org/un-population-2024-revision (accessed April 05, 2025).
Robbins J. *Diet for a New America: How Your Food Choices Affect Your Health, Your Happiness, and the Future of Life on Earth.* H J Kramer, Tiburon, California, 2012.
Salam GP, Wang LT, Zanderighi G. The Higgs boson turns ten. *Nature.* 2022;607(7917):41–47.
Shishido H, Sato R, Ieki D, Matsuo G, Saito K, Konagai M, Ishikawa R. High-efficiency perovskite/silicon tandem solar cells with flexibility. *Solar RRL.* 2025;9:2400899.
OECD. *Artificial Intelligence in Science: Challenges, Opportunities and the Future of Research,* OECD Publishing, Paris, 2023. https://www.oecd.org/en/publications/artificial-intelligence-in-science_a8d820bd-en.html (accessed May 07, 2025).
Smith A. U.S. Views of Technology and the Future – Science in the Next 50 Years, Pew Research Internet Project, April 17, 2014. http://www.pewinternet.org/2014/04/17/us-views-of-technology-and-the-future/
Stocker TF, Qin D, Plattner GK, Tignor M, Allen SK, Boschung J, et al. *Climate Change 2013: The Physical Science Basis. Contribution of Working Group I to the Fifth Assessment Report of the Intergovernmental Panel on Climate Change.* Cambridge University Press, Cambridge, United Kingdom and New York. 2013;1535 pp., IPCC.
Wang ZL, Song J. Piezoelectric nanogenerators based on zinc oxide nanowire arrays. *Science.* 2006 Apr 14;312(5771):242–246.
Wheeler J. ANI vs. AGI vs. ASI: What They Mean and Why 2025 Could Be the Year of AGI. January 06, 2025. *Virtually Caffeinated – A Double-Shot of Knowledge.* https://www.virtuallycaffeinated.com/2025/01/06/ani-vs-agi-vs-asi-what-they-mean-and-why-2025-could-be-the-year-of-agi/ (accessed May 07, 2025).
Wilson L. Average Household Electricity Consumption – 2025. Shrink That Footprint. https://shrinkthatfootprint.com/average-household-electricity-consumption/ (accessed April 10, 2025).
Womersley J. Beyond the Standard Model. *Symmetry Magazine.* February 2005. https://www.symmetrymagazine.org/article/february-2005/beyond-standard-model?language_content_entity=und (accessed April 05, 2025).
Zax D. Envisioning the Omnipresent, Benevolent Internet of the Future, Fast Company, May 26, 2011. http://www.fastcompany.com/1755500/envisioning-omnipresent-benevolent-internet-future (accessed April 10, 2025).
Zolman ON. Longevity escape velocity medicine: A new medical specialty for longevity? *Rejuvenation Res.* 2018 Feb;21(1):1–2.

2 Factory@Home
Democratizing the Coming Industrial Revolution

Margaret Morris and Frank J. Boehm

CONTENTS

2.1 Introduction 13
2.2 Advantages and Utility 14
2.3 Survey of 3D Printing Technologies: Prelude to F@H 16
2.3.1 Jet Engines, Space Wrench, and Cameras 18
2.3.2 Energy 19
2.3.3 Transportation 20
2.3.4 Consumer Goods 20
2.3.5 Robotics 21
2.3.6 Construction 23
2.3.7 Tissue Engineering 23
2.4 Differentiating 3D Printing and F@H 24
2.5 Micro/Nano-Electro-Mechanical Systems (MEMS) 24
2.6 Molecular Manufacturing 25
2.7 Societal Implications 26
2.7.1 Eradication of Poverty 26
2.7.2 Decentralized Electricity and the Environment 27
2.7.3 Dimensional Constraints 28
2.7.4 Future of Work 28
2.8 Potential for Abuse and Regulatory Prudence 29
2.9 Price:performance Ratio 30
2.10 First-Generation F@H 30
2.11 Second-Generation F@H 31
2.12 Third-Generation F@H 32
2.13 Conclusion 32
References 33

2.1 INTRODUCTION

At first glance, the phrase "Factory-at-Home" (F@H) might evoke the drudgery of either bringing factory work to one's home or involve the transformation of one's home into some type of cottage industry. In stark contrast, envisioned future F@H systems will

DOI: 10.1201/9781003724896-2

function as indispensable, omnipresent, on-site cornucopia desktop/countertop (and larger) factories, realized through the integration of highly sophisticated technologies. These systems might achieve the capacity to fabricate practically any type of consumer item, including nutritious gourmet foods, and via mature specialized units, autonomous nanomedical devices that address all that ails us, including the disease of aging.

To be practical, they will operate not only using commercially available feedstocks (e.g., smart molecular/atomic feedstocks) but (downstream) also with all manner of atomically recycled materials and renewable feedstocks. Unlike today's factories that manufacture a plethora of identical objects and their variants, the versatility of F@H systems will enable common, customized, or one-of-a-kind products on demand. Indeed, we may someday take for granted these F@H systems as they sustain our businesses, homes, and personal needs, while we, due to the likely emergence of significant paradigm shifts, in terms of the desuetude of the notion that humanity must labor by rote to obtain money to survive (Albanesi et al., 2023; Crisanto et al., 2024; Uzougbo et al., 2024; AZAI, Research Team 2025; Stiefenhofer, 2025) brought on by robotics, ANI (Artificial Narrow Intelligence)/AGI (Artificial General Intelligence)/ ASI (Artificial Superintelligence) agents, perhaps in conjunction with integrated B/ CI cognitive prosthetics (Martins et al., 2019), set our focus on the creative endeavors and activities to which we enthusiastically devote our passions, and from which we derive our best fulfillment and joy. Here, we consider the potentially radical social paradigm shifts that F@H systems might bring to bear (in conjunction with mechanistic strategies and processes) that will likely make them increasingly versatile, as the scientific and technological disciplines that enable them continue to advance.

2.2 ADVANTAGES AND UTILITY

As its prerequisite technologies are perfected, routine F@H products will encompass almost any typical consumer item, including common household and decorative items and customized fashion and practical clothing (e.g., extreme weather gear and ultra-comfortable shoes that are made to fit perfectly, imbue posture-correcting structuring, or satisfy other concerns). Functional clothing that is completely water and strain resistant (superhydrophic) (Ahmad & Kan, 2016) self-cleaning and deodorizing (Wu & Long, 2011; Manna et al., 2015) and/or embedded with nanometric piezogenerators that generate electricity through user movements to power portable or wearable digital devices (including an array of diagnostic nanosensors) (Kim et al., 2017; Lu & Skorobogatiy, 2017) may also become customized and commonplace. Advancing F@H will have the ability to manufacture increasingly complex and multifaceted personalized products, such as wearable medical devices equipped with communications that interact with healthcare providers and databases to help safeguard individual health.

Even more sophisticated iterations will enable the generation of personalized medicines and nutritional formulas that are tailored to individual requirements. With further insights into epigenetics (e.g., gene expression through chemical modifications, typically favorably using functional foods and specific nutrients in therapeutic doses) gleaned by medical science, customized formulations might induce specific genes to switch on or off, thereby promoting improved gene expression and greater potential for extreme longevity in good health. In due course, physician-prescribed autonomous nanomedical devices may be fabricated to atomic scale precision, which will play a prominent role in the establishment and worldwide proliferation of an envisaged Global Health Care Equivalency (GHCE) platform (Boehm, 2013; Boehm, 2025). Entrepreneurs will be able to rapidly prototype and test experimental F@H built designs and manufacture short runs of their products. Artists, crafters, and designers will employ F@H to generate their unique creations (e.g., one-of-a-kind jewelry pieces) that might be virtually impossible to fabricate by hand.

When parallel computational power and robotic technologies advance sufficiently, F@H might be utilized to fabricate components of various superintelligent robots and other machines, including shape-shifting robots that repair, maintain, and efficiently run homes and associated gardens, orchards, properties, and related infrastructures. Such robots might have the capacity to perform domestic and professional tasks as they manage land and resources, as well as collect, test, and optimize soil and water, grow, tend, harvest, dry, can, store, inventory, and prepare food via highly calibrated F@H-built robotics and other equipment.

Singularity University lists robotic development as an exponentially growing technology, as well as technologies that expedite robotic evolution, encompassing high performance computing, improved algorithms (e.g., employing big data), digital fabrication, nanotechnology, and others. For exponential growth, only 30 doublings are equivalent to a billion-fold increase in performance. One of the ultimate achievements of advanced F@H may be the fabrication of sophisticated autonomous nanomedical robots (Figure 2.1) that are able to remain within

FIGURE 2.1 Conceptual autonomous nanomedical robot. (By FotoRichter via Pixabay. Creative Commons Zero (CC0) license.)

the body for extended periods to attain, secure, and preserve health, even under extremely adverse conditions. A further critical accomplishment will be to significantly equip space colonies and help build wealth on Earth when humankind is able to inhabit Mars, customize our solar system, and send spacecraft and humans to the stars.

Robert A. Freitas Jr., an early pioneer of nanomedicine and winner of the 2009 Feynman Prize for Theory and innovator in nanomanufacturing, provides the following description (in collaboration with Ralph Merkle) of a conceptual nanofactory, and why diamondoid may be a preferred material.

> The nanofactory is a proposed compact molecular manufacturing system, possibly small enough to sit on a desktop, that could build a diverse selection of large-scale atomically precise diamondoid products. The nanofactory is potentially a high quality, extremely low cost, and very flexible manufacturing system... Our nanofactory will be constructed from diamondoid components of the same sort that it can itself manufacture. While molecular manufacturing systems made from DNA, other biopolymers, or even biological organisms are possible, such systems would be unable to build products that approach the remarkable strength, stiffness, temperature range, lightness, electrical, optical and other properties that can achieved with diamondoid materials.
>
> **Freitas & Merkle, 2024**

As articulated further below, while highly advanced F@H will be designed to have the capacity to integrate molecular manufacturing, first-generation F@H will not rely on such technology, and instead will incorporate AI-driven 3D printing and shape-changing, miniaturized robotic factory configurations, and certain nanotechnology innovations that do not necessitate nanomanufacturing per se.

2.3 SURVEY OF 3D PRINTING TECHNOLOGIES: PRELUDE TO F@H

Although AI/computer-guided 3D printing (also referred to as additive manufacturing) may be considered a forerunner of F@H (and might well evolve to it), there are clear distinctions between the two. A brief survey of the progress in 3D printing can illustrate some of the promise of desktop manufacturing generally. Thousands of product manufacturing specifications are freely available as open-source software for 3D printers (Figure 2.2) (Laureto & Pearce, 2017; Sander et al., 2017), and ongoing progress illustrates just how rapidly advances can emerge (Antezana et al., 2023; Belgin et al., 2025; Ling et al., 2025; Sun et al., 2025; Wu et al., 2025).

3D printing involves the AI-controlled layer-by-layer application of various types of inks (comprised of polymers or other materials) to form 3D objects with almost any conceivable geometries. Printed objects may range from being fully transparent to completely opaque, while their structures can be internally and externally customized according to design. For example, complex entities can be fabricated

FIGURE 2.2 3D printers operating in parallel. (Reproduced from Wikimedia Commons, public domain https://www.cdc.gov/niosh/hhe/reports/pdfs/2017-0059-3291.pdf.)

with dissolvable support materials when one or more of the printer materials used is degradable (Hildreth et al., 2016; Farzin et al., 2018; Zammit et al., 2020; Zhang et al., 2021; Chaurasia et al., 2024; Du et al., 2024; Rashad et al., 2024; Korkeamäki et al., 2025). The 3D printed objects may have internal cavities or networks that are left hollow or filled with a different type of substance or device, either during or subsequent to printing. An object can also be 3D printed to contain single or multiple moving parts.

Further, advanced 3D printer systems incorporate multiple print heads that utilize a range of different inks and processing methods. Myriad classes of materials include various liquid polymers (Li et al., 2019; Xie et al., 2023); melted wax (Gu et al., 2023); metal powders (Svozilova et al., 2024; Li et al., 2025); graphene oxide (Fu et al., 2016; Zhou et al., 2022; Cernencu et al., 2023); elemental organic materials (e.g., for printing designer molecules) (Service, 2015;

Zhu et al., 2019; Zhang et al., 2022); and living biological cells for fabricating skin (Pourchet et al., 2017; Balavigneswaran et al., 2023; Ghobadi et al., 2025); and other tissues (Zhu et al., 2016; Tan et al., 2017; Feng et al., 2021; Davoodi et al., 2025; Rajzer et al., 2025); and organs (Ji & Guvendiren, 2017; Pedde et al., 2017; Dey & Ozbolat, 2020) for the treatment of an extensive range of medical conditions.

One of the processes involved is the use of a hot laser, employed to fuse inks made with resins or metal powders. The 3D printer concept has also been employed to create a 3D welder, wherein a spool of wire is fed to the welder, such that the wire is applied layer-by-layer to create 3D metal objects.

2.3.1 Jet Engines, Space Wrench, and Cameras

In 2015, a research team at the Monash Centre for Additive Manufacturing (MCAM) (Monash University, Australia), 3D-printed the world's first jet engine by employing selective laser melting (SLM). Using an older gas turbine engine as a model, the printed engine was comprised of approximately 24 components that were scanned and then 3D printed using various materials, such as alloys of nickel, aluminum, and titanium (Science in Public, 2015). In the following year, the GE Aviation Additive Development Center (Cincinnati, OH) 3D printed a complete mini jet engine using melted metal powder applied layer upon layer. Subsequently, the engine was tested to run at 33,000 RPM (Keller, 2016). The same GE facility also produced the world's largest commercial jet engine, the GE9X (designed for the Boeing 777X) with metal parts that were 3D-printed and then manually assembled. This engine exhibited greater efficiency, power (100,000 pounds/45 metric tons of thrust), and heat resistance than its predecessors (Coren, 2016).

Recently, Rocket Lab (Long Beach, CA) 3D printed a Rutherford engine (the core element of its Electron rocket), which is the first such engine to reach space. The printer used to fabricate it applied finely powdered (Ø45 μm) super alloy metals including aluminum, copper, nickel, or titanium, layer by layer, which robustly tolerate high temperature rocket blasts. These alloys were applied to a printer bed and fused via laser-sintering (Zaleski, 2025) (Figure 2.3).

In 2014, the Additive Manufacturing Facility on board the International Space Station (ISS) 3D printed its first tool (a ratchet wrench), which was also the first tool to ever be printed in space. Designed to operate under microgravity, this 3D printer is anticipated to have the capacity to manufacture 30% of the replacement parts required for the ISS (Harbaugh, 2014).

Further, a complete 35 mm camera, with aperture, lens, and other moving parts, was entirely printed by 3D designer Amos Dudley in 2016 (Scott, 2016). At a far more diminutive scale, Simon Thiele and associates at the University of Stuttgart (Germany) 3D printed a miniaturized camera comprised of four different objective lenses on a semiconductor image-sensor. The lenses were positioned such that each had a different field of view, which enabled the generation of high-resolution images at its center (similar to how eagle and human eyes function). A computer was employed to merge the data from the lenses to form a sharp central image (Thiele et al., 2017).

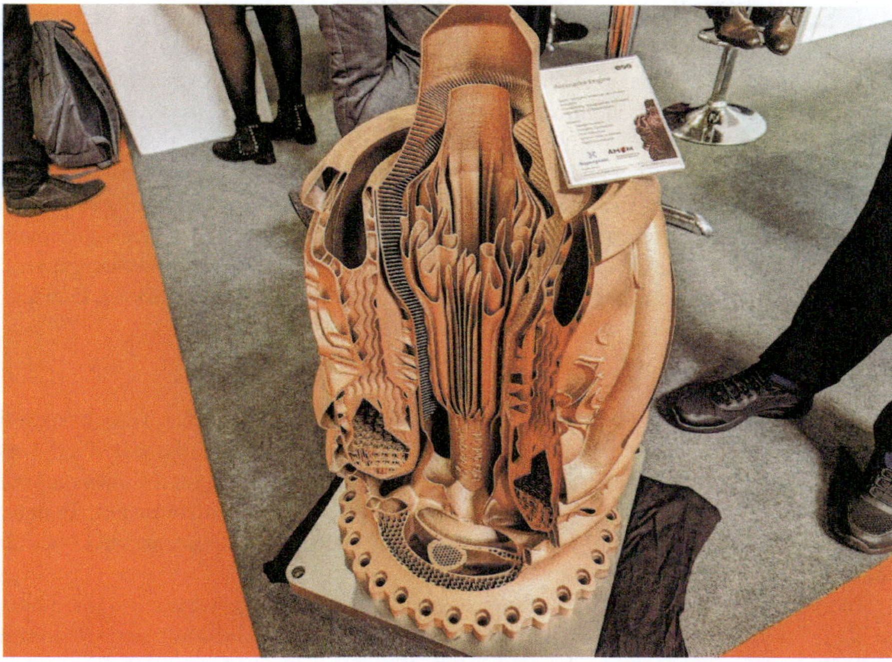

FIGURE 2.3 3D printed aerospike engine by AMCM and hyperganic. Photo by Michael Petch. (Reproduced from Alex Tyrer-Jones' article, "3D Printed Aerospike Engine Tested on MIRA II Spaceplane" for *3D Printing Industry*, December 12, 2024. https://3dprintingindustry.com/news/3d-printed-aerospike-engine-tested-on-mira-ii-spaceplane-235115/ (accessed May 19, 2025).)

2.3.2 Energy

The manufacturing of components to advance the energy sector may also be revolutionized by 3D printing. Researchers at the Flexible Electronics Laboratory (Clayton, Australia) demonstrated the first roll-to-roll printing of hybrid serially interconnected perovskite solar cell modules under ambient room temperatures/pressures. These cells exhibited conversion efficacies of "up to 15.5% for individual small-area cells and 11.0% for serially interconnected cells in large-area modules." (Weerasinghe et al., 2024). These types of photovoltaic cells might provide reliable energy to the estimated ~750 million people currently without power in developing countries (IEA, 2024). This system may also provide backup energy in developed countries, particularly during power grid failures.

Semi-transparent (Pompilio et al., 2022; Yu et al., 2023) and fully transparent (Liu et al., 2024; Jowett, 2025) photovoltaic materials will enable roofs, windows, and glass doors to be covered with solar panels that may generate between 10 and 50 W/m^2 at 1–5% efficiencies. To put this in perspective, consider that a mobile phone operates at a power level of 0.6 watts. Lighting is more energy intensive, where ten 100-watt light bulbs lit for an hour consume one kilowatt of power. Owing to their large surface areas, tall office buildings may be transformed to energy towers.

In another energy-related application, stainless steel powder was employed for the 3D printing of clean, efficient, micro fuel cells (Scotti et al., 2014; Jin et al., 2022; Khalifa et al., 2024; Elhadad et al., 2025), which can power portable devices with electric current generated by chemical reactions. To create the fuel cells, a computer program generates 3D models for the printer, with the advantage being that the resulting 3D printed lattice and other stainless-steel components cannot be fabricated using conventional manufacturing techniques at the required dimensions (internal geometries may be at micron or even nanometric scales). This work provides excellent exemplars of miniaturized stainless-steel components (e.g., pins, springs, gears, screws, gaskets, etc.), which might inspire the manufacturing processes utilized for a desktop F@H.

2.3.3 Transportation

Designed to be environmentally compatible, all components of a vehicle called Urbee were manufactured using 3D printed components, while the car itself was manually assembled. Urbee was to have been driven across the United States on only ten gallons of gas with the capacity to accommodate two people and one pet. To date, it has not been independently confirmed that the Urbee or its successor, Urbee 2, made the trip (King, 2013).

In 2015, Divergent Technologies used metal powder and aerospace carbon fibers to 3D print the components for its Blade supercar prototype, the company's first eco-friendly proof-of-concept model. The Blade was comprised of modular 3D-printed structures that were fitted together by hand. The company demonstrated a radical shift in the automotive industry in terms of greatly reducing manufacturing pollution and construction complexity. Blade's 3D printed aerospace carbon fiber 102 lb chassis (Marchese, 2019) possessed the same industrial strength and safety protection as a steel chassis weighing up to 1000 lbs. Owing to its lightweight interior, the two-seater offered an improved power-to-weight ratio than present-day Formula 1 racing cars. Among others in the growing list of 3D printed cars, the Blade project proved that a massive factory assembly line could be replaced by strategic 3D printing in conjunction with advanced design concepts (Figure 2.4).

In the aerospace domain, experimental drones have been progressing facilitated by 3D printing (Darsin et al., 2025). The U.S. Marine Corps recognized the economical and other advantages of using 3D printing to create drones, and now designs and tests prototype models for various applications (Atherton, 2021). Currently, myriad 3D printed drone applications include LIDAR mapping (Bierlich et al., 2024; Tang et al., 2025), medical supply transiting to remote locations (Dhakal & Karunakaran, 2023; Charan et al., 2024), assisting with wildfire containment (Saffre et al., 2022; Ramadan et al., 2024) and rescue missions (Holzmann et al., 2021; Nozica et al., 2023), military reconnaissance (Petrovski & Radovanovic, 2021) commercial delivery services (Mitchell et al., 2023; Charan et al., 2024), etc.

2.3.4 Consumer Goods

The fashion industry is driving the 3D printed design of clothing (Wu et al., 2022; Yang et al., 2022; Zhao et al., 2025), including dresses fabricated as a single piece (Kalenderian,

FIGURE 2.4 The Blade 3D printed supercar. (Reproduced from adlughmin article, World's First 3D Printed Supercar is Unveiled – 0–60 in 2.2 seconds, 700 HP motor – built from Unique Node System. *3DPrint.com*. June 24, 2015. https://3dprint.com/74810/3d-printed-supercar-blade/ (accessed May 19, 2025).)

2014), shoes (Baranowski et al., 2024; Lee & Chun, 2025), and avant-garde accessories (Pasricha & Greeninger, 2018; Unionfab, 2023). Rather than traditional 2D woven cloth, 3D interlocking weaves are sometimes featured. Industry goals include producing pieces that cannot be made by traditional methods, custom fashions that eliminate the need for tailoring, and personalized clothing that is customized for individual body shapes.

Professional confectioners find 3D printing ideal for creating unique designer frosting decorations (Santhoshkumar & Moses, 2025) and novel candy designs (Qiu et al., 2024). These 3D printers employ food-grade inks made of ingredients such as sugar, flour, chocolate, peanut butter, and natural coloring.

In terms of 3D printed foods, NASA sponsored a project that printed and baked pizzas, creating a system designed for future astronauts (Chan, 2017). The novelty of 3D printing also spawned an international restaurant chain called Food Ink (Wang, 2016), which serves gourmet printed food (Domżalska & Jakubczyk, 2025) including cakes and dishes printed with hummus, cheese, chocolate, etc.).

Several 3D printing kitchen appliances are sold domestically for the creation of uniquely shaped pasta (Siniauer, 2017), pancakes (O'Neal, 2015), fancy deserts (infused innovations, 2022), and other items that would be difficult and time-consuming to handcraft (D'Monte, 2020).

2.3.5 Robotics

The Spanish telecommunications company BQ unveiled a 3D printed programmable toy mascot robot in 2015, called Zowie, that walks, dances, stands on its tiptoes (or on elevated feet through up and down maneuvering), as it avoids obstacles with sonic sensors. Zowie is microphone equipped and works with a mobile device using Bluetooth and a downloadable app (Montez, 2015). France's Inria's 3D printed robot

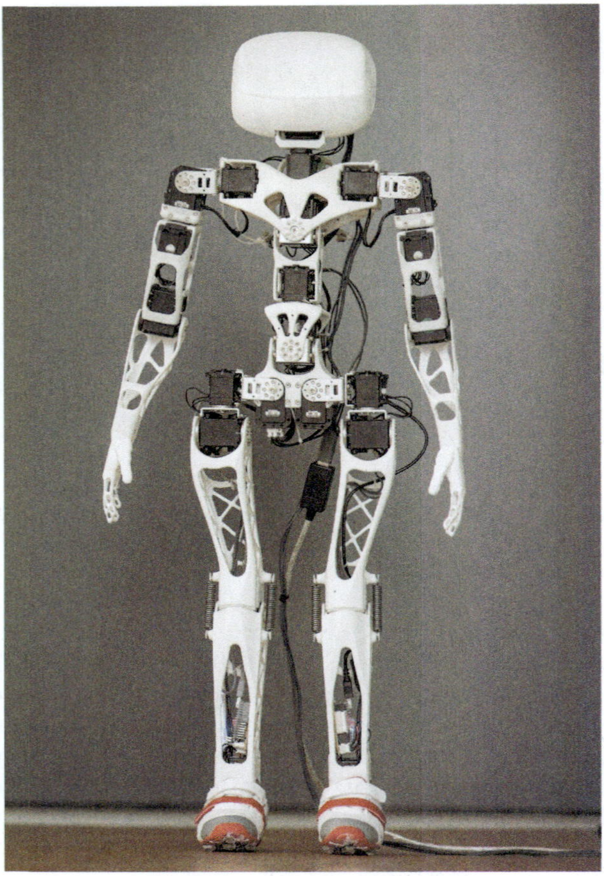

FIGURE 2.5 Poppy Humanoid 3D printed robot. (Reproduced from Inria/Poppy-project.org/Photo H. Raguet. Creative Commons Attribution-Share Alike 4.0 International license.)

(assembled from printed parts) called Poppy Humanoid (Figure 2.5) was made using open-source hardware and software, which was adopted by universities as a research platform for experimentation. It operates on Odroid XU4 with Ubuntu 14.04, has 25 actuators, a substantial FOV, and an HD camera (Champeau, 2013). Such interactive, Internet-connected robots can provide up-to-the-minute storm and other emergency warnings, selective or general news, financial information, and can schedule reminders, provide around-the-clock home security watch with alerts when danger is perceived, and on-demand Googling to answer queries.

A growing list of 3D-printed robots and their applications feature increasing sophistication due to continuing developments in the miniaturization of microelectronics and other required technologies, which them sufficiently economical to build. Exemplars of current (2025) humanoid robots include the PIB (printable intelligent bot), which is completely 3D-printed and recently won the 2025

German Design Award (Gartner, 2025). The PIB is comprised of totally digitally fabricated structural components, which demonstrates the capacities of additive manufacturing to generate joints with functional complexity, while producing a convincing aesthetically humanlike exterior. This customizable humanoid robot can be adapted for research, education, and artistic applications (isento robotics UG, 2025).

A cost effective ($70) open-source 3D-printed humanoid robot (ALANA) was developed by independent designer, Shashwat Batish in 2025. ALANA's realistic arms operate with six degrees of freedom, representing how 3D printing can democratize hardware innovations to facilitate further experimentation and progress in humanoid robotics (Malayil, 2025).

The customizable Berkeley Humanoid Lite was developed by workers at UC Berkeley as an educational and experimental platform mainly aimed at researchers, educators, and hobbyists toward the advancement of humanoid robotics. At a cost of under $5000 the Berkeley Humanoid Lite can be assembled using standard parts and desktop 3D printers (McFadden, 2025).

2.3.6 Construction

Using large 3D printers, ten tiny homes devoid of basements were constructed within 24 hours, at a cost of under $5000. The primary material typically used is quick drying concrete; however, some projects having added sawdust, recycled materials from earlier constructions, and/or polystyrene. The automation of construction work has the advantages of preventing many injuries and lost lives (being one of the most dangerous professions), keeping projects on schedule and budget, while streamlining the entire production process, including wiring, plumbing, and structural reinforcement. Additional benefits encompass lower housing prices (via preventing material waste and reducing construction costs), less construction waste, and expedited relief in areas devastated by natural disasters. Further, any 3D printed building can be customized to order by modifying the architectural design program sent to the printer (Plaid Zebra, 2015).

2.3.7 Tissue Engineering

Biomedical researchers now have the capacity to bioprint various biological tissues with "inks" comprised of cultured biological cells and biodegradable gels (Ze et al., 2022; Karamchand et al., 2023). This process involves scanning to obtain a computer-generated 3D model for printing a correctly formed scaffolding, to which the cells adhere. Similarly, natural organ growth requires an appropriate substrate (generated by the body) on which cells grow, which makes 3D printing well suited for mimicking this natural process. Among the benefits of 3D printing human tissues and organs is that the body does not reject them (as they are comprised of its own cells), and patients do not have to endure the excruciating wait for donors.

Researchers at the University of California (San Diego, La Jolla) overcame one of the largest obstacles associated with the use of 3D printed organs in 2017, by successfully 3D printing the vascular system (Choi et al., 2023) required for supplying nutrients and oxygen, while removing metabolic wastes. The challenge for the development of 3D-bioprinted blood vessels will be to match or exceed the flexibility and burst pressures of natural human blood vessels (Chen et al., 2021; Wang et al., 2022; Jiang et al., 2023). In the next decade or two we may witness the emergence of 3D-printing technologies that have the capacity to fabricate most human organs from the patient's own stem cells, including the heart (Freeman, 2019; Rosellini et al., 2023; Bhandari et al., 2023), lungs (Ng et al., 2021; Choi et al., 2023), kidney (Homan et al., 2019; Chaudhary & Chakraborty, 2022), and liver (Ma et al., 2020; Sun et al., 2023), etc. Progress in 3D printing exemplifies an evolving web of converging technologies, processes, materials, and increasing capabilities (3DSourced, 2024).

2.4 DIFFERENTIATING 3D PRINTING AND F@H

Although both 3D printing systems do, and prospective F@H systems will, facilitate the production of 3D objects, they will operate in fundamentally different ways. Among the similarities are that the user provides a command (or series of orders) to the device for the desired product/s, and production requirements are calculated once an object is either scanned (for replication) or is fabricated based on internal memory or through interactions with a database.

Among the differences are that while 3D printers employ various inks, F@H would formulate its own inks when it opts to incorporate 3D printing as a facet of a particular manufacturing strategy. Further, F@H will utilize sealed cartridges loaded with pristine molecular and atomic feedstocks, and/or various preassembled smart or raw material components. Subsequent to the emergence and potentially ubiquitous implementation of F@H, an accessory appliance referred to as Recykl@ Home (R@H) would have the capacity to completely disassemble all manner of waste and materials to their molecular and atomic constituents. These elements would then be sorted, purified, and repackaged as pristine feedstocks for reuse in F@H, which would close the loop in terms of the democratization of domestic molecular manufacturing.

Another distinction is that advanced F@H models, endowed with molecular manufacturing, might have the option of being partially supplemented by bioengineered components and synthetic biology, enabling myriad hybrid products or systems. As such, there will be a stark distinction between how conventional 3D printed objects are fabricated and, for example, fully operational (~Ø1 μm) autonomous nanomedical devices are molecularly manufactured (atomic layer by atomic layer) via F@H. Thus, the end products generated by F@H will typically be imbued with vastly deeper sophistication, technological dynamics, dimensional resolution, and diversification than any of the 3D printing technologies that exist today.

2.5 MICRO/NANO-ELECTRO-MECHANICAL SYSTEMS (MEMS)

Micro-Electro-Mechanical Systems (MEMS) technologies enable the production of micron scale (~Ø1 μm–Ø100 μm) devices and systems, which are typically integrated with mechanical and electronic sub-systems, which facilitate myriad applications

(Toshiyoshi et al., 2019; Bayer, 2022; Chircov & Grumezescu, 2022; Aronne et al., 2024; Fan et al., 2024). Typical materials used for the fabrication of MEMS include metals, organic polymers, silicon, and ceramics (Faraji Rad et al., 2021). Examples of MEMS devices include the integration of AI with MEMS sensors to enable a variety of smart sensors (Chen et al., 2022; Stella et al., 2023; Wu et al., 2024); MEMS accelerometers used to measure acceleration, activate automotive airbags, and work with various electronics (Mohammed et al., 2018; Rasras et al., 2019; Ahmed et al., 2023); MEMS inertial measurement units that sense both rotation and acceleration to facilitate robotic navigation (Sun et al., 2024; Xuan et al., 2024); MEMS gyroscopes, installed in drones and other vehicles (Gill et al., 2022; Xu et al., 2023); and MEMS microphones for mobile phones and laptops (Zhang et al., 2018; Zawawi et al., 2020; Gemelli et al., 2023).

Diminutive arrays of nanomotors (<Ø1 μm) were developed using Nano-Electro-Mechanical Systems (NEMS) and operated continuously for 15 hours at 18,000 RPM over 240,000 cycles (Kim et al., 2014). These types of nanomotors are among many (Rahman et al., 2017; Guo et al., 2018), which demonstrated that devices can currently be miniaturized even smaller than might be required for a range of F@H operations.

2.6 MOLECULAR MANUFACTURING

The emergence of a countertop molecular assembler has long been inching its way into existence. The fictional Star Trek Replicator instantly produced (and recycled) items on demand (Drum & Gordon, 2003; Hollow, 2013). The popularity of the TV show solidified the assembler concept firmly into the modern imagination. In 1959, physicist Richard Feynman presented a lecture at Caltech entitled "There's Plenty of Room at the Bottom," which entertains, among other things, the possibility of making nanoscale machines that arrange atoms by mechanical manipulation to build complex objects, thereby eliminating the inexact processes of traditional chemistry experimentation (Feynman, 1960).

Engineer K. Eric Drexler initially proposed a molecular assembler in his 1991 MIT doctoral thesis (Drexler, 1981, 1986, 1991), which operated on the principle of directing chemical reactions by mechanically situating reactive molecules with atomic-scale precision. In principle, the concept is feasible in that the process exists in nature in the example of ribosomes, which assemble the proteins that make up our bodies (Wilson et al., 2012; Ni & Buszczak, 2023; Vanden et al., 2024). Ribosomes accept DNA instructions and messenger RNA (mRNA molecules) signals and organize them along with other molecules, as well as collect and assemble amino acid sequences (e.g., from the food we eat) to synthesize specifically arranged protein molecules.

A team at MIT achieved a milestone in molecular manufacturing via biological processing by bioengineering a virus (harmless to humans and other animals) to facilitate the fabrication of a lithium-ion battery. As a facet of their operation, specific organisms (M-13 bacteriophages) were employed to grasp conductive metals (cobalt oxide and gold) from solution. As the viruses clung to the metals and changed positions, well-arranged wires were formed. The goal was to arrange billions of such organic machines to work in parallel, and function as molecular foundries to create specific components, products, or layers thereof (Lee et al., 2009).

The first synthetic life form was created in 2010 by geneticist Craig Venter and coworkers, which added to the potential for programmed organic nanometric assembly (Gibson et al., 2010).

The Venter team's strategy enabled materials to be computationally grown, which created further possibilities for a combination of various biophysics approaches to manufacturing. There are several ways to manipulate (apply gentle force to and guide) the behaviors of organisms with dimensions on the order of average body cells and smaller.

For example, among thousands of other applications, lasers can be employed for materials processing and operate optical tools. For example, a laser wrench was developed that could trap (hold stable in three dimensions), spin, and twist microscale objects in any direction along any axis. This optical wrench was made possible through the use of flexible optical fibers that could manipulate laser light (Black & Mohanty, 2012). The operating principle here was that the force of the laser light was sufficiently robust to rotate microscopic particles. Since the optical wrench was precise enough to twist and untwist DNA strands, it had the capacity to manipulate microscopic motors and dynamos. The predecessors of optical wrenches were optical tweezers, which can immobilize and relocate microscopic objects (Favre-Bulle & Scott, 2022; Berns, 2020; Sparkes, 2018). Further, microfluidic operations may also be utilized to apply forces to diminutive entities such as micro/nanoparticles and cells (Manshadi et al., 2020; Abedini-Nassab et al., 2021; Tong et al., 2024).

2.7 SOCIETAL IMPLICATIONS

Since decentralized mechanisms for the generation of material abundance are typically derived only through monetary resources, the F@H concept presents a significant democratizing shift for humanity. Envisaged F@H and other systems that might produce similar results can be expected to dramatically ease global shortages and supply chain/distribution constraints for goods and services. Moreover, these systems may initiate the systematic increase of programmed abundance.

2.7.1 Eradication of Poverty

Currently (2025) at least ~8.5% of the global population of ~8.23 billion, or ~700 million people worldwide, dwell in extreme poverty, who attempt to survive on the equivalent of under ~$2.15 USD per day (Mahapatra, 2025). These individuals would benefit greatly from even very rudimentary iterations of F@H that are equipped with a rechargeable power pack attachment, which uses recyclables to help provide for their basic needs. Interactive (globally linguistically fluent) F@H would be able to make systematic verbal recommendations for those who need even the most fundamental items for survival (e.g., water purification materials/devices, nutrient/protein-rich foods, sanitation provisions, alternative power, lighting, heating, and cooling devices, first-aid preparations, rudimentary tools, and the like).

Interactive units might instruct users on widely available feedstocks that may be suitable for the manufacturing of various products. Incorporated lab-on-a-chip technologies could recommend against or reject unsafe feedstocks. Ideally, with a nascent portable F@H, even the homeless could survive at the level of a well-accommodated camper until strong, comfortable facilities (made with modular F@H-built components), furnishings, and other goods might eventually be provided. That is, F@H could allow those with virtually nothing to build and collect necessities in the prioritized order of strategic importance in the most systematic, pragmatic, and feasible way.

2.7.2 Decentralized Electricity and the Environment

In 2025, approximately 750 million people lives without electricity globally. F@H could be utilized to produce simple environmentally compatible power devices of several kinds, such that virtually everyone will have access to electrical power, including fail-safes for power failures. Alternative clean energy devices and carbon conversion devices (e.g., that can convert CO_2 to either sugar or electricity) could be used to regulate the carbon footprint of each dwelling and industry. This, so CO_2 and other carbon compounds (typically generated due to the combustion of fossil fuels) are geared toward ameliorating environmental imbalances.

Several strategies may be employed through which F@H might facilitate and contribute to the creation of a sustainable future to support the survival of humanity. Greenhouse gas mitigation alone may provide sufficient incentives to garner support from willing governments, corporations, philanthropists, and other sponsors to expedite F@H development and implementation. With informed intelligent planning, collective efforts should be possible without users having to sacrifice the major benefits of F@H.

The provision of electric power would also support access to global communications, which might facilitate Internet-enhanced F@H. Based on real-time data and statistics, F@H could provide autonomous recommendations for cooperative teamwork. The utilization of household trash for recycling into useful new products could dramatically reduce growing burdens on landfills and contamination from trash incinerators. Further, F@H could serve as an expansively ecologically clean technology when enabled to sanitize, sterilize, and detoxify and put to good purpose much of the rubbish that would otherwise overflow landfills or litter the environment. A reduction in throwaway articles would be natural transition as people constantly seek free-of-charge recyclables for F@H reprocessing and manufacturing.

Some level of revenues or bonuses (e.g., tax or debt reductions) could be offered by regional or national governments to households that contribute to such environmental efforts. F@H cooperatives could produce robotic teams that monitor and clean up bodies of water, and process the incoming plastics or other wastes according to government regulations. Monetary points or assets might be awarded to those businesses that create biodegradable and/or otherwise environmentally friendly products with F@H devices. In these ways, the health of forests and other ecosystems could be improved with conscientious, cooperative F@H use.

2.7.3 Dimensional Constraints

Early F@H desktop units could have the dimensions of today's typical home 3D printers, which can measure ~24″ × 24″ × 24″. Nascent F@H might accommodate internal components that reconfigure to shape-shift (e.g., able to be rolled up or folded within the appliance), which enables them to unfold into larger components outside of the unit, for the manufacture of articles that may be much larger than the machine itself. For the attainment of self-sufficiency, which is of particular importance for those who prefer to live off grid in mobile homes or tiny homes, a F@H of this size may suffice, especially when F@H-built robotics can be manufactured as assistants. Homes that accommodate larger appliances (such as standard-sized water heaters and furnaces) can benefit from larger F@H devices, especially if at least one advanced F@H-fabricated humanoid robot organizes household spaces for maximum efficacy. Consequently, F@H units may be a good investment for homeowners and small businesses, which use their machines to earn passive income. Even when individuals have the capacity to build everything they need, income will still be required for feedstocks, and for as long as taxation systems are enforced.

2.7.4 Future of Work

Presently (2025), there are deepening concerns as relates to the future of work; the replacement of human labor by STARA (smart technology, artificial intelligence, robotics, and algorithms). Industrial psychologists forecast that "85 million jobs may be displaced by a shift in the division of labor between humans and machines, while 97 million new roles may emerge that are more adapted to the new division of labor between humans, machines and algorithms" (Oosthuizen, 2022). The Future of Jobs Report 2025 by the World Economic Forum estimates that "AI and automation will transform 86% of businesses by 2030, with research drawn from a survey of 1,000 companies across 22 industries and 55 economies, representing more than 14 million workers" (Wheeler, 2025).

From another perspective, "the impact of automation on jobs is not as straightforward as simply 'robots replacing workers.' While automation may eliminate some jobs, it will also create new ones. As machines take over mundane tasks, humans will be freed up to focus on more complex and creative work that requires critical thinking, emotional intelligence, and problem-solving."

Using F@H systems, individuals would have the capacity to fabricate their own special-purpose or dynamic humanoid robots that perform myriad tasks, which could engage in building or enhancing careers for the continuous generation of revenue. Robots will be able to undertake a great variety of skill sets to undertake manual labor jobs, white collar positions that require high intellectual capacity and learning, while filling several other functional domains. The issue may not be that robots will eliminate jobs, but that job-filling robots are not currently available to be owned by those who are losing jobs. As such, there is an urgent need for F@H (or similar systems) that can prevent loss of income for the masses from the rapid transitions to come due to STARA (Babamiri et al., 2022).

An additional aspect of F@H will involve capacities for the massively distributed and decentralized fabrication of all manner of consumer goods, including nutritious gourmet foods. This paradigm shift will likely translate to an age of abundance, which will certainly alter myriad occupational landscapes. For example, the necessity to transport consumer goods on a global scale will decrease dramatically when they can be produced domestically on site. Further, utilities infrastructures and energy grids will have far less demands placed on them, with the emergence of highly efficient domestic water recycling/purification systems that require only occasional water intakes from municipal water systems to keep them topped up, in conjunction with power generating devices/systems can be fabricated in one's own home, leading to most homes becoming completely energy self-sufficient. The downstream domestic generation of autonomous nanomedical devices by F@H with the capacity to address all that ails us (including aging) will relegate most large hospitals and other centralized medical infrastructures to obsolescence.

2.8 POTENTIAL FOR ABUSE AND REGULATORY PRUDENCE

Any technologies as potentially powerful as F@H may be abused. The potential fabrication of devices that can harm others or the environment underscores the need for stringent laws and regulations, as well as the incorporation of strong incentives as preventative measures against nefarious activities. While it is conceivable that advanced F@H might be made to exponentially self-replicate (to a certain extent) for good cause, such a system or its offspring could do significant global damage if created and replicated for careless/malicious purposes. Items such as military assault weapons and drones have already been 3D printed (Panella, 2025; Velo3D, Inc., 2025), which has raised serious concerns regarding the urgent need for legislation. By the time F@H is available (in ~10–15 years), it is logical to expect that stringent regulations will have been instituted due to the concerns already raised by 3D printing.

Appropriate and fair stopgaps should also be integrated into F@H devices and their accessories. Anything as pervasive as F@H might become required to be manufactured such that it cannot be employed to abuse environmental or other regulations. Further, the use of F@H to fabricate foods (such as artificial meats) at the cellular level is likely to raise ethical concerns centering around purpose, usage, and what exactly will be created as science and technologies continue to progress (Handral et al., 2022; Tejada-Ortigoza & Cuan-Urquizo, 2022; Wen et al., 2024). Today, knowledgeable individuals can manipulate genes and cells in Petri dishes with readily available devices (Benedetto et al., 2014; Yadav et al., 2024); thus, the same may be expected for F@H as it becomes increasingly more technologically enabled.

For example, today, as mentioned, several food items are already being manufactured by 3D printers, and when the recipe calls for an egg, a natural egg is used as part of the ink. The same could be done with F@H by strictly using its printing abilities to produce food items. However, when F@H/molecular manufacturing is technologically sophisticated enough to fabricate a compositionally indistinguishable synthetic egg at atomic resolution, a potential Pandora's box of nefarious uses can open. Consequently, bioethical oversight bodies must pre-emptively resolve such issues to prevent the runaway creation of harmful synthetic pathogens and other life forms. It will be critical to thoroughly elucidate dangers that can result from molecular manufacturing toward the development of appropriate regulations. Progress in this regard was made by the Center for Responsible Nanotechnology (Phoenix, 2008) and Foresight Institute (Jacobstein, 2006), both of which suggest ethical guidelines for the responsible development of nanotechnology.

2.9 PRICE:PERFORMANCE RATIO

As relates to engineered products such as computers, the price:performance ratio refers to the efficacy of a given product to deliver one or more types of work correlated with its cost. Factors that will boost the price performance ratio of incrementally more sophisticated F@H iterations will include progressive improvements based on developments in existing and emerging technologies and sciences. For instance, when laboratory-on-a-chip technologies (Liu & Zhu, 2005; Pumera, 2006; Xing et al., 2021; Ardila, 2025) are well-developed, which are important for in-home use, as they can detect chemical contaminants, mold counts and their toxins, microbe types and counts in water and food. Lab-on-a-chip technologies can be utilized to prevent food poisoning, as well as air and waterborne diseases, and will be essential for assuring that recycled feedstocks are safe for the fabrication of new products by F@H. They will also be vital as more mature F@H iterations incorporate multifaceted productivity improvements, so that it is enabled to operate with bacteria-powered nanomotors and virus-like structures that manipulate molecules. It must be assured that the organisms used in F@H are inert and properly contained and never allowed to evolve in a way that could be harmful.

Price performance value may also be added through the incorporation of low wear/friction-free materials in F@H units that virtually never wear out. In addition to greatly increasing the precision manufacturing capability of the F@H internal components, the absence of friction reduces vibration and noise levels, and thermal distortion in materials such as silicon and plastics when they are being rapidly processed. Generative engineering algorithms and optical computing will be able to sort and sift through the deluge of data and test for engineering parameters in prototypes to be manufactured and determine optimal strategies for resolving any manufacturing issues that may arise.

Price performance can also be increased through the development of F@H models that can fabricate their own replacement parts, as well as those for robots and other devices, self-diagnose, and either self-repair or build robots that can perform repairs. Further enhancement can be realized when new products are made from recycled materials or when truly valuable products are fabricated. As mentioned above, light industrial robotic systems may be manufactured and contracted out to earn passive income for F@H users. High performance capabilities can allow F@H to pay for itself many times over for homeowners, industries, and space ventures.

2.10 FIRST-GENERATION F@H

Early iterations of F@H (Figure 2.6) might likely incorporate advanced 3D printings technologies that can draw on a wide variety of plugin cartridge or filament type feedstock materials (solid and liquid), electromechanical assembly systems with interchangeable accessories, as well as integrated MEMS, NEMS, and microfluidics capacities. It may be advantageous in terms of operation and safety to have distinct fabrication

FIGURE 2.6 Conceptual first-generation Factory@Home system (author-generated image via Shutterstock AI).

This conceptual illustration does not literally represent the actual F@H design.

compartments (e.g., side by side, or front and back) with dedicated doors. One compartment would be dedicated to the fabrication of foods and health supplements, including synthesized vitamins and health rejuvenating peptides (Zhang et al., 2022).

A display and integrated AI agent will assist users with learning to operate/troubleshoot the system in any language. The F@H appliance may typically function in conjunction with the Industrial Internet of Things (IIOT) (Salih et al., 2022; Stadnicka et al., 2022; Meira et al., 2023) to enhance operational efficiencies via engaging with open-source software, networking, and a host of applications, allowing users to consider various feedstock options.

2.11 SECOND-GENERATION F@H

Subsequent second-generation F@H iterations would integrate more advanced versions of the technologies used in preceding iterations, which would be supplemented by nanotechnology-based fabrication capacities, including AI-guided electrospraying and electrospinning (Feng et al., 2023; Nosoudi et al., 2023; Pires et al., 2023) to produce the charged complex molecules, polymers, aerogels, nanoparticles, nanofibers, meta-materials, and other micro/nanomaterials and objects that comprise desired end products. These F@H would also have capacities for self-diagnostic, self-repair, and the fabrication of replacement parts. Although F@H might generally fabricate modular robotic components, which would be subsequently manually assembled by the user, a dedicated accessory might work specifically with the latest advancements for building robotics domestically (Shepheard, 2024). This may include humanoid robots that can be either upgraded or recycled as they become obsolete. Further, it would function with generative engineering algorithms, enabling users to input requirements rather than specific standardized engineering designs. The F@H could also provide options for the fabrication and testing of robotic prototypes.

Further to first-generation functionality, enhanced optical capacities will translate to more rapid fabrication and overall performance. Technological developments will also be aimed at employing light energy instead of electrical energy for various processes. A more extensive range of feedstocks will complement advances in MEMS, NEMS, microarray and nanoscale assembly systems, material compositions, and database upgrades resulting in a significantly broader array of products and systems that can be created. Internal F@H components will be more robust, faster, and more diversified, with an increased ability to create materials at the molecular level by exploiting the latest scientific and technological advances.

2.12 THIRD-GENERATION F@H

Third-generation F@H systems would (in addition to the capabilities mentioned above) begin to incorporate early iterations of molecular manufacturing, once it becomes a reality. This will enable the molecular manufacturing of practically any consumer product imaginable (within the safety, security, and dimensional constraints of the appliances), nutritious gourmet foods that are indistinguishable in every respect from the real thing, biological processing (within the safety, security, and dimensional constraints of the appliances), as well as the fabrication of medically certified passive, semi-autonomous, and fully autonomous nanomedical devices to address most injuries and illnesses, including aging. These F@H would also be equipped to engage in whole system replication, thereby opening the possibility for multiple F@H that can operate in parallel.

2.13 CONCLUSION

It is reasonable to posit that F@H or other emerging systems with similar capabilities will dramatically shift paradigms across virtually every aspect of society while fostering well-being and general abundance for humanity. It is reasonable to expect that F@H models will increasingly progress in ways we have yet to imagine, driven by rapid advances in nanotechnology, nanomedicine, AI/AGI/ASI, and quantum computation. Given the amazing achievements that can be realized by teams of like-minded, passionate, ambitious, productive, enthusiastic, and loving individuals when working collectively to improve humanity and the planet, the time and timeliness could not be better for the emergence of F@H.

REFERENCES

3DSourced. 6 Exciting 3D Printed Organs & 3D Bioprinting Projects. *3DSOURCED*. February 20, 2024. https://www.3dsourced.com/guides/3d-printed-organs-bioprinting/ (accessed May 12, 2025).

Abedini-Nassab R, Pouryosef Miandoab M, Şaşmaz M. Microfluidic synthesis, control, and sensing of magnetic nanoparticles: a review. *Micromachines (Basel)*. 2021 Jun 29;12(7):768.

Ahmad I, Kan CW. A review on development and applications of bio-inspired superhydrophobic textiles. *Materials (Basel)*. 2016 Nov 3;9(11):892.

Ahmed HE, Sahandabadi S, Bhawya, Ahamed MJ. Application of MEMS accelerometers in dynamic vibration monitoring of a vehicle. *Micromachines (Basel)*. 2023 Apr 24;14(5):923.

Albanesi S, Dias da Silva A, Jimeno JF, Lamo A, Wabitsch A. Reports of AI ending human labour may be greatly exaggerated. *European Central Bank Research Bulletin No. 113*. November 28, 2023. https://www.ecb.europa.eu/press/researchpublications/resbull/2023/html/ecb.rb231128~0a16e73d87.en.html (accessed May 08, 2025).

Antezana PE, Municoy S, Ostapchuk G, Catalano PN, Hardy JG, Evelson PA, et al. 4D printing: the development of responsive materials using 3D-printing technology. *Pharmaceutics*. 2023 Dec 7;15(12):2743.

Ardila CM. Advancing healthcare through laboratory on a chip technology: transforming microorganism identification and diagnostics. *World J Clin Cases*. 2025 Jan 26;13(3):97737.

Aronne M, Bertana V, Schimmenti F, Roppolo I, Chiappone A, Cocuzza M, Marasso SL, Scaltrito L, Ferrero S. 3D-printed MEMS in Italy. *Micromachines (Basel)*. 2024 May. 22;15(6):678.

Atherton KD. The Marine Corps wants to 3D print cheaper drones. *Popular Science*. May 18, 2021. https://www.popsci.com/marine-corps-3d-printed-drones/ (accessed May 11, 2025).

AZAI Research Team. Future-Proofing the Economy: How UBI and AGI Can Redefine Economic Models by 2040. *Deepest Research*. 2025. https://www.deepestresearch.com/2025/02/ubi-and-agi-redefine-economic-models.html (accessed May 08, 2025).

Babamiri M, Heidarimoghadam R, Ghasemi F, Tapak L, Mortezapour A. Insights into the relationship between usability and willingness to use a robot in the future workplaces: studying the mediating role of trust and the moderating roles of age and STARA. *PLoS One*. 2022 Jun 3;17(6):e0268942.

Balavigneswaran CK, Selvaraj S, Vasudha TK, Iniyan S, Muthuvijayan V. Tissue engineered skin substitutes: a comprehensive review of basic design, fabrication using 3D printing, recent advances and challenges. *Biomater Adv*. 2023 Oct;153:213570.

Baranowski P, Kapusta A, Płatek P, Sarzyński M. Influence of 3D-printed cellular shoe soles on plantar pressure during running – experimental and numerical studies. *Biocybern Biomed Eng*. 2024;44(4):858–873.

Bayer IS. MEMS-based tactile sensors: materials, processes and applications in robotics. *Micromachines (Basel)*. 2022 Nov 23;13(12):2051.

Belgin Paul DL, Susila PA, Karthick M. Exploring manufacturing techniques in bioceramic scaffold fabrication with a focus on DIW 3D printing for tissue engineering applications. *Ann Biomed Eng*. 2025;53:1525–1542.

Benedetto A, Accetta G, Fujita Y, Charras G. Spatiotemporal control of gene expression using microfluidics. *Lab Chip*. 2014 Apr 7;14(7):1336–1347.

Berns MW. Laser scissors and tweezers to study chromosomes: a review. *Front Bioeng Biotechnol*. 2020 Jul 16;8:721.

Bhandari S, Yadav V, Ishaq A, Sanipini S, Ekhator C, Khleif R, et al. Trends and challenges in the development of 3D-printed heart valves and other cardiac implants: a review of current advances. *Cureus*. 2023 Aug 9;15(8):e43204.

Bierlich KC, Wengrove D, Bird CN, Davidson R, Chandler T, Torres LG, Cantor M. LidarBoX: a 3D-printed, open-source altimeter system to improve photogrammetric accuracy for off-the-shelf drones. *Drone Syst Appl*. 2024;12:1–10.

Black BJ, Mohanty SK. Fiber-optic spanner. *Opt Lett*. 2012 Dec 15;37(24):5030–5032.

Boehm FJ. *Nanomedical Device and Systems Design – Challenges, Possibilities, Visions*. CRC Press, Taylor & Francis, Boca Raton, FL, USA. 2013.

Boehm *FJ. Nanotechnology, Nanomedicine, and AI: Toward the Dream of Global Health Care Equivalency*. CRC Press, Taylor and Francis, Boca Raton, FL, USA. 2025.

Cernencu AI, Vlasceanu GM, Serafim A, Pircalabioru G, Ionita M. 3D double-reinforced graphene oxide – nanocellulose biomaterial inks for tissue engineered constructs. *RSC Adv*. 2023 Aug 10;13(34):24053–24063.

Champeau G. Poppy, an open-source humanoid robot made in France. *numerama*. November 18, 2013. https://www.numerama.com/sciences/27530-poppy-un-robot-humanoiumlde-open-source-made-in-france.html (accessed May 12, 2025).

Chan R. This Robot Could 3D Print Pizzas in Space for Hungry Astronauts. *Inverse*. March 7, 2017. https://www.inverse.com/article/28758-pizza-3d-printer-space-travel-beehex-nasa (accessed May 12, 2025).

Charan AK, Abhishek R, Navya R, Sathwika C. "Design and fabrication of an automated delivery drone using 3D printing," in *Recent Advances in Mechanical Engineering. Volume 1 (Lecture Notes in Mechanical Engineering)*, Springer. 2024.

Chaudhary S, Chakraborty E. Hydrogel based tissue engineering and its future applications in personalized disease modeling and regenerative therapy. *Beni Suef Univ J Basic Appl Sci*. 2022;11(1):3.

Chaurasia P, Singh R, Mahto SK. FRESH-based 3D bioprinting of complex biological geometries using chitosan bioink. *Biofabrication*. 2024 Jul 16;16(4):045007.

Chen EP, Toksoy Z, Davis BA, Geibel JP. 3D bioprinting of vascularized tissues for *in vitro* and *in vivo* applications. *Front Bioeng Biotechnol*. 2021 May 13;9:664188.

Chen L, Wei L, Wang Y, Wang J, Li W. Monitoring and predictive maintenance of centrifugal pumps based on smart sensors. *Sensors (Basel)*. 2022 Mar 9;22(6):2106.

Chircov C, Grumezescu AM. Microelectromechanical systems (MEMS) for biomedical applications. *Micromachines (Basel)*. 2022 Jan 22;13(2):164.

Choi YM, Lee H, Ann M, Song M, Rheey J, Jang J. 3D bioprinted vascularized lung cancer organoid models with underlying disease capable of more precise drug evaluation. *Biofabrication*. 2023 Jun 6;15(3):034104.

Choi J, Lee EJ, Jang WB, Kwon SM. Development of biocompatible 3D-printed artificial blood vessels through multidimensional approaches. *J Funct Biomater*. 2023 Oct 8;14(10):497.

Coren MJ. GE fired up world's largest commercial jet engine using 3D-printed metal parts. *3D Print.com*. April 25, 2016. https://3dprint.com/131192/ge-aviation-big-jet-engine/ (accessed May 10, 2025).

Crisanto CC, Leuterio, CB, Prenio J, Yong J Regulating AI in the financial sector: recent developments and main challenges. FSI Insights on policy implementation No 63. *Financial Stability Institute, Bank for International Settlements*. December 2024. https://www.bis.org/fsi/publ/insights63.pdf (accessed May 08, 2025).

D'Monte L. These 3D Prints Not Only Look Good – They're Delicious! *engineering.com*. June 12, 2020. https://mobile.engineering.com/amp/20391.html (accessed May 12, 2025).

Darsin M, Pamungkas W, Syuhri SNH, Wibowo RK, Basuki HA, Djumharianto D. Advanced optimization of drone frame design through the application of generative design techniques and 3D printing technology. *J Mech Sci Technol*. 2025;39:119–128.

Davoodi E, Li J, Ma X, Najafabadi AH, Yoo J, Lu G, et al. 3D printing and electrospinning of drug- and graphene-enhanced polycaprolactone scaffolds for osteochondral nasal repair. *Materials (Basel)*. 2025 Apr 16;18(8):1826.

Dey M, Ozbolat IT. 3D bioprinting of cells, tissues and organs. *Sci Rep*. 2020 Aug 18;10(1):14023.

Dhakal S, Karunakaran P. Design and implementation of drone technology for medical supplement delivery services in rural regions. *Int. Res. J. Eng. Technol*. 2023 Dec;10(12):545–550.

Domżalska Z, Jakubczyk E. Characteristics of food printing inks and their impact on selected product properties. *Foods*. 2025 Jan 25;14(3):393.

Drexler KE. Molecular engineering: an approach to the development of general capabilities for molecular manipulation. *Proc Natl Acad Sci USA*. 1981;78:5275–5278.

Drexler KE. *Engines of Creation: The Coming Era of Nanotechnology*. Anchor/Doubleday, New York. 1986.

Drexler KE. *Molecular machinery and manufacturing with applications to computation*. Thesis (Ph.D.). Massachusetts Institute of Technology, Dept. of Architecture, 1991.

Drum RW, Gordon R. Star Trek replicators and diatom nanotechnology. *Trends Biotechnol*. 2003 Aug;21(8):325–328.

Du S, Huynh T, Lu YZ, Parker BJ, Tham SK, Thissen H, et al. Bioactive polymer composite scaffolds fabricated from 3D printed negative molds enable bone formation and vascularization. *Acta Biomatters*. 2024 Sep 15;186:260–274.

Elhadad A, Li G, Yang J, Liu D, Choi S. 3D-Printed Stainless Steel Electrodes for Advancing Mems Microbial Fuel Cells toward Sustainable On-Chip Energy. *2025 IEEE 38th International Conference on Micro Electromechanical Systems (MEMS)*, Kaohsiung, Taiwan. 2025;643–646.

Fan X, He C, Ding J, Gao Q, Ma H, Lemme MC, Zhang W. Graphene MEMS and NEMS. *Microsyst Nanoeng*. 2024 Oct 28;10(1):154.

Faraji, Rad Z, Prewett PD, Davies GJ. An overview of microneedle applications, materials, and fabrication methods. *Beilstein J Nanotechnol*. 2021 Sep 13;12:1034–1046.

Farzin A, Miri AK, Sharifi F, Faramarzi N, Jaberi A, Mostafavi A, et al. 3D-printed sugar-based stents facilitating vascular anastomosis. *Adv Healthc Mater*. 2018 Dec;7(24):e1800702.

Favre-Bulle IA, Scott EK. Optical tweezers across scales in cell biology. *Trends Cell Biol*. 2022 Nov;32(11):932–946.

Feng K, Huangfu L, Liu C, Bonfili L, Xiang Q, Wu H, Bai Y. Electrospinning and electrospraying: emerging techniques for probiotic stabilization and application. *Polymers (Basel)*. 2023 May 22;15(10):2402.

Feng Y, Zhu S, Mei D, Li J, Zhang J, Yang S, Guan S. Application of 3D printing technology in bone tissue engineering: a review. *Curr Drug Deliv*. 2021;18(7):847–861.

Feynman RP. There's plenty of room at the bottom. *Caltech Eng Sci*. 1960;23:22–36.

Freeman D. Israeli scientists create world's first 3D-printed heart using human cells. NBC News. April 19, 2019. https://www.nbcnews.com/mach/science/israeli-scientists-create-world-s-first-3d-printed-heart-using-ncna996031 (accessed May 12, 2025).

Freitas Jr RA, Merkle R. What Is a Nanofactory? Nanofactory Collaboration. 2024. https://www.molecularassembler.com/Nanofactory/ (accessed May 09, 2025).

Fu K, Wang Y, Yan C, Yao Y, Chen Y, Dai J, et al. Graphene oxide-based electrode inks for 3D-printed lithium-ion batteries. *Adv Mater*. 2016 Apr 6;28(13):2587–2594.

Gartner L. 3D-printed robot pib wins German Design Award 2025. *3Printr.com*. Feb 11, 2025. https://www.3printr.com/3d-printed-robot-pib-wins-german-design-award-2025-0978035/ (accessed May 12, 2025).

Gemelli A, Tambussi M, Fusetto S, Aprile A, Moisello E, Bonizzoni E, Malcovati P. Recent trends in structures and interfaces of MEMS transducers for audio applications: a review. *Micromachines (Basel)*. 2023 Apr 14;14(4):847.

Ghobadi F, Kalantarzadeh R, Ashrafnia Menarbazari A, Salehi G, Fatahi Y, Simorgh S, et al. Innovating chitosan-based bioinks for dermal wound healing: current progress and future prospects. *Int J Biol Macromol*. 2025 Apr;298:140013.

Gibson DG, Glass JI, Lartigue C, Noskov VN, Chuang RY, Algire MA, et al. Creation of a bacterial cell controlled by a chemically synthesized genome. *Science*. 2010 Jul 2;329(5987):52–56.

Gill WA, Howard I, Mazhar I, McKee K. A review of MEMS vibrating gyroscopes and their reliability issues in harsh environments. *Sensors (Basel)*. 2022 Sep 29;22(19):7405.

Gu X, Cui L, Meng Z. Differences of wax-based emulsion gel in 3D printing performance: crystal distribution and droplet stability. *Food Chem*. 2023 Dec 1;428:136760.

Guo J, Gallegos JJ, Tom AR, Fan D. Electric-field-guided precision manipulation of catalytic nanomotors for cargo delivery and powering nanoelectromechanical devices. *ACS Nano*. 2018 Feb 27;12(2):1179–1187.

Handral HK, Hua, Tay S, Wan Chan W, Choudhury D. 3D printing of cultured meat products. *Crit Rev Food Sci Nutr*. 2022;62(1):272–281.

Harbaugh JA. Space Station 3-D Printer Builds Ratchet Wrench to Complete First Phase of Operations. December 22, 2014. https://www.nasa.gov/missions/station/space-station-3-d-printer-builds-ratchet-wrench-to-complete-first-phase-of-operations/ (accessed May 09, 2025).

Hildreth OJ, Nassar AR, Chasse KR, Simpson TW. Dissolvable metal supports for 3D direct metal printing. *3D Print Addit Manuf*. 2016 Jun;3(2):90–99.

Hollow M (2013). Confronting a new era of duplication? 3D printing, replicating technology and the search for authenticity in George O. Smith's Venus Equilateral series. *SSRN Electronic Journal*. https://doi.org/10.2139/SSRN.2333496 (accessed May 16, 2025).

Holzmann P, Wankmüller C, Globocnik D, Schwarz EJ. Drones to the rescue? Exploring rescue workers' behavioral intention to adopt drones in mountain rescue missions. *Int J Phys Distrib Logist Manag*. 2021;51(4):381–402.

Homan KA, Gupta N, Kroll KT, Kolesky DB, Skylar-Scott M, Miyoshi T, et al. Flow-enhanced vascularization and maturation of kidney organoids in vitro. *Nat Methods*. 2019;16(3):255–262.

IEA. Electricity access continues to improve in 2024 – after first global setback in decades. Summit on the Future of Energy Security. International Energy Agency. November 15, 2024. https://www.iea.org/commentaries/electricity-access-continues-to-improve-in-2024-after-first-global-setback-in-decades (accessed May 11, 2025).

Infused Innovations. Innovation Spotlight: Ukrainian Pastry Chef Dinara Kasko's 3D-Printed Cake Designs. *Infused Innovations*. May 9, 2022. https://infusedinnovations.com/blog/secure-intelligent-workplace/innovation-spotlight-pastry-chef-dinara-kaskos-3d-printed-cake-designs (accessed May 12, 2025).

isento robotics UG. pib Turns 3: From 3D-Printed Prototype to a Moving, Thinking Humanoid Robot. *isento GmbH*. April 4, 2025. https://pib.rocks/pib-turns-3-from-3d-printed-prototype-to-a-moving-thinking-humanoid-robot/ (accessed May 12, 2025).

Jacobstein N. Foresight Guidelines for Responsible Nanotechnology Development. *Foresight Institute*. April 06, 2006. https://legacy.foresight.org/guidelines/current.php (accessed May 17, 2025).

Ji S, Guvendiren M. Recent advances in bioink design for 3D bioprinting of tissues and organs. *Front Bioeng Biotechnol*. 2017 Apr 5;5:23.

Jiang H, Li X, Chen T, Liu Y, Wang Q, Wang Z, Jia J. Bioprinted vascular tissue: Assessing functions from cellular, tissue to organ levels. *Mater Today Bio*. 2023 Oct 28;23:100846.

Jin CK, Kim JH, Lee B-S. Powder bed fusion 3D printing and performance of stainless-steel bipolar plate with rectangular microchannels and microribs. *Energies*. 2022;15(22):8463.

Jowett P. Transparent tandem solar cell hits record 12.3% efficiency. *PV Magazine*. March 21, 2025. https://www.pv-magazine.com/2025/03/21/transparent-tandem-solar-cell-hits-record-12-3-efficiency/ (accessed May 11, 2025).

Kalenderian M This dress was 3D printed in one piece. *PLAIN Magazine*. December 23, 2014. https://plainmagazine.com/this-dress-was-3d-printed-in-one-piece/ (accessed May 12, 2025).

Karamchand L, Makeiff D, Gao Y, Azyat K, Serpe MJ, Kulka M. Biomaterial inks and bioinks for fabricating 3D biomimetic lung tissue: a delicate balancing act between biocompatibility and mechanical printability. *Bioprinting*. 2023;29:e00255.

Keller M. These Engineers 3D Printed a Mini Jet Engine, Then Took It to 33,000 RPM GE Reports, Sep 5, 2016. https://www.ge.com/news/reports/these-engineers-3d-printed-a-mini-jet-engine-then?pStoreID=massmutual%27A=0 (accessed May 11, 2025).

Khalifa RE, Shalaby AA, Špitalský Z. Toward revolutionizing PEMFC manufacturing for clean energy conversion: a review on the innovative contribution of 3D printing techniques. *Int J Adv Manuf Technol*. 2024;135:4119–4146.

Kim SH, Haines CS, Li N, Kim KJ, Mun TJ, Choi C, et al. Harvesting electrical energy from carbon nanotube yarn twist. *Science*. 2017 Aug 25;357(6353):773–778.

Kim K, Xu X, Guo J, Fan DL. Ultrahigh-speed rotating nanoelectromechanical system devices assembled from nanoscale building blocks. *Nat Commun*. 2014 Apr 7;5:3632.

King D. 3D-printed Urbee 2 ready to cross US on 10 gallons of fuel. *autoblog*. Nov 27, 2013. https://www.autoblog.com/news/3d-printed-urbee-2-ready-to-cross-us-on-10-gallons-of-fuel (accessed May 11, 2025).

Korkeamäki JT, Rashad A, Ojansivu M, Karvinen J, Koivisto JT, Syverud K, et al. Systematic development and bioprinting of novel nanostructured multi-material bioinks for bone tissue engineering. *Biofabrication*. 2025 Jan 28;17(2),025005.

Laureto JJ, Pearce JM. Open source multi-head 3D printer for polymer-metal composite component manufacturing. *Technologies*. 2017;5(2):36.

Lee J, Chun J. Development of personalized cast shoes using 3D printing technology. *Fash Text*. 2025;12(7).

Lee YJ, Yi H, Kim WJ, Kang K, Yun DS, Strano MS, et al. Fabricating genetically engineered high-power lithium-ion batteries using multiple virus genes. *Science*. 2009 May 22;324(5930):1051–1055.

Li J, Jiang W, Lei J, Xing X. Research on plasma arc flame length detection technology based on region of interest. *PLoS One*. 2025 Apr 3;20(4):e0321110.

Li L, Lin Q, Tang M, Duncan AJE, Ke C. Advanced polymer designs for direct-ink-write 3D printing. *Chemistry*. 2019 Aug 14;25(46):10768–10781.

Ling K, Wang W, Liu J. Current developments in 3D printing technology for orthopedic trauma: a review. *Medicine (Baltimore)*. 2025 Mar 21;104(12):e41946.

Liu T, Almutairi MMS, Ma J, Stewart A, Xing Z, Liu M, Hou B, Cho Y. Solution-processed thin film transparent photovoltaics: present challenges and future development. *Nanomicro Lett*. 2024 Oct 23;17(1):49.

Liu WT, Zhu L. Environmental microbiology-on-a-chip and its future impacts. *Trends Biotechnol*. 2005 Apr;23(4):174–179.

Lu X, Qu H, Skorobogatiy M. Piezoelectric micro- and nanostructured fibers fabricated from thermoplastic nanocomposites using a fiber drawing technique: comparative study and potential applications. *ACS Nano*. 2017 Feb 28;11(2):2103–2114.

Ma L, Wu Y, Li Y, Aazmi A, Zhou H, Zhang B, Yang H. Current advances on 3D-bioprinted liver tissue models. *Adv Healthc Mater*. 2020;9(24):e2001517.

Mahapatra R. Trends for 2025: Burden of poverty. *DownToEarth*. January 02, 2025. https://www.downtoearth.org.in/governance/trends-for-2025-burden-of-poverty (accessed May 16, 2025).

Malayil J. Video: 3D-printed humanoid robot made in just $70 with lifelike arms, chatbot brain. *Interesting Engineering*. May 09, 2025. https://interestingengineering.com/innovation/alana-3d-printed-humanoid-robot (accessed May 12, 2025).

Manna J, Goswami S, Shilpa N, Sahu N, Rana RK. Biomimetic method to assemble nanostructured Ag@ZnO on cotton fabrics: Application as self-cleaning flexible materials with visible-light photocatalysis and antibacterial activities. *ACS Appl Mater Interfaces*. 2015 Apr 22;7(15):8076–8082.

Manshadi MKD, Mohammadi M, Monfared LK, Sanati-Nezhad A. Manipulation of micro- and nanoparticles in viscoelastic fluid flows within microfluid systems. *Biotechnol Bioeng*. 2020 Feb;117(2):580–592.

Marchese K. Divergent debuts "blade," the world's first 3D printed hypercar. *designboom*. May 16, 2019. https://www.designboom.com/technology/divergent-the-blade-3d-printed-hypercar-05-16-2019/ (accessed May 11, 2025).

Martins NRB, Angelica A, Chakravarthy K, Svidinenko Y, Boehm FJ, Opris I, Lebedev MA, Swan M, Garan SA, Rosenfeld JV, Hogg T, Freitas RA Jr. Human brain/cloud interface. *Front. Neurosci*. 2019;13:112.

McFadden C. US scientists develop 3D-printable open-source humanoid robot that costs just $5K. *Interesting Engineering*. https://interestingengineering.com/innovation/us-scientists-3d-printable-open-source-robot-5k (accessed May 12, 2025).

Meira J, Matos G, Perdigão A, Cação J, Resende C, Moreira W, et al. Industrial internet of things over 5G: A practical implementation. *Sensors (Basel)*. 2023 May 30;23(11):5199.

Mitchell S, Steinbach J, Flanagan T, Ghabezi P, Harrison N, O'Reilly S, et al. Evaluating the sustainability of lightweight drones for delivery: towards a suitable methodology for assessment. *Functional Composite Mater*. 2023;4(1):4.

Mohammed Z, Elfadel IAM, Rasras M. Monolithic multi degree of freedom (MDoF) capacitive MEMS accelerometers. *Micromachines (Basel)*. 2018 Nov 16;9(11):602.

Montez J. BQ promotes Open Day in Madrid and presents news. *TecheNet*. October 14, 2015. https://www.techenet.com/2015/10/bq-promove-open-day-em-madrid-e-apresenta-novidades/ (accessed May 12, 2025).

Ng WL, Ayi TC, Liu YC, Sing SL, Yeong WY, Tan BH. Fabrication and characterization of 3D bioprinted triple-layered human alveolar lung models. *Int J Bioprint*. 2021 Apr 9;7(2):332.

Ni C, Buszczak M. Ribosome biogenesis and function in development and disease. *Development*. 2023 Mar 1;150(5):dev201187.

Nosoudi N, Hasanzadeh A, Hart M, Weaver B. Advancements and future perspectives in cell electrospinning and bio-electrospraying. *Adv Biol (Weinh)*. 2023 Oct;7(10):e2300213.

Nozica D, Blazevic D, Keser T. Communication convergence for improvement of the unmanned aerial search and rescue missions. *Applied Engineering Letters*. 2023;8(2):80–90.

O'Neal B. 3D Printing in the Kitchen: Pancakes Featuring Your Own Face. *3Dprint.com*. February 17, 2015 https://3dprint.com/45222/3d-print-pancake-w-your-face/ (accessed May 12, 2025).

Oosthuizen RM. The fourth industrial revolution – smart technology, artificial intelligence, robotics and algorithms: industrial psychologists in future workplaces. *Front Artif Intell*. 2022 Jul 6;5:913168.

Panella C. Check out Hellhound, a 375-mph, 3D printed, turbojet-powered exploding drone competing for a spot in the US Army's arsenal. *Business Insider*. May 13, 2025. https://www.msn.com/en-us/money/other/check-out-hellhound-a-375-mph-3d-printed-turbojet-powered-exploding-drone-competing-for-a-spot-in-the-us-armys-arsenal/ar-AA1EHTYx (accessed May 17, 2025).

Pasricha A, Greeninger R. Exploration of 3D printing to create zero-waste sustainable fashion notions and jewelry. *Fash Text.* 2018;5:30.

Pedde RD, Mirani B, Navaei A, Styan T, Wong S, Mehrali M, et al. Emerging biofabrication strategies for engineering complex tissue constructs. *Adv Mater.* 2017:29:1606061.

Petrovski A, Radovanovic M. Application of detection and reconnaissance technologies used by drones in collaboration with C4ISR for military purposes. *Contemp Maced Def.* 2021;117:23–35.

Phoenix C. Dangers of Molecular Manufacturing. *Center for Responsible Nanotechnology.* 2008. http://crnano.org/dangers.htm (accessed May17, 2025).

Pires JB, Santos FND, Costa IHL, Kringel DH, Zavareze EDR, Dias ARG. Essential oil encapsulation by electrospinning and electrospraying using food proteins: a review. *Food Res Int.* 2023;170:112970.

Plaid Zebra. With this 3-D printer you can build 10 tiny houses in under 24 hrs for less than $5,000 each. The Plaid Zebra, September 22, 2015. https://theplaidzebra.com/with-this-3-d-printer-you-can-build-10-tiny-houses-in-under-24-hrs-for-less-than-5000/ (accessed May 12, 2025).

Pompilio M, Ierides I, Cacialli F. Biomimetic approaches to "Transparent." *Molecules.* 2022 Dec 25;28(1):180.

Pourchet LJ, Thepot A, Albouy M, Courtial EJ, Boher A, Blum LJ, Marquette CA. Human skin 3D bioprinting using scaffold-free approach. *Adv Healthc Mater.* 2017;6:1601101.

Pumera M. Analysis of nerve agents using capillary electrophoresis and laboratory-on-a-chip technology. *J Chromatogr A.* 2006 Apr 28;1113(1–2):5–13.

Qiu L, Zhang M, Ghazal AF, Chu Z, Luo Z. Development of 3D printed k-carrageenan-based gummy candies modified by fenugreek gum: correlating 3D printing performance with sol-gel transition. *Int J Biol Macromol.* 2024;265, Part 1: 130865.

Rahman MM, Chowdhury MM, Alam MK. Rotating-electric-field-induced carbon-nanotube-based nanomotor in water: a molecular dynamics study. *Small.* 2017;13:1603978.

Ramadan NA, Basmaji T, Gad A, Hamdan H, Akgün BT, Ali MAH, et al. Towards early forest fire detection and prevention using AI-powered drones and the IoT. *Internet Things.* 2024;(2): 101248.

Rashad A, Gomez A, Gangrade A, Zehtabi F, Mandal K, Maity S, et al. Effect of viscosity of gelatin methacryloyl-based bioinks on bone cells. *Biofabrication.* 2024 Sep 3;16(4):10.1088/1758-5090/ad6d91.

Rasras M, Elfadel IAM, Ngo HD. Editorial for the special issue on MEMS accelerometers. *Micromachines (Basel).* 2019 Apr 29;10(5):290.

Rosellini E, Cascone MG, Guidi L, Schubert DW, Roether JA, Boccaccini AR. Mending a broken heart by biomimetic 3D printed natural biomaterial-based cardiac patches: a review. *Front Bioeng Biotechnol.* 2023 Nov 16;11:1254739.

Saffre F, Hildmann H, Karvonen H, Lind T. Monitoring and cordoning wildfires with an autonomous swarm of unmanned aerial vehicles. *Drones.* 2022;6(10):301.

Salih KOM, Rashid TA, Radovanovic D, Bacanin N. A comprehensive survey on the internet of things with the industrial marketplace. *Sensors (Basel).* 2022 Jan 19;22(3):730.

Sander IM, McGoldrick MT, Helms MN, Betts A, van Avermaete A, Owers E, et al. Three-dimensional printing of X-ray computed tomography datasets with multiple materials using open-source data processing. *Anat Sci Educ.* 2017 Jul;10(4):383–391.

Santhoshkumar P, Moses JA. Impact of konjac glucomannan on hot extrusion 3D printability of cake gel. *Int J Biol Macromol.* 2025 May;305(Pt 2):141162.

Science in Public. The world's first printed jet engine. *Phys.org.* February 25, 2015. https://phys.org/news/2015-02-world-jet.html (accessed May 11, 2025).

Scott C. From Shutter to Lens, This 35mm Camera is Entirely 3D Printed. *3DPrint.com.* September 9, 2016. https://3dprint.com/148836/3d-printed-35mm-camera/ (accessed May 11, 2025).

Scotti G, Matilainen V, Kanninen P, Piili H, Salminen A, Kallio T, et al. Laser additive manufacturing of stainless steel micro fuel cells. *J Power Sources*. 2014 Dec 25;272:356–361.

Service RF. The synthesis machine. *Science*. 2015Mar;347(6227):1190–1193.

Shepheard S. 22+ Coolest 3D Printed Robotics Projects (2024 Update). 3DSourced. https://www.3dsourced.com/feature-stories/3d-printed-robotics-robots/ (accessed May 17, 2025).

Siniauer P. 3D Printers Make Incredible Pastas Your Nonna Could Only Dream About. *SAVEUR*. November 9, 2017. https://www.saveur.com/3d-printers-pasta-barilla/ (accessed May 12, 2025).

Sparkes I. Lessons from optical tweezers: quantifying organelle interactions, dynamics and modelling subcellular events. *Curr Opin Plant Biol*. 2018 Dec;46:55–61.

Stadnicka D, Sęp J, Amadio R, Mazzei D, Tyrovolas M, Stylios C, et al. Industrial needs in the fields of artificial intelligence, internet of things and edge computing. *Sensors (Basel)*. 2022 Jun 14;22(12):4501.

Stella GM, Lettieri S, Piloni D, Ferrarotti I, Perrotta F, Corsico AG, Bortolotto C. Smart sensors and microtechnologies in the precision medicine approach against lung cancer. *Pharmaceuticals (Basel)*. 2023 Jul 22;16(7):1042.

Stiefenhofer P. Artificial General Intelligence and the End of Human Employment: The Need to Renegotiate the Social Contract. *arXiv*. February 10, 2025. https://arxiv.org/abs/2502.07050. (accessed May 08, 2025).

Sun J, Chen C, Zhang B, Yao C, Zhang Y. Advances in 3D-printed scaffold technologies for bone defect repair: materials, biomechanics, and clinical prospects. *Biomed Eng Online*. 2025 Apr 30;24(1):51.

Sun W, Huang H, Sun P, Ding W. MEMS strapdown inertial attitude measurement system using rotational modulation technology. *PLoS One*. 2024 Feb 13;19(2):e0298168.

Sun L, Wang Y, Zhang S, Yang H, Mao Y. 3D bioprinted liver tissue and disease models: current advances and future perspectives. *Biomater Adv*. 2023 Sep;152:21349.

Svozilova S, Zetková I, Marin JFS, Garay JAT. Evaluation of recycled and reused metal powders for DMLS 3D printing. *Materials (Basel)*. 2024 Dec 18;17(24):6184.

Tan Z, Liu T, Zhong J, Yang Y, Tan W. Control of cell growth on 3D-printed cell culture platforms for tissue engineering. *J Biomed Mater Res A*. 2017 Dec;105(12):3281–3292.

Tang J, Gao Y, Yang D, Yan L, Yue Y, Yang Y. DroneSplat: 3D Gaussian Splatting for Robust 3D Reconstruction from in-the-Wild Drone Imagery. *arXiv preprint*. 2025;2503.16964.

Tejada-Ortigoza V, Cuan-Urquizo E. Towards the development of 3D- printed food: a rheological and mechanical approach. *Foods*. 2022 Apr 20;11(9):1191.

Thiele S, Arzenbacher K, Gissibl T, Giessen H, Herkommer AM. 3D-printed eagle eye: compound microlens system for foveated imaging. *Sci Adv*. 2017 Feb 15;3(2):e1602655.

Tong S, Niu J, Wang Z, Jiao Y, Fu Y, Li D, et al. The evolution of microfluidic-based drug-loading techniques for cells and their derivatives. *Small*. 2024 Nov;20(47):e2403422.

Toshiyoshi H, Ju S, Honma H, Ji CH, Fujita H. MEMS vibrational energy harvesters. *Sci Technol Adv Mater*. 2019 Feb 18;20(1):124–143.

Unionfab. A New Dimension of Luxury: 3D Printing's Impact on Fashion. August 21, 2023. https://www.unionfab.com/blog/2023/08/3d-printing-luxury-fashion (accessed May 12, 2025).

Uzougbo NS, Ikegwu CG, Adewusi AO. Regulatory frameworks for Decentralized Finance (DeFi): challenges and opportunities. *GSC Adv Res Rev*. 2024;19:116–129.

Vanden, Broeck A, Klinge S. Eukaryotic ribosome assembly. *Annu Rev Biochem*. 2024;93(1):189–210.

Velo3D, Inc. Velo3D Announces Agreement with Ohio Ordnance Works to Support 3D Printed Military Weapons Development Initiative as Part of Rapid Production Services. *PR Newswire*. May 5, 2025. Velo3D Announces Agreement with Ohio Ordnance Works to Support 3D Printed Military Weapons Development Initiative as Part of Rapid Production Services (accessed May 17, 2025).

Wang L. World's first 3D-printing restaurant may serve the future of sustainable food. *Inhabitat.* November 07, 2016. https://inhabitat.com/worlds-first-3d-printing-restaurant-may-serve-the-future-of-sustainable-food/ (accessed May 12, 2025).

Wang D, Maharjan S, Kuang X, Wang Z, Mille LS, Tao M, et al. Microfluidic bioprinting of tough hydrogel-based vascular conduits for functional blood vessels. *Sci Adv.* 2022 Oct 28;8(43):eabq6900.

Weerasinghe HC, Macadam N, Kim JE, Sutherland LJ, Angmo D, Ng LWT, et al. The first demonstration of entirely roll-to-roll fabricated perovskite solar cell modules under ambient room conditions. *Nat Commun.* 2024 Mar 12;15(1):1656.

Wen Y, Che QT, Wang S, Park HJ, Kim HW. Elaboration of dimensional quality in 3D-printed food: key factors in process steps. *Compr Rev Food Sci Food Saf.* 2024 Jan;23(1):e13267.

Wheeler K. WEF: How AI Will Reshape 86% of Businesses by 2030. *Technology Magazine.* January 14, 2025. https://technologymagazine.com/articles/wef-report-the-impact-of-ai-driving-170m-new-jobs-by-2030 (accessed May 16, 2025).

Wilson DN, Doudna Cate JH. The structure and function of the eukaryotic ribosome. *Cold Spring Harb Perspect Biol.* 2012 May 1;4(5):a011536.

Wu H, Diao J, Li X, Yue D, He G, Jiang X, Li P. Hydrogel-based 3D printing technology: from interfacial engineering to precision medicine. *Adv Colloid Interface Sci.* 2025 Mar 19;341:103481.

Wu Y, Lei M, Xia X. Research progress of MEMS gas sensors: a comprehensive review of sensing materials. *Sensors (Basel).* 2024 Dec 19;24(24):8125.

Wu D, Long M. Realizing visible-light-induced self-cleaning property of cotton through coating N-TiO_2 film and loading AgI particles. *ACS Appl Mater Interfaces.* 2011Dec;3(12):4770–4774.

Wu S, Zeng T, Liu Z, Ma G, Xiong Z, Zuo L, Zhou Z. 3D printing technology for smart clothing: a topic review. *Materials (Basel).* 2022 Oct 21;15(20):7391.

Xie X, Xu Z, Yu X, Jiang H, Li H, Feng W. Liquid-in-liquid printing of 3D and mechanically tunable conductive hydrogels. *Nat Commun.* 2023 Jul 18;14(1):4289.

Xing W, Wang J, Zhao C, Wang H, Bai L, Pan L, et al. A highly automated mobile laboratory for on-site molecular diagnostics in the COVID-19 pandemic. *Clin Chem.* 2021 Mar 31;67(4):672–683.

Xu Y, Liu S, He C, Wu H, Cheng L, Huang Q, Yan G. Research on packaging reliability and quality factor degradation model for wafer-level vacuum sealing MEMS gyroscopes. *Micromachines (Basel).* 2023 Oct 20;14(10):1956.

Xuan J, Zhu T, Peng G, Sun F, Dong D. A review on the inertial measurement unit array of microelectromechanical systems. *Sensors (Basel).* 2024 Nov 6;24(22):7140.

Yadav R, Mahajan S, Singh H, Mehra NK, Madan J, Doijad N, et al. Emerging in vitro and in vivo models: hope for the better understanding of cancer progression and treatment. *Adv Biol (Weinh).* 2024;8(6):e2300487.

Yang Z, Ma Y, Jia S, Zhang C, Li P, Zhang Y, Li Q. 3D-printed flexible phase-change nonwoven fabrics toward multifunctional clothing. *ACS Appl Mater Interfaces.* 2022 Feb 9;14(5):7283–7291.

Yu H, Wang J, Zhou Q, Qin J, Wang Y, Lu X, Cheng P. Semi-transparent organic photovoltaics. *Chem Soc Rev.* 2023 Jul 3;52(13):4132–4148.

Zaleski A. 3D printing shakes up the space industry. *The Washington Post.* May 06, 2025. https://www.washingtonpost.com/business/2025/05/06/rocket-engines-3d-printing-space/ (accessed May 10, 2025).

Zammit D, Safran T, Ponnudurai N, Jaberi M, Chen L, Noel G, Gilardino MS. Step-specific simulation: the utility of 3D printing for the fabrication of a low-cost, learning needs-based rhinoplasty simulator. *Aesthet Surg J.* 2020 May 16;40(6):NP340–NP345.

Zawawi SA, Hamzah AA, Majlis BY, Mohd-Yasin F. A review of MEMS capacitive microphones. *Micromachines (Basel).* 2020 May 8;11(5):484.

Ze Y, Li Y, Huang L, Shi Y, Li P, Gong P, Lin J, Yao Y. Biodegradable inks in indirect three-dimensional bioprinting for tissue vascularization. *Front Bioeng Biotechnol.* 2022 Mar 25;10:856398.

Zhang R, Deng L, Guo J, Yang H, Zhang L, Cao X, et al. Solvent mediating the *in situ* self-assembly of polysaccharides for 3D printing biomimetic tissue scaffolds. *ACS Nano.* 2021 Nov 23;15(11):17790–17803.

Zhang L, Forgham H, Shen A, Wang J, Zhu J, Huang X, et al. Nanomaterial integrated 3D printing for biomedical applications. *J Mater Chem B.* 2022 Sep 28;10(37):7473–7490.

Zhang Y, Reid A, Windmill JFC. Insect-inspired acoustic micro-sensors. *Curr Opin Insect Sci.* 2018 Dec;30:33–38.

Zhang J, Wang Y, Rodriguez BJ, Yang R, Yu B, Mei D, Li J, Tao K, Gazit E. Microfabrication of peptide self-assemblies: inspired by nature towards applications. *Chem Soc Rev.* 2022 Aug 15;51(16):6936–6947.

Zhao Z, Liu W, Liu H. Flexible and durable direct ink writing 3D-printed conductive fabrics for smart wearables. *ACS Omega.* 2025 Apr 1;10(14):14138–14149.

Zhou GX, Yu YG, Yang ZH, Jia DC, Poulin P, Zhou Y, Zhong J. 3D printing graphene oxide soft robotics. *ACS Nano.* 2022 Mar 22;16(3):3664–3673.

Zhu W, Ma X, Gou M, Mei D, Zhang K, Chen S. 3D printing of functional biomaterials for tissue engineering. *Curr Opin Biotechnol.* 2016 Aug;40:103–112.

Zhu C, Pascall AJ, Dudukovic N, Worsley MA, Kuntz JD, Duoss EB, Spadaccini CM. Colloidal materials for 3D printing. *Annu Rev Chem Biomol Eng.* 2019 Jun 7;10:17–42.

3 Quantum Nanoscience
Atomically Precise Manufacturing Meets Quantum Computation

Melanie Swan

CONTENTS

3.1 Introduction 45
3.1.1 The Scale of Molecular Manufacturing 45
3.2 Hardware: Atomic Scale Fabrication 47
3.2.1 Atomic Scale Fabrication 48
3.2.1.1 Atomically Precise Manufacturing 48
3.2.1.2 Atomic Precision Advanced Manufacturing 49
3.2.1.3 Digital Methods and Autonomous Robotic Nanofabrication 50
3.2.2 Assembly 51
3.2.2.1 MEMS/NEMS and Bio-MEMS/NEMS 51
3.2.2.2 Nanoparticles 51
3.2.2.3 Molecular Assemblers 52
3.2.2.4 Biomolecular Condensates 52
3.2.2.5 Self-Assembled Drug Delivery Platforms 53
3.2.3 DNA Nanotechnology 53
3.2.3.1 Topological Molecules 54
3.2.3.1.1 Molecular Knots 54
3.2.3.1.2 Molecular Weaving 55
3.2.3.2 Biological Liquid Crystals and DNA Chirality Inversion 55
3.2.3.2.1 Virtual Self-Assembly Liquid 56
3.3 Hardware: Quantum Nanoscience Technologies 56
3.3.1 Quantum Dots (Semiconductor Nanoparticles) 56
3.3.2 Spin Engineering (Magnetic Moments) 57
3.3.2.1 Spintronics 57
3.3.2.2 Spin Chains 57
3.3.3 Electron Engineering 58
3.3.4 Artificial Lattices (Spin and Electron) 58
3.3.5 Floquet (Periodic Time) Engineering 59
3.3.5.1 Discrete Time Crystals 59
3.3.5.2 Floquet Circuits: Discrete Time Crystals NISQ and Classical Hardware 59
3.3.5.3 Floquet Tri-unitary Time Circuits 60

DOI: 10.1201/9781003724896-3

3.3.6 Quantum Matter (Topological Materials) 60
3.3.6.1 Quantum Spin Liquids and Topological Insulators 61
3.3.6.2 Long-Range Entanglement 62
3.3.6.2.1 Quantum Hall States 62
3.3.6.2.2 Quantum Spin Liquids 62
3.3.6.2.3 Topological Entanglement Entropy 63
3.3.6.3 Short-Range Protected Topological Order 63
3.3.6.3.1 Symmetry-Protected Topological Insulators 63
3.3.7 Molecular Coding 64
3.3.7.1 Quantum Error Correction 64
3.3.7.2 Stabilizer Codes 65
3.3.7.3 Bosonic Codes 65
3.3.7.4 Molecular Codes 65
3.4 Software 65
3.4.1 Virtual Reality Prototyping 66
3.4.1.1 Digital Twins and Programmable Smart Matter 66
3.4.1.1.1 Smart Programmable Matter (4D Printing) 66
3.4.1.1.2 Levels of Smart Matter 67
3.4.1.2 Game Engine Platforms 67
3.4.1.2.1 The Asset Store 68
3.4.1.2.2 Gaming Engine Headsets: Cognitive Neuroscience Consumer BCIs 68
3.4.2 Machine Learning 68
3.4.2.1 Classical Machine Learning 68
3.4.2.2 Quantum Machine Learning 69
3.4.2.3 Quantum Machine Learning and Quantum Nanoscience 69
3.4.3 Blockchain 69
3.4.3.1 Blockchain Definition and Status 69
3.4.3.1.1 Blockchain Mining Nodes 70
3.4.3.1.2 Cryptoeconomics 70
3.4.3.1.3 Status of Blockchain Adoption 71
3.4.3.1.4 Non-Fungible Token (NFT) Intellectual Property Registries 71
3.4.3.2 Blockchain Science 72
3.4.3.2.1 Smart Contract Molecular Dynamics Simulations 72
3.4.3.2.2 BioBlockchain 72
3.4.3.3 Applications of Blockchain in Manufacturing Industries 72
3.4.3.3.1 Blockchain Manufacturing, Clinical Trials, and Genomics 73
3.4.3.3.2 Pharmaceutical Manufacturing and Chemical Industry 73
3.4.3.4 Molecular Manufacturing Economy 74
3.4.3.4.1 Participant-Provided Infrastructure and Financing 74
3.5 Applications 75
3.5.1 Biology 75
3.5.1.1 The Human Brain 75
3.5.1.2 Neuroscience Physics 76
3.5.1.3 Chern-Simons/Biology 76

3.5.2 Space 77
3.5.2.1 Off-World Manufacturing 77
3.6 Conclusion 78
3.6.1 Summary 78
3.6.2 Risks and Limitations 78
3.6.3 Farther Future 79
References 79
Glossary 92

3.1 INTRODUCTION

"Our ability to arrange atoms lies at the foundation of technology."

Drexler, *Engines of Creation*, 1986

Our current comprehension of the physical world is astounding. At the very large scale, the Milky Way galaxy (a small portion of the Virgo Supercluster) is now seen within the much larger Laniakea Supercluster (using a new method for analyzing the relative velocities of galaxies as being similar to watershed divides) (Tully et al., 2014). The James Webb Space Telescope was launched in December 2021, with the aim of being able to "see" much farther back toward the Big Bang by targeting the infrared spectrum and having a 6.25× larger light collecting surface area than the Hubble Space Telescope (NASA, 2021). Over 5,000 exoplanet discoveries were confirmed by NASA's Transiting Exoplanet Survey Satellite (TESS) in January 2022, with over 800 having more than one planet (MIT, 2022). At the diminutive scale, scientists have been able to reverse the orbital moment of a single electron (Rejali et al., 2020) and "weigh" frontier electron orbitals via optical microscopy (Tan et al., 2021). Novel topological quantum matter phases at zero temperature have been discovered, such as discrete time crystals, fractional quantum Hall states, and quantum spin liquids, with direct portability to quantum computation (Giustino et al., 2021). The implication of this slate of findings is that the more that is known about matter at all scales, the easier it may be to create and ultimately manipulate matter at these scales as well.

This chapter discusses the current status of molecular manufacturing (MM) (the controlled molecularly and/or atomically precise fabrication of physical objects) via a three-tiered model of hardware, software, and applications (Figure 3.1). Not only is MM transitioning to scientific quantum information techniques along with many other fields, but also leading the effort given its core capacities for molecular precision. The early scientific visions proposed by Feynman (a universal quantum simulator, 1982), Drexler (a molecular assembler, 1986), and Merkle (self-replicating systems, 1992) are moving closer to realization, facilitated, in particular, by quantum nanoscience (fabricated nanostructures that exploit quantum properties).

3.1.1 The Scale of Molecular Manufacturing

Physical matter is comprised of atoms and molecules (Tables 3.1 and 3.2), with atoms as the smallest discrete units of matter. Molecules are aggregates of atoms that are

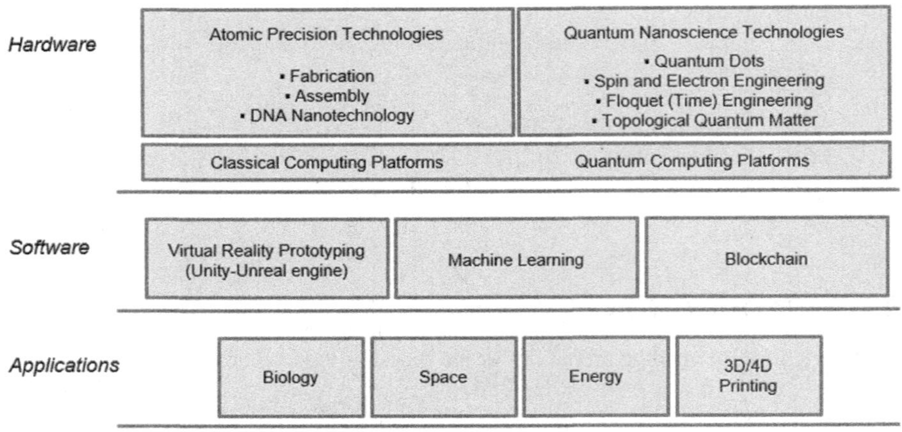

FIGURE 3.1 Molecular manufacturing stack: hardware, software, applications.

TABLE 3.1
Exemplars of Entities at the Nanometer Scale

Size	Entity Description
100,000 nm	Thickness of a sheet of paper or human hair.
100 nm	Nanoparticles (10 nm: colloidal nanoparticle, 1,000 nm: nanorobots).
2 nm	Width of DNA (20 nm: synaptic cleft).
	DNA: like a human hair the length of a football field.
0.27 nm	Width of a water molecule (H_2O; 2 H, 1 O).
	Atom: nucleus like a ping pong ball, electrons circling 1 km away.

TABLE 3.2
Orders of Magnitude Scales of Atoms and Molecules

Entity	Size	Scale	Notable Ranges	
Human	>1 m	Meters		
Cell	10^{-3} m	Millimeters		
Molecule	10^{-6} m	Micrometers	Mesoscopic scale	
Atom	10^{-9} m	Nanometers	Mesoscopic scale	Quantum scale
Ions and photons	10^{-12} m	Picometers		Quantum scale
Atomic nucleus	10^{-15} m	Femtometers		Quantum scale
Electron	10^{-18} m	Attometers		
Planck length	10^{-35} m			

bound together by covalent (electron sharing), ionic (electrostatic forces), or metallic bonds. The outer dimensions of atoms (electron shells) are typically ~100,000 times larger than their nuclei (if the atomic nucleus were a ping pong ball, its electrons would be circling in orbits about a kilometer away (Smith, 1999)).

The dimensions of molecules can vary significantly, with water and carbon dioxide being some of the smallest and proteins some of the largest. A water molecule is only ~0.27 nm across, whereas a DNA duplex is ~2 nm wide and can be meters in length. The width-to-length ratio of human DNA is analogous to a hair that is as long as a football field (Rathburn, 2013). A typical virus is Ø100 nm (a bacterium is Ø1,000 nm), while the nanoparticles/nanocarriers employed in nanomedicine are generally Ø100 nm; however, they may range in size from Ø4 nm to Ø200 nm (Nance, 2019). In contrast, the thickness of a piece of paper or human hair is about 100,000 nm, and a red blood cell is Ø7,000 nm. Envisaged nanomedical robots may typically range in size from ~Ø1,000 to Ø3,000 nm for respirocytes (artificial red blood cells), clottocytes (artificial platelets), and microbivores (artificial phagocytes) (Freitas, 2012). The upper size limit for circulating nanorobots (recommended by Freitas) is ~Ø1,000 nm (Ø1 μm), as this is close to the size of the internal lumen of the smallest capillaries in humans (~Ø4 μm) (Eugene & Renkin, 1989; Gaehtgens, 1980; Skalak & Branemark, 1969). In the brain, medical neuralnanorobots would need to be on the order of ~Ø1,000 nm for axonal endoneurobots and gliabots, and perhaps from ~Ø30 to 300 nm in size for synaptobots (Martins et al., 2019).

Some of the relevant scale tiers for MM include the mesoscopic scale (10^{-6} to 10^{-9}) and the quantum scale (10^{-9} to 10^{-15}). Mesoscopic physics is the intermediate scale between microscopic and macroscopic domains, which generally relates to materials between the micrometer (10^{-6}) and nanometer (10^{-9}) scale, where quantum effects start to become important. For example, at the macroscopic level, the conductance of a wire increases continuously with diameter, whereas at the mesoscopic level, its conductance is quantized and increases in discrete steps. The word "quantum" refers to the scale of atoms and molecules (10^{-9} to 10^{-15}), meaning atoms at the nanometer scale (10^{-9}), ions and photons at the picometer scale (10^{-12}), and subatomic particles at the femtometer scale (10^{-15}).

Quantum effects may be observed at different scales in various systems through the primary quantum properties of superposition, entanglement, and interference (Swan et al., 2022a). Superposition is the property of an unobserved particle that exists in all possible states simultaneously, until collapsing to only a single one when measured. Entanglement is the property of physical attributes being correlated between particles (e.g., position, momentum, polarization, and spin), even when separated by a distance (if one particle is in a spin-up state, the other is in a spin-down state; a "heads-tails" relationship). Interference is the property of waves that reinforce or cancel each other out (coherence or decoherence); the observed pattern of phase difference between two wavefunction amplitudes.

3.2 HARDWARE: ATOMIC SCALE FABRICATION

Considering hardware platforms, one way of categorizing MM activities is between atomic precision and quantum nanoscience technologies to reflect the distinction between strategies that employ a combination of classical and quantum methods, and

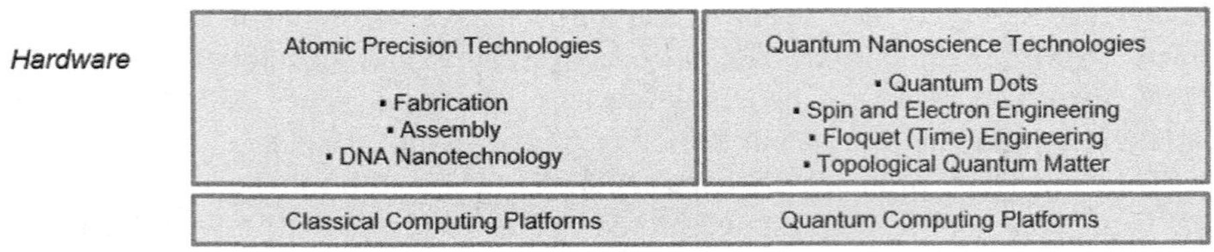

FIGURE 3.2 Molecular manufacturing hardware platforms.

those that explicitly target quantum properties through quantum information science methods (Figure 3.2). Atomic precision technologies consist primarily of atomic force microscopy (AFM) and scanning tunneling microscopy (STM)-based fabrication (Bouju et al., 1994; Fishlock et al., 2000), biobased assembly, and self-assembly techniques (Jungck et al., 2022; Thomas & Schwartz, 2017), and DNA nanotechnology, which entails the use of oligonucleotides as fabrication "biobots" and nucleic acids as building materials (Pu et al., 2014; Xia et al., 2020) rather than carriers of genetic information. Quantum nanoscience technologies focus on the explicit construction and harnessing of quantum effects in molecular devices, encompassing quantum dots (semiconductor nanoparticles), quantum spin, and electron engineering, Floquet time engineering, and quantum matter (matter phases at zero temperature with topological properties). The digital methods that support the orchestration of MM processes are also expanding from primarily classical to quantum computational platforms.

3.2.1 Atomic Scale Fabrication

3.2.1.1 Atomically Precise Manufacturing

Clear examples toward the realization of MM have been demonstrated by atomically precise manufacturing strategies, which are fabrication techniques in which single atoms and molecules can be precisely positioned to create macroscopic objects (Chen et al., 2023; Ding et al., 2023). The real-world operation of this process (Figure 3.3) illustrates the ZyVector, which is an STM control system for atomically precise lithography (Randall, 2021), which includes four steps. The first step outlines the structure of the design, while the second specifies the crystal lattice vector layout. The third step involves the writing (deposition) of atoms with the STM tip, whereas the fourth completes an atomically precise pattern. Various applications of atomically precise lithography encompass integrated circuits, thin films, etch masks, and molecular electronics. Renewable energy initiatives are supported by the potentially reduced energy use and increased performance capacities of materials, structures, devices, and finished goods (Kaarsberg, 2019).

Various microscopy methods may be employed in the precise fabrication of atomic scale structures, most notably STM, AFM, and electron beam-based scanning transmission electron microscopy (STEM), with various levels of resolution

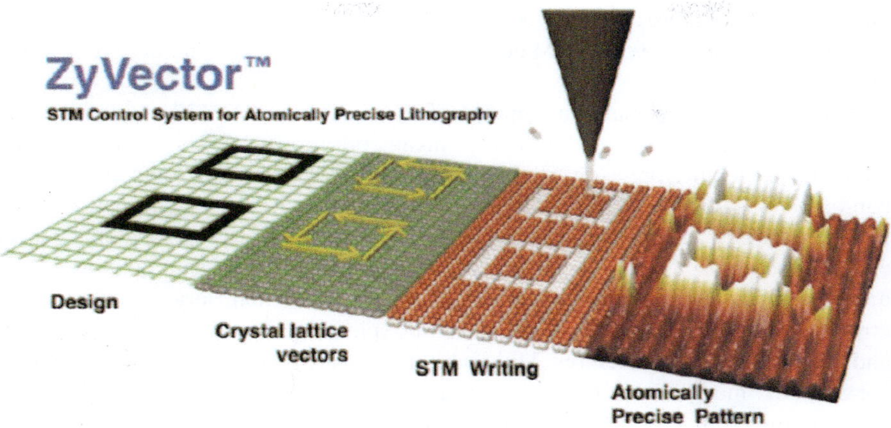

FIGURE 3.3 Atomically precise writing (deposition) with a scanning tunneling microscope. ZyVector: STM Control System for Atomically Precise Lithography. https://www.zyvexlabs.com/apm/products/zyvector/ (Used with permission from Zyvex Labs.)

relevant to different applications (Dyck et al., 2019). The microscopes themselves serve as fabrication tools, not merely for inspecting and identifying samples, but also as probes and construction platforms. The workhorse microscope of MM is the STM, which is employed to study the three-dimensional structures of samples at the atomic level by scanning the surface with a sharp, metallic, conductive tip in close proximity to the sample. Individual atoms are routinely imaged and manipulated via a single-atom-wide tip that is slowly moved across the sample surface at a distance of one atomic diameter. The tip moves up and down according to the vertical structures of the sample and measures the results utilizing the quantum tunneling current that is created. Quantum tunneling is a quantum mechanical effect that occurs when electrons traverse a barrier due to their wave-like properties to create a measurable current across the tunnel according to the thickness of the barrier.

3.2.1.2 Atomic Precision Advanced Manufacturing

A next-generation version of atomically precise manufacturing (APM) is atomic precision advanced manufacturing (APAM). Atomic precision advanced manufacturing is a materials synthesis strategy that involves the doping (addition of impurities to modulate electrical, optical, or structural properties) of an atomically thin layer of material. Specifically, APAM may include the atomically precise ultra-doping of silicon with atomic resists, area-selective surface chemistry, and molecules (such as hydride and halide precursor molecules and aluminum alkyls) (Owen et al., 2021). One research aim beyond Moore's law is to investigate digital electronics at the atomic limit and integrate APAM with the fabrication of computer chips (Ward et al., 2020). This challenge involves shrinking the smallest features from nanometers to atoms (10 nm corresponds to 30 silicon atoms), while targeting the absolute physical limits of atoms themselves to shift the paradigm, from classical to quantum computation. A key use of APAM could be for the fabrication of multiqubit quantum processing chips.

The techniques for quantum chip manufacturing include APAM and ion implantation; however, APAM may be the best means of tailoring silicon nanoelectronics with full 3D atomic precision (Bussmann et al., 2021). The development of a silicon-based quantum chip manufacturing industry is desirable, as fabrication could proceed utilizing established silicon foundry techniques and existing global supply chains. A current research focus is set on quantum chips that employ atomic spin qubits as a possibility for fault-tolerant quantum computation (large-scale universal error-corrected quantum computation). Atomic spin qubits have long dephasing times and high-fidelity gate structures. Multiqubit devices are challenging to fabricate with conventional methods due to tight interqubit pitches forced by the short-range spin interactions; however, APAM offers the required molecular scale precision to systematically address these issues. The creation of high-fidelity two-qubit gates is a current focus in dopant spin qubit research (in both APAM and ion implantation). It must be acknowledged, however, that significant developments in techniques and know-how are required to apply APAM for the fabrication of circuitry with increasing numbers of qubits. For example, with a focus on developing digital strategies such as technology-based computer-aided design (TCAD) methods (Bussmann et al., 2021).

3.2.1.3 Digital Methods and Autonomous Robotic Nanofabrication

To facilitate a global scale classical-quantum transition, digital fabrication methods will be essential. This, as stepwise improvements in manufacturing precision enabled by Moore's Law, cannot be sustained much longer when approaching the limits imposed by the quantized nature of matter (Randall et al., 2018). Shifting MM from analog to digital fabrication techniques might provide the same types of benefits that digital information technologies have had over analog methods. One example of digital fabrication includes that digital processes involve the making and breaking of chemical bonds, according to a digital address grid that is spatial in nature. Error detection and correction processes are applied according to the analyses of digital spatial grids being applied to physical domains. In atomic layer deposition and di-block copolymer lithography, for instance, processes occur at the scale of atoms and molecules that have not yet been fully developed as digital fabrication technologies. Such fabrication methods will necessitate digitally managed precision control of the thickness of a physical vapor deposited film entailing a uniform flux of the material across an entire substrate. This would involve a highly accurate flux monitor, metrology to calibrate the flux monitor, and multi-directional flux detection contingent on the topography of the substrate.

One way that digital fabrication processes are being deployed is through the autonomous robotic nanofabrication of supramolecular structures from single molecules (Leinen et al., 2020). This process consists of controlling single molecules via the machine learning agent-based manipulation of scanning probe microscope (SPM) actuators (using reinforcement learning/goal-directed updating) to autonomously remove molecules from a supramolecular structure.

3.2.2 Assembly

Approaches for the synthesis of atomically precise materials include both the externally controlled microscopy-based placement of atoms and molecules, and the intrinsically inherent principles of assembly; the natural processes through which atoms and molecules coalesce into larger structures. In particular, self-assembly is of great interest for the organization of individual units of material into arranged structures or patterns at atomic resolution. The ultimate accomplishment will be to biomimetically harness nature's own highly efficient molecular self-assembly processes, which join molecules together in an assembly-line like fashion (Lo et al., 2010; Wilson et al., 2012). In biological environments, self-assembled structures often arise due to the spontaneous and reversible non-covalent localized interactions of molecules. An extensive range of both organic and inorganic self-assembled nanomaterials is being applied in nanotechnology, imaging techniques, biosensors, and medicine with the advantageous attributes of simplicity, spontaneity, scalability, versatility, and cost-effectiveness. The principles of self-assembly might be applicable beyond the molecular domain across a wide range of scales (Whitesides & Grzybowski, 2002), including for the construction of macroscale structures (Gross & Dorigo, 2008).

3.2.2.1 MEMS/NEMS and Bio-MEMS/NEMS

As the miniaturization of electronics shrinks to ever smaller scales, fabrication strategies aim to emulate natural assembly principles that are thought to be inherent to matter at these diminutive domains. Microelectromechanical systems (MEMS) and nanoelectromechanical systems (NEMS) are devices/systems fabricated at millimeter- and nanometer-scales, which operate and are endowed with functionality in these domains, while bio-MEMS/NEMS are deployed from the biological context. MEMS are millimeter scale machines with moving parts (20–1000 μm devices with components at 1–100 μm), which typically consist of a central processing unit combined with sensors that interact with the environment. MEMS are comprised of materials that include silicon, polymers, metals, and ceramics. These miniature systems integrate electrical and mechanical functionality through the use of transistor-like electronics that control mechanical actuators, pumps, and motors. Their smaller size gives rise to the prevalence of alternative properties such as high surface-to-volume ratios that influence electromagnetism (electrostatic charges and magnetic moments), fluid dynamics (surface tension and viscosity), and surface chemistry.

3.2.2.2 Nanoparticles

Inorganic nanoparticles have initially been applied for the potential clinical diagnosis and treatment of disease. However, the introduction of microscopic inorganic matter into the complex organochemical environment of the human body is a challenging proposition, as the renal accumulation of nanoparticles may become toxic if they are not cleared in a timely manner. However, their immediate wholesale excretion must also be avoided to facilitate the optimization of their beneficial effects (Choi et al., 2007). Many of the initially developed nanoparticles were too large, which precluded their efficient clearance from the body as intact nanoparticles via renal filtration and urinary excretion processes. Without such nanoparticle clearance mechanisms or their degradation to biologically benign components, toxicity is potentially amplified and radiological imaging is hindered. Choi et al. found that a diameter of <5.5 nm resulted in rapid and efficient urinary excretion of nanoparticles, whereas the addition of organic coatings to particles increased their diameter to >15 nm and prevented renal excretion. Improvements in the elucidation of the behavior of nanoparticles in the body have suggested their appropriate dimensions for various interventions (Kumar et al., 2021; Qi et al., 2023).

More recent nanoparticle delivery platforms incorporate FDA-approved biodegradable polymers (e.g., polylactic-co-glycolic acid (PLGA)) (Ding & Zhu, 2018). One experimental setup featured nanoparticles with surface chemistry coatings that maximized their traversal of the blood-brain barrier (BBB) (Li et al., 2021). Small interfering RNA (siRNA) molecules were delivered to the brain (in a mouse model) to treat secondary injuries associated with traumatic brain injury that can lead to Alzheimer's disease, Parkinson's disease, and other neurodegenerative conditions. The results indicated a 50% reduction in the expression of the tau protein, which is implicated in these pathologies.

3.2.2.3 Molecular Assemblers

Optimal assembly systems that emulate the assembly-line like production of molecules utilized by nature are envisioned as self-replicating platforms in the MM environment (Freitas & Merkle, 2004; Merkle, 1992). A contemporary project developed a synthetic molecular assembler that produced (disulfide) polymers (large molecules of repeating subunits) (Engwerda & Fletcher, 2020). The work demonstrated how to create primitive assemblers from simple phase-separated reactant processes. There are two phases in which reactants initially self-reproduce until they reach a critical concentration, to then begin to produce polymers. The polymer size can be controlled by adjusting the temperature and reaction time, or through the introduction of a capping agent. Traditionally, diffusion and Brownian motion have limited the feasibility of molecular assembly-line processes, which were overcome here by the molecular assembler operating within a confined space, which increased contact between the reactants. This research has opened new directions for the development of functional supramolecular systems via molecular assembly.

3.2.2.4 Biomolecular Condensates

Self-assembly mechanisms are leading to an improved understanding of intracellular communication through biomolecular condensates, which are organelles in living cells (not bounded by membranes) that spontaneously self-assemble to carry out various tasks (Bracha et al., 2019). Biocondensates are formed by the phase separation of proteins, RNA, and other biopolymers into either colloidal emulsions, liquid crystals, solid crystals, or cellular aggregates. These biomolecular condensates in membraneless compartments appear to be involved in critical biological functions such as stress response, intracellular signaling, and transcriptional control. As such, the biomolecular condensates self-assemble in these compartments as proteins and nucleic acids separate to form droplets. Self-assembly is driven by the net attractive interactions between disordered polypeptides and specific binding interactions between proteins, DNA, RNA, and other molecules. Researchers have investigated how the self-assembly of a large number of condensates may be temporally and spatially coordinated within a single intracellular compartment (Jacobs, 2021). This work found that the number of distinct condensates that can be reliably assembled grows to the extent that condensates are allowed to share components (up to a certain threshold) (Ramella et al., 2022).

3.2.2.5 Self-Assembled Drug Delivery Platforms

One of the biggest use cases for biological self-assembly is drug delivery through the use of platforms that are prepared via self-assembly, which offer a scalable and tunable platform. Self-assembled drug delivery platforms include micelles, hydrogels, and vesicles (e.g., liposomes), which are comprised of amphiphilic molecules (both water-loving and fat-loving), designed to avoid renal excretion, and facilitate targeted drug delivery to specific tissues or cells through surface functionalization (Erdoğar et al., 2018; Osorno et al., 2021). Micelles are compact spheres (Ø10–100 nm) that are externally hydrophilic with hydrophobic cores. Hydrogels are cross-linked structures of polymer networks that can absorb significant amounts of water without dissolving, similar to natural tissue. Vesicles are larger hollow versions of micelles; small spheres that can deliver both hydrophilic compounds encapsulated within the hollow core, and hydrophobic compounds loaded between the bilayer. There are several FDA-approved biopolymers employed in the formation of self-assembled structures for controlled drug delivery via micelles (Ghezzi et al., 2021; Hwang et al., 2020), hydrogels (Bernhard & Tibbitt, 2021; Zielińska et al., 2022), and vesicles (Guimarães et al., 2021; Herrmann et al., 2021). The majority of these entities are employed for oncology (cancer drug delivery), while others address bacterial infections, glucose control, ocular conditions, and enzyme replacement. Specific formulations that are approved or in clinical trials include the cancer drugs Paclitaxel, Doxorubicin, and Irinotecan; the leukemia drug Vincristine; and cystic fibrosis drug Ciprofloxacin.

Smart hydrogels (nanogels) are a research focus as they are responsive to external stimuli such as pH, temperature, ionic strength, and electric fields. They serve as an exemplar of macromolecular self-assembly and more precise drug action through controlled mechanical properties and tunable drug release. The nanomechanical properties of hydrogels (Khalili et al., 2023; Zaguri et al., 2020) prepared through self-assembly may be enhanced by integrating self-assembly motifs with polymeric networks (interlaced but not covalently bonded networks) to create "double-network hydrogels" (Webber & Pashuck, 2021; Yin et al., 2023).

3.2.3 DNA Nanotechnology

DNA nanotechnology entails the design and synthesis of artificial nucleic acid structures for technological applications (He et al., 2023; Zhan et al., 2023) using nucleic acids as non-biological engineering materials for nanotechnology (see Chapter 5) (Seeman, 1998, 2020), not as carriers of genetic formation in living cells. High-profile developments include the creation of topological structures such as molecular knots and chirality reversibility. The recent combination of advanced AI with CRISPR will likely expedite the emergence of highly precise DNA-based MM (Noshay et al., 2023).

3.2.3.1 Topological Molecules

3.2.3.1.1 Molecular Knots

Molecular knotting is a potentially important technology for MM, as unique tasks can be undertaken at the nanoscale, such as increasing the local bulk and introducing chirality as strands are knotted (Schaufelberger, 2020). Knots and entanglements are common topological features of practical and mathematical interest, which are observed at both the macroscopic scale and the molecular level. In particular, molecular knots are mechanically interlocked molecular architectures analogous to macroscopic knots that are found in natural organic molecules such as DNA, RNA, and proteins, and synthesized in labs. Knotting is implicated in various DNA-related processes such as transcription, replication, compaction, and recombination. Further, enzymes (topoisomerases) can resolve errors by catalyzing the unknotting of DNA duplexes via a "cut and paste" mechanism, which breaks the chain into small segments and rearranges them to eliminate the knot (Sun et al., 2022).

The detection of knots in topologically complex systems requires mathematical methods to classify the knot type. The Alexander polynomial is a first-line algorithm that is applied to classify a knot according to the minimum number of crossings it has in a projection of the chain onto a plane. Knots are labelled with Alexander-Briggs notation in which the first number is the crossing number (a measure of knot complexity) and the subscripted index number denotes the knot's order among all knots with that crossing number. A simple ring with zero crossings is referred to as the unknot (0_1), while the simplest non-trivial knot type is the trefoil knot (3_1) with three crossings, and the figure-of-eight knot (4_1) has four crossings. The Jones and HOMFLY polynomials are more advanced algorithms that are used to discriminate knot types of higher complexity. DNA comprises a long biopolymer that forms various chiral, achiral, torus, and twist knots ranging from simple trefoil (3_1) knots to those that are more complex (e.g., with nine crossings). Viral genomic DNA has been found to exhibit chiral and torus knots (Lim & Jackson, 2015).

Research has extended the techniques for tying different classes of knots in a synthetic nanoscale strand to form a molecular endless (7_4) knot (Leigh et al., 2021). Strategies for the synthesis of molecular knots typically focus on the twisting, folding, and threading of molecular building blocks. Instead, Leigh and workers used zinc and iron ions to weave ligand strands to form a woven 3×3 molecular grid. The grid was joined to form a 7_4 knot topology that corresponded to that of an endless knot (a recognized macroscale motif in Celtic and Chinese traditions). A related project prepared synthetic knots from a molecular strand by using transition metal and lanthanide ions to guide chain folding in a manner reminiscent of the action of protein chaperones (Leigh et al., 2021). Recent advances have enabled the creation and modification of increasingly complex synthetic molecular topologies for the design and fabrication of sophisticated nanomaterials (Ashbridge et al., 2022).

3.2.3.1.2 *Molecular Weaving*

Fabrics (materials consisting of layers of woven fibers) are essential materials for everyday life. In a significant advance, researchers have booted up their chemical looms to create a completely organic nanoweave for the first time (Lewandowska et al., 2017). The macroscale principle of threads woven at 90° angles was applied to molecular systems to produce stable nanomaterials (woven nanofabrics comprised weft threads and warp threads aligned at right angles to each other). The resulting nanoweave was composed of three strands made purely from organic molecules using a helical-shaped peptide (an oligoproline helix), related to collagen. Two perylene-monoimides were attached to each end of the helical-shaped peptide (disc-shaped molecules that adhere to each other through so-called p-stacking). These components arranged into long threads and formed a kagome (triangle-hexagon) weave when left to cool in solution.

The trend in woven nanofabrics has continued through work that demonstrated the self-assembly of layered two-dimensional molecularly woven fabrics via the direct, bottom-up assembly of molecular building blocks into linear organic polymer chains woven in two dimensions (August et al., 2020; Ding et al., 2020). Connections between slowly precipitating pre-woven grids, followed by the removal of an ion template, resulted in a wholly organic molecular material that formed as stacks and clusters of thin sheets (each sheet was up to hundreds of µm long and wide but only about four nm thick) in which warp and weft single-chain polymer strands remained associated through periodic mechanical entanglements within each sheet.

A team of scientists from the University of Manchester was honored with a Guinness World Record for weaving threads of individual molecules together to create the world's finest fabric, overtaking fine Egyptian linen. The team developed a strategy to weave molecular threads in two-dimensional layers. In doing so they produced a 2D molecularly woven fabric with a thread count of 40–60 million (in contrast, the finest Egyptian linen has a thread count of ~1,500) (thread count is the number of strands per inch) (University of Manchester, 2021). More recent work in molecular weaving considers polymer networks and molecular nanotopology. The signature of the periodic nanoscale weaving of polymer chains is a foundational structure that is distinct from other structures, such as randomly entangled polymer networks and rigid crystalline frameworks (Zhang et al., 2022).

3.2.3.2 Biological Liquid Crystals and DNA Chirality Inversion

Liquid crystals inhabit a state of matter that resides between a liquid and a crystal, which flows like a liquid, albeit contains molecules that are arranged as a crystal (positioned in ordered lattices and directionally oriented). Materials can undergo liquid-crystal phase transition based on temperature, solvent concentration, and inorganic-organic composition. Liquid crystal phases are abundant in biology (Jia et al., 2022; Safinya et al., 2013) in membranes (phospholipid bilayers), DNA, RNA, and proteins (spider silk is a liquid crystal phase, as a concentrated protein solution with molecules in specific arrangements).

Biological liquid crystals (short strands of liquid-crystal DNA in solution) have been employed to investigate the transfer of structural information from the nanoscale to the macroscale, as a strategy for the development of dynamic and adaptive materials (Wang et al., 2020; Zhao et al., 2019). Research has demonstrated that the knotting and unknotting of molecular strands may be utilized to control and invert the handedness of helical organization within liquid crystals (Katsonis et al., 2020). A liquid crystal DNA strand folds into an overhand knot of single-handedness upon the introduction of a chiral dopant (lanthanide ions), which results in opposite handedness (Ru et al., 2022). When the dopant molecules are removed, the molecules dynamically switch back to the original chirality. The result is a lower-cost method for shifting handedness (chirality) that avoids costly covalent bond breaking (Winogradoff et al., 2021).

3.2.3.2.1 Virtual Self-Assembly Liquid

A self-replication and self-assembly model for MM has been proposed in the form of a virtual liquid platform (Ewaschuk & Turney, 2006), which consists of a computational simulation of nanoscale machines floating in a virtual liquid. The molecular machines bond together to form strands (chains) that self-assemble and self-replicate into user-specified meshes. Based on von Neumann's outline for self-replicating machines (von Neumann, 1966), the JohnnyVon 2.0 simulation includes four types of machines drifting in a two-dimensional continuous space with Brownian motion and viscosity (a simulated liquid) (Ewaschuk & Turney, 2005). The machines exert attractive and repulsive forces on each other, with internal states governing bond formation or rupture. As the bonding forces change, the bonding angles alter and the strand folds, in an analogue to protein-folding. The strands assemble into polygonal meshes, a large variety of which can be manufactured (in the simulation) by altering the sequence of input machines (Jurková & Zámečník, 2023).

3.3 HARDWARE: QUANTUM NANOSCIENCE TECHNOLOGIES

> "The past few decades have seen a revolution in the ability to structure matter at the nanoscale."
>
> ***Roadmap on Quantum Nanotechnologies*** **(Laucht et al., 2021)**

3.3.1 Quantum Dots (Semiconductor Nanoparticles)

The first standardized building blocks for nanotechnology that came into use in the 1980s were quantum dots, which are semiconductor nanoparticles a few nm in size. These entities possess quantum mechanical properties (photonic and electronic) that are not available in larger particles (Ekimov & Onushchenko, 1982; Gowtham et al., 2023; Nabil & Megahed, 2023). Quantum dots have intermediate properties that reside between those of bulk semiconductors and discrete atoms or molecules. These nanoscale semiconductors have the capacity to confine electrons and electron holes, thereby exciting electrons to high-energy states for controlled transition between two semiconducting bands (valence and conductance). Some of the use cases for these nanometric entities include imaging, microscopy, communications, quantum computation, and biology (Bao et al., 2023; Barroso, 2011; Chakraborty et al., 2022;

Gil et al., 2021; Pleskova et al., 2018). One deployment of quantum dots in biology is as alternative imaging agents, as they are brighter and more stable than the traditional fluorescent dyes used in organic staining (e.g., to image real-time single-cell migration in vitro) (Xu et al., 2021). However, deployment may be restricted depending on constituent materials such as cadmium, as toxicity may be implicated for in vivo use (Le et al., 2022).

3.3.2 Spin Engineering (Magnetic Moments)

Spin engineering involves the control and manipulation of quantum spin systems via spintronics and spin chains. Nanoparticles have various physical attributes that may be exploited to encode them as qubits (quantum data) and control them in the nanoscale manufacturing environment. These properties include spin, charge, angular momentum, and polarization (Flamini et al., 2019). Electronic charge and magnetic moment (spin) are the properties most frequently manipulated in computational devices, where spin describes the intrinsic angular momentum of a particle. Intrinsic angular momentum (spin), magnitude (speed), and direction (projection) of orbital angular momentum also exist. Bosons (force particles) have integer-based spin values (spin-1), while fermions (matter particles) have half-odd integer spin values (spin-1/2, 3/2, etc.). Nanoparticles with spin possess a magnetic dipole moment that can be manipulated in electronic devices, with directional pointing designated as spin-up or spin-down. Although spin may be imagined/depicted as an axis of rotation, the actual nanoparticles do not rotate.

3.3.2.1 Spintronics

Spintronics (spin electronics) is the study of the intrinsic spins of electrons correlated with their magnetic moments, in addition to their fundamental electronic charges (the typical focus in electronics) (Han et al., 2023; Hu & Xiang, 2020; Kim et al., 2022). Thus, spintronics differs from traditional electronics, as beyond the charge state, electron spins (magnetic moments) are exploited as a further degree of freedom. This has implications for the efficiency of data storage and transfer in quantum devices (spintronics encompasses "two-for-one" properties, both electronic and magnetic).

3.3.2.2 Spin Chains

Spin chains are collections of linearly arranged magnetic moments that interact via spin-spin coupling. Creating spin chains with spin coupling in engineered atomic structures is further foundational building block in designed quantum systems (Gopalakrishnan & Vasseur, 2023; Hess et al., 2017). Scanning tunneling microscopes are utilized to generate and probe the spin interactions of distinct atomic-scale magnetic structures. In an early demonstration, linear chains of up to 10 manganese atoms were assembled one atom at a time on a thin insulating layer (Hirjibehedin et al., 2006). The spin excitation spectra of the structures were measured using electron tunneling spectroscopy. The observed excitations of the coupled atomic spins had the capacity to alter both the total spin and its orientation. Contemporary spin chain engineering projects, using optics and quantum machine learning, have been able to achieve a simulation of 135 spins (Cong et al., 2019).

3.3.3 Electron Engineering

Electron engineering is the characterization and control of the electron-resolution behaviors of atoms and molecules, which has been enabled by advances in microscopy. Whereas atomic motion is on the scale of femtoseconds (10^{-15} seconds), electron activities are on the scale of attoseconds (10^{-18} seconds). Recent progress in ultrafast (attosecond) microscopy enables of atoms and molecules (Chen & Tao, 2009; Sessa & Rahm, 2022).

Monitoring the effects of local excitation in real-time is an important priority. One experiment combined tomographic photoemission imaging with a femtosecond pump-probe method to trace the excited state molecular orbitals of surface-adsorbed molecules with both spatial and temporal resolution. This, to investigate the ultrafast (attosecond scale 10^{-18}) electron transfer dynamics of such processes as chemical reactions on surfaces and intermolecular charge transfers (Wallauer et al., 2021). Another investigation used STM to trace the coherent evolution of a pair of coupled atomic spins, which also employed a pump-probe scheme to detect localized magnetization following current-induced excitation performed on one of the spins (electron spin resonance performed locally at the probe tip) (Veldman et al., 2021).

Likewise, the elucidation and manipulation of electron orbitals is an important research topic.

Research in this area aims to independently access both the spin and orbital degrees of freedom of single atoms, to incite and probe the excitations of each moment (Rejali et al., 2020). Another team optically "weighed" the contributions of different imaged electron orbitals in strong field tunneling ionization experiments (Tan et al., 2021). Using digital methods, a third experiment simulated the electron dynamics of molecular systems by parametrizing the wavefunction as a matrix product state. This enabled the dynamics simulation of systems that contained up to 20 electrons and 32 orbitals (with a time-dependent Density Matrix Renormalization Group (DMRG) algorithm) (Baiardi, 2021).

3.3.4 Artificial Lattices (Spin and Electron)

Artificial lattices are applied in both spin and electron engineering, where engineered spin chains are utilized as inputs for artificial lattices. These are atomic-scale structures designed to confine electrons or spins (magnetic moments) within a specific lattice (grid) structure (Freeney et al., 2022; Polini et al., 2013), the properties of which (such as spacing) may be precisely controlled. Specifically, artificial electronic lattices are used to study the effects of coupling, topology, strain, periodicity, and dimensionality, while artificial spin lattices are employed to investigate complex magnetic structures and spin. Artificial lattices (both electron and spin) can be made by engineering either localized (bound) or delocalized (free) interactions (Leis et al., 2021). A tight-binding approach uses a nearest-neighbor orbital overlap or vacancy states method, whereas a nearly free electron strategy utilizes confined two-dimensional electron gases as a scattering center. Artificial lattices can be fabricated through atomically precise microscopy. In particular, the advent of STM techniques has made it possible to engineer spin chains via atom-by-atom manipulation, while accessing their spin states at the atomic scale (Choi et al., 2019).

3.3.5 Floquet (Periodic Time) Engineering

Further to spin (magnetic moment) and electron (charge) engineering, the temporal control of quantum systems as they evolve is another important aspect of quantum nanoscience. Quantum time evolution is primarily orchestrated through Floquet engineering, which is the management of periodically driven behavior in quantum systems (Hübener et al., 2023; Ji et al., 2022; Kennes et al., 2018). Floquet methods are known in condensed matter physics and are gaining new traction in quantum nanoscience. A central formulation in quantum mechanics is the Schrödinger wave equation, which quantifies the positions or speeds (momenta) of complete system configurations. However, it is generally intractable as it is comprised of complex-valued probability amplitudes (with real and imaginary wave-shaped components). This situation may be remedied in certain cases, as the Bloch theorem is a solution to the Schrödinger equation for periodic systems, while the Floquet theorem transforms periodic systems (Bloch theorem) to solvable linear differential equations. Thus, Floquet methods are emerging as a standard first-line strategy for the treatment of time cycle behaviors in quantum systems.

3.3.5.1 Discrete Time Crystals

Crystals comprised of repeating structures are important architectures found in the natural world. Although crystals have been traditionally understood as repeating spatial motifs, time crystals have also been discovered with repeating temporal motifs (Scheie et al., 2022). A notable structure is the discrete time crystal, which is a novel form of matter that does not attain thermal equilibrium (Else et al., 2020; Peng, 2022; Zhang et al., 2017). Floquet methods are used to characterize discrete time crystals and implementing the discrete time crystal in quantum computation circuits is somewhat of a "Hello, World" task (i.e., a standard demonstration). Notably, time entanglement has a different (more multiplicity-oriented) structure than space entanglement (Ringbauer et al., 2018), which has implications for quantum cryptography based on information hidden in spacetime (Hayden & May, 2019; Neumann et al., 2022; Yin et al., 2020).

3.3.5.2 Floquet Circuits: Discrete Time Crystals NISQ and Classical Hardware

Floquet circuits (the simulation of discrete time crystals) have been implemented using both classical and quantum platforms (currently available NISQ (noisy intermediate-scale quantum) devices that do not require error correction). At the quantum scale, a Sycamore 53-qubit quantum computer (a solid-state, Josephson junction-based platform) was programmed to realize the discrete time crystal phase and experimentally detect its dynamical properties (Ippoliti et al., 2021). On the classical side, a TPU architecture (Tensor Processing Unit, linear algebra acceleration ASICs) applied a two-layer Floquet unitary circuit to wavefunctions (simulating the dynamics of 34 qubits for over 105 Floquet periods with 128 TPU cores) (Morningstar et al., 2021). These results established digital computational methods for the realization of a discrete time crystal as a standard quantum component.

3.3.5.3 Floquet Tri-unitary Time Circuits

More theoretically oriented research introduced quantum circuits that enabled greater control over temporal parameters. The Floquet tri-unitary time circuits are tri-unitary in time in the sense of being unitary (evolving in directed lock-step units) along three distinct "arrows of time" (two light cone edges and the worldline) (Jonay et al., 2021). When computing the dynamics of a quantum many-body system, the identification of spatiotemporal correlations between local operators is critical for the physical characterization of the system, albeit they are difficult to obtain. The ability to model a quantum matter system by constraining its structure allows spatiotemporal correlations to be computed more exactly, namely with random circuit models that allow dynamics to be constrained to those that are unitary in the space and time directions. The result is that system correlations are pinned to three potential directions in spacetime rather than just two, as in dual-unitary circuits (Claeys & Lamacraft, 2021). With tri-unitary circuits, correlation functions are obtained exactly in terms of quantum channels built from the individual gates that make up the circuit, which enables entanglement management and other aspects of quantum matter system control. Recently, ternary unitary quantum circuits have been described, with the authors stating that, "When used as building blocks of lattice models with periodic boundary conditions in time and space (corresponding to infinite temperature states), dynamical correlation functions exhibit a light ray structure" (Milbradt et al., 2023).

3.3.6 Quantum Matter (Topological Materials)

Quantum matter (topological materials) is a term that refers to novel phases of matter that arise at a temperature of absolute zero (0 kelvin) with emergent orders and exotic properties. Prior to quantum matter, symmetry-based formulations (namely, Landau symmetry-breaking) were sufficient for the characterization of materials. However, quantum matter requires topological and possibly also geometry-based descriptions; hence, the name topological materials, or topological quantum matter (Thedford et al., 2023).

In the fast-paced development of quantum matter, topology, and symmetry (including symmetry-breaking), and entanglement are the primary tools for describing, creating, and manipulating quantum matter phases. Other attributes of quantum matter systems may include the emergence of quasiparticles (or collective excitations)

(Wölfle, 2018) with anyonic exchange statistics (identical particles that swap places by altering their wavefunctions in a process called "braiding"), quantum phase transitions, and gauge theory. Known examples of quantum matter phases include quantum spin liquids, fractional quantum Hall states, topological insulators, and strongly correlated quantum liquid states (and condensates) (Ko et al., 2023).

Quantum matter enables the treatment of quantum many-body systems (systems with three or more elements), whose collective behavior may give rise to quasiparticles (fermions) and collective excitations (bosons). One collective behavior of interest is many-body localization (MBL), the situation in which collective interactions cause quantum particles to be localized and maintained in an in- and out-of-equilibrium state (De Roeck & Imbrie, 2017).

Fermions and bosons are the two elementary types of particles. Fermions are matter particles (electrons, protons, neutrons, quarks) with half-odd integer (1/2, 3/2, etc.) spins that do not like to clump together and keep themselves spread out (e.g., electron shells of atoms) (per the Pauli exclusion principle) (Yazdani et al., 2023). Bosons are force particles (photons, gluons, gravitons) with integer spins that prefer to coalesce (e.g., to form a beam of light) (Ma et al., 2021). While fermions and bosons are foundational particles, anyons are a third type ("any" particle between fermions and bosons) that arise through the behaviors of other particles and exhibit statistical signatures between those of bosonic bunching and fermionic exclusion (Feldman & Halperin, 2021).

3.3.6.1 Quantum Spin Liquids and Topological Insulators

In conjunction with entanglement, topology and symmetry are valuable measurement and control parameters for quantum matter phases. Entanglement is becoming increasingly important as a means to describe interactions and correlations in topologically ordered many-body systems across distances (Cirac et al., 2021). Quantum matter activities may be distinguished as long-range entanglement and short-range protection. Matter phases with long-range entanglement include quantum Hall states and quantum spin liquids (allowing quantum error correction through non-locality). Short-range protection consists of symmetry-protected topological phases and topological insulators (allowing for the controlled symmetry-breaking of local topological phases) (Table 3.3) (Lodge et al., 2021; Loidl et al., 2021). The key benefit of long-range entanglement is being able to perform quantum error correction (through non-local measurements) (Sivak et al., 2023), particularly via the "fractional quantum Hall effect," for which the 1998 Nobel prize was received (Ball, 2019).

TABLE 3.3
Quantum Matter (Topological Matter Phases, Quantum Materials with Topological Properties)

Long-Range Entanglement	Short-Range Protection
Quantum Hall states	Symmetry-protected topological order
Quantum spin liquids	Topological insulators

3.3.6.2 Long-Range Entanglement

3.3.6.2.1 Quantum Hall States

Quantum Hall states are a form of topological order that cannot be described by the conventional methods of Landau symmetry breaking. The Hall effect (discovered by Edwin Hall in 1879) occurs in a sheet of metal in a perpendicular magnetic field (Hall, 1879). An electric voltage perpendicular to the charge current is induced by the magnetic field, due to the Lorentz force acting on electrons (Liu et al., 2016). In a two-dimensional electron gas under low temperature and a strong magnetic field, the Hall conductance becomes quantized as the integer quantum Hall effect. At lower temperatures, the fractional quantum Hall effect occurs, as strongly correlated electron states with fractionally quantized Hall conductivity. The fractional quantum Hall effect can be obtained by applying a strong magnetic field perpendicular to a two-dimensional electron system at low temperature (Sharma et al., 2023).

3.3.6.2.2 Quantum Spin Liquids

Quantum Hall states and also quantum spin liquids are research priorities in long-range entangled quantum matter systems. A quantum spin liquid is a magnetic system that does not settle into a long-range ordered configuration, even at zero temperature (Broholm et al., 2020; Chamorro et al., 2021). There are three different strategies for characterizing classical magnetic systems.

1. Structure in which all of the spins point in the same direction (e.g., standard refrigerator ferromagnets).
2. Structure in which the spins point in opposite directions; one spin is disordered as relates to those of its neighbors (e.g., antiferromagnet).
3. Combination of both ordered and disordered spins (e.g., "frustrated order" of a spin liquid or a spin glass), which can be manipulated as they do not settle into a fixed configuration.

As the name suggests, quantum spin liquids are the quantum version of a spin liquid. When ordered magnets (ferromagnets) are cooled to low temperatures, the electrons stabilize and form large-scale patterns (e.g., checkerboards or stripes) that have magnetic properties. However, in a quantum spin liquid, the electrons do not stabilize when cooled and preserve their disorder similar to the way that liquid water exists in a disordered state. The electrons in a quantum spin liquid continue to fluctuate in a highly entangled quantum state and do not settle into a regular lattice as they would in a solid. This property makes quantum spin liquids conducive for use in electronics as they produce a magnetic system that does not stabilize in a large-scale ordered configuration, but rather persists in a somewhat disordered ground state, which can be operationally manipulated.

Quantum spin liquids were discovered in nature in the form of a mineral called herbertsmithite in Chile (1972), and subsequently in other locations (Iran, Chile, Arizona, and Greece). The magnetic particles of herbertsmithite indicate constantly fluctuating scattered orientations in a kagome (triangle-hexagon) lattice. Named after the mineralogist Herbert Smith, the substance is comprised of zinc, copper, oxygen, hydrogen, and carbon. Quantum spin liquids have also been demonstrated in the laboratory (Huang et al., 2021; Pozo et al., 2020; Shaginyan et al., 2019). In 2021, two experiments undertook to create quantum spin liquids from scratch to specifically demonstrate long-range entanglement properties in quantum computational platforms, namely superconducting circuits (Jiang, 2021) and optical arrays (Semeghini et al., 2021).

A team from Google Quantum AI used a 32-qubit quantum processor to study the ground state and excitations of the toric code (Satzinger et al., 2021). A team from the Lukin laboratory at Harvard detected signatures of a toric code-type quantum spin liquid in a two-dimensional array of Rydberg atoms held in optical tweezers (lasers) (Semeghini et al., 2021). The central achievement for both projects was engineering the topological order known as the toric code. This is an archetypical two-dimensional lattice model that exhibits the exotic properties of topologically ordered states, which have been proposed for quantum error correction (Kitaev, 2003; Kubica & Vasmer, 2022).

3.3.6.2.3 Topological Entanglement Entropy

Topological entanglement entropy is a topology-based measure of entanglement entropy that is specific to phase transitions in quantum systems. An issue that arises is that the phases on the two sides of a quantum critical point may have different types of topological order. Since it may not be possible to distinguish the phases based on a localized order parameter, a non-local parameter is required such as a topology-based measure of long-range system entanglement. Topological entanglement entropy is a formulation beyond conventional quantum entropy measures (e.g., von Neumann entropy and Rényi entropy) that allows entanglement entropy to be measured in quantum many-body states with topological order (Kitaev & Preskill, 2006; Sankar et al., 2023). (Specifically, topological entanglement entropy is computed as the logarithm of the total quantum dimension of the quasiparticle excitations of the state, or by comparing the von Neumann entropy between a spatial block and the rest of the system.) Topological entanglement entropy is further instantiated in the form of a topological quantum field theory with tripartite information (equations involving two space and one-time dimensions) (Hosur et al., 2016).

3.3.6.3 Short-Range Protected Topological Order

3.3.6.3.1 Symmetry-Protected Topological Insulators

Both symmetry and topology are important characterization and control parameters for quantum matter. Designed quantum matter systems engage short-range properties through symmetry-protected topological phases such as topological insulators. Symmetry-protected topological order refers to quantum matter phases that have, on one hand, trivial topological order (entanglement), but on the other hand, symmetry protection (symmetry-based properties that can be manipulated). Since short-range states have only trivial topological order (entanglement properties), they are referred to as symmetry-protected order. Symmetry-protected phases are often realized through topological insulators; materials that have a conducting surface and an insulating interior, whose surface states are symmetry-protected by time-reversal, particle-hole, or chiral symmetry. In particular, statistical topological insulators are employed to characterize matter phases that remain stable against disorder, while preserving average symmetry (Poyhonen et al., 2021). With symmetry-protected states, topological insulators may also be employed as controlled probes to test long-range entangled systems such as quantum spin Hall states. Digital methods are proving indispensable, as algorithms (Qsymm; the quantum symmetry algorithm) are deployed to find all the symmetries of a given family of Hamiltonians and predict where symmetry protection will break down (Varjas et al., 2018).

3.3.7 Molecular Coding

3.3.7.1 Quantum Error Correction

Quantum systems are typically more sensitive to environmental noise than classical systems, and particularly in computation, which introduces additional errors that need to be corrected. Quantum systems are thus designed with both the physical elements (e.g., atoms, molecules, spins) that conduct the computational processes in question, and a much larger set of elements (e.g., entangled ancilla of ancillary bits) to correct the errors. Classical error correction methods, such as making redundant copies or checking information integrity prior to transmission, are not possible in quantum systems, as information cannot be copied or inspected (due to the no-cloning and no-measurement principles of quantum mechanics). Consequently, the primary mode of quantum error correction thus far incorporates entanglement rather than redundancy (Pereira & Mancini, 2022). The quantum state to be protected is entangled with a much larger group of states (the ancilla) from which it can be indirectly corrected (Swan et al., 2020).

The standard errors encountered in quantum systems are bit flips, sign flips (sign of the phase), or both. Basic codes are applied to diagnose errors, which correspond to Pauli matrices for controlling the qubit (or spin) in the X, Y, and Z dimensions (Khalifa et al., 2021). The error is expressed as a superposition of basis operations given by the Pauli matrices. If there is an error, the same Pauli operator is applied to act again on the corrupt qubit to reverse the error effect. The unitary correction returns the state to its initial condition without measuring the qubit directly. The physical state of the qubit (e.g., spin, atom, photon, molecule) is controlled by using its dynamic properties (e.g., rotational state) to realign it to its initial configuration (Swan et al., 2022a). Various quantum error correction codes are listed in Table 3.4.

TABLE 3.4
Quantum Error Correction (QEC) Codes for Quantum Systems

Class	Code	Description
1	Stabilizer codes	Topology-based Pauli operators (X, Y, Z) correct a bit-flip or a spin flip.
1a	Toric code	Stabilizer operators defined on a 2D torus-shaped spin lattice.
1b	Surface code	Stabilizer operators defined on a 2D spin lattice in any shape.
2	Bosonic codes	Self-contained photon-based oscillator system with bosonic modes.
2a	GKP code	Squeezed states protect position and amplitude shifts with rotations.
2b	Molecular code	Rotations performed on any asymmetric body (molecule) in free space.

3.3.7.2 Stabilizer Codes

Stabilizer codes (quantum analogues of the linear codes applied in classical systems) are the most basic forms of quantum error-correcting codes. The stabilizer code is a technique based on topology that treats particle movements and their corrections through the structures of lattices (multi-dimensional grids) (Pereira et al., 2023). Two of the main implementations of stabilizer codes are toric and surface codes. The canonical approach is toric codes, which are stabilizer codes instantiated on a two-dimensional lattice with periodic boundary conditions (which suggest the shape of a torus). There are stabilizer operators on the spins around each vertex and plaquette (face). The other method is surface codes, which are a more generic formulation of stabilizer codes. Surface codes are also defined on two-dimensional spin lattices but can take various shapes that are not necessarily toroidal. Quantum systems may be engineered to switch between error-coding mechanisms on the fly, using techniques called "lattice surgery" (Erhard et al., 2021).

3.3.7.3 Bosonic Codes

Bosonic codes extend stabilized codes as the conventional approach to quantum error correction (Del Pino & Zilberberg, 2023). A typical strategy involves the use of multiple qubits in two-tier registers; however, this format is not conducive to system scaling. Thus, an improved technique has been proposed that integrates encoding qubits into a self-contained system such as bosonic modes (photon states), within the structure of a harmonic oscillator (Royer et al., 2020). An even more generalized form of bosonic codes is provided by GKP (Gottesman, Kitaev, Preskill) codes, which perform error-corrected qubit encoding in an oscillator using superpositions of squeezed states (quantum noise-reduced oscillatory states) (Noh et al., 2020). The application of GKP codes allows for changes in molecular position and amplitude to be controlled, and is applied on the basis of grid states in an oscillator (Gottesman et al., 2001). The standardized term for superpositioned qubits (used in error correction and other applications) is (Schrödinger) cat codes or cat states (Li et al., 2017; Mirrahimi, 2016).

3.3.7.4 Molecular Codes

GKP codes represent a significant advance toward the correction of errors (molecular displacement) by manipulating the position and momentum of an oscillator with known symmetric rotations. A subsequent advance is molecular codes, which extend the functionality of GKP codes by allowing not only known symmetry rotations, but any type of rotation to be performed on asymmetric rigid bodies in free space. Molecular codes apply to a broader range of quantum systems that extend not only from oscillators, but to molecules themselves such as diatomic and polyatomic molecules (Albert et al., 2020). The classes of molecular control methods afforded by molecular codes more firmly link quantum information science and digital computational methods with MM.

3.4 SOFTWARE

A theme already evident in the hardware discussion of MM is the increasing use of digital strategies to orchestrate processes. This theme continues to be prevalent in the software domain, as advanced methods allow progress in the realization of MM

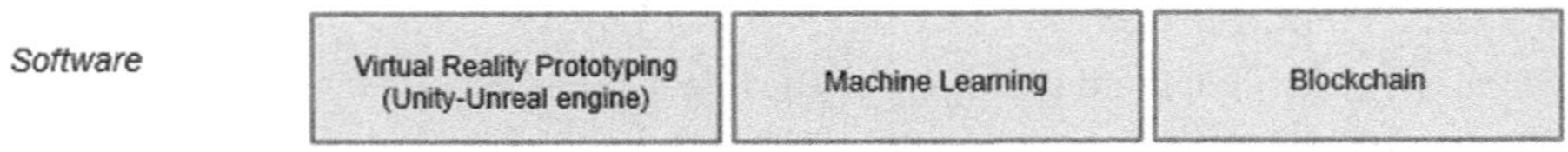

FIGURE 3.4 Molecular manufacturing software platforms.

activities. Three areas of software applications are prominent in the MM program (virtual reality prototyping with game engine platforms, machine learning algorithms, and blockchain economic automation) (Figure 3.4).

3.4.1 Virtual Reality Prototyping

3.4.1.1 Digital Twins and Programmable Smart Matter

A related cluster of terms is used to designate the idea of running parallel physical and virtual entities, including those that increasingly communicate to update each other. These terms encompass digital twins, mirror worlds, and digital materialization to generally refer to the virtual-physical existence of objects and smart programmable matter (4D printing), to mean the manipulation of physical-world objects (once instantiated) by virtual or other means.

Central to virtual prototyping is the notion of digital twins; virtual representations serving as real-time digital counterparts to physical objects or processes, which might potentially be in constant communication and updating via Internet of Things sensors (Grieves, 2019; Mahmood et al., 2023). Application domains are extensive and include manufacturing, space, energy, and healthcare. In manufacturing, digital twins are the result of constant improvements in the creation of product designs and engineering activities. Product drawings and engineering specifications have progressed from manual drafting to computer-aided drafting and computer-aided design to 3D model-based systems engineering. The physical manufacturing objects are virtualized and represented as digital twin models that are closely integrated between the physical and virtual spaces. Digital materialization more explicitly calls out a two-way communication capability between the virtual and physical objects, which is envisioned to be necessary for universal desktop fabrication (Vilbrandt et al., 2008).

3.4.1.1.1 Smart Programmable Matter (4D Printing)

"Hydrogen … given enough time, turns into people."

British cosmologist Ted Harrison (2015)

Smart programmable matter (4D printing) refers to the notion of being able to modify an object subsequent to its generation. This type of programmable matter has the capacity to programmably alter its physical attributes (e.g., shape, density, moduli, conductivity, optical properties) in response to user inputs or autonomous sensing (Goldstein et al., 2005). The concept of information processing is thus tethered to both the digital representation and physical matter (Damioli et al., 2022; Toffoli & Margolus, 1991).

3.4.1.1.2 *Levels of Smart Matter*

The development of variously tiered smart matter models has been proposed (Levine & Goldman, 2023; Pishvar & Harne, 2020). One smart matter "IQ" system consists of three levels of activity based on the responsiveness and agility of materials in the deployment of smart structures fabricated as an integration of sensors, actuators, and control systems (Cao et al., 1999). Another hierarchy proposes four levels of material qualities that are structural, responsive, adaptive, and intelligent (meaning that they can store information and learn) (Kaspar et al., 2021; Shao et al., 2022). Intelligent matter is defined by four functional elements and their interplay for data processing, sensors (data receptors), actuators (properties that can change the material), memory (that can store data), and network connectivity (through wireless protocols such as Bluetooth). Intelligent matter structures could be activated via neuromorphic processing models similar to the brain's own spiking.

Tegmark (2017) segregates life into three phases (biology, culture, and technology) based on their computational sophistication. Life 1.0 (biology) evolves its hardware and its software, Life 2.0 (culture) evolves its hardware and designs much of its software, while Life 3.0 (technology) designs both its hardware and software. One of the most startling developments in the 13.8 billion years of the universe's history is that dumb and lifeless matter has become intelligent. In principle, Tegmark argues that any matter can be imbued with the necessary information processing features to enable it to remember, compute, and learn, namely commencing with a NAND gate. NAND gates and neurons are two important examples of universal "computational atoms" (Mougkogiannis & Adamatzky, 2023). The implication is that any chunk of matter may serve as a substrate for memory and computation (computronium) with the proviso that it contains many different stable states as building blocks, which can be combined to implement computational functions. Further, AI is a natural substrate for the development of smart matter platforms due to its ability to learn.

3.4.1.2 Game Engine Platforms

Virtual prototyping is often driven by gaming engines such as Unity, Unreal, and Outerra. The two main platforms, Unity and Unreal, are 2D- or 3D-based game engine software platforms used for the rapid coding of video games, digital assets, and virtual prototypes. The standard features of these game engines integrate animation, sound, scripting, artificial intelligence, collision detection, and memory management in virtual environments. Virtual asset development is streamlined through existing templates that can roll out to a variety of desktop, mobile, console, and virtual reality platforms. For beginners, Unity may be more appropriate as it has a simpler user interface, many tutorials and examples, and available asset templates (Christopoulou & Xinogalos, 2017). High-performance hardware is not required, but C# programming is (as opposed to C++). For experienced users, the Unreal Engine may be preferable as it supports Visual Scripting (El-Wajeh et al., 2022). This program is more difficult to use but offers a more complex graphical environment and higher graphics quality, which requires high-performance hardware. The third program, Outerra, is a 3D planetary graphics engine for space-based prototyping such as spacecraft, asteroid, and planet modeling.

3.4.1.2.1 The Asset Store

3D game engines are a general form of software infrastructure and a virtual prototyping tool that includes both video games and opportunities in automotive, aerospace, construction, and medical industries (Corke, 2019). Virtual prototyping users in aerospace include CERN, JPL, and CalTech, and in the automotive industry, Volkswagen, Audi, BMW, Volvo, Lexus, Lincoln, and Toyota. These gaming engines can be used to create high-quality visualizations of complex models in real-time. Designers can make changes in computer-aided design models, transfer the model into the gaming engine, render it instantly with realistic materials, and then explore it on an interactive and immersive basis on many different devices. Prototype designers can register, share, and merchandize their created designs and objects through the Asset Store (the analogue to the App Store for smartphone applications). Software plug-ins may offer the ability to mint blockchain-based non-fungible tokens (NFTs) when registering assets to the Asset Store. NFTs are unique identifiers and intellectual property trackers for the assets described in more detail in the blockchain section.

3.4.1.2.2 Gaming Engine Headsets: Cognitive Neuroscience Consumer BCIs

Consumer-grade virtual reality headsets (Oculus Rift, Valve Index, HTC Vive) are a form of consumer brain-computer interface (BCI) technology that may make cognitive neuroscience studies more feasible to conduct (Yang et al., 2023). A research project investigated the precision and accuracy of the timing mechanisms of the Unreal Engine 4 and the SteamVR platform in combination with the HTC Vive VR system and found that the timing of stimulus and reaction response was appropriate to the rigors of cognitive neuroscience research (Wiesing et al., 2020). The work concluded that the experimental setup might be used in subsequent patient-based studies as virtual reality is increasingly seen as a research tool in behavioral neuroscience and intervention, for example, in investigating the attentional capacities of persons with ADHD in a realistic, well-controlled environment (Paschall et al., 2022). As applied to MM, virtual reality headsets may enable researchers to monitor and fine-tune simulated molecular/atomic scale fabrication processes toward the development and optimization of real-world mature MM systems (Müller et al., 2018).

3.4.2 Machine Learning

3.4.2.1 Classical Machine Learning

Machine learning is a routinely applied artificial intelligence technique through which a computer algorithm writes an equation (function) to describe inputted data that it iteratively improves, toward the aim of correctly identifying new data (Choi et al., 2020). A related method, deep learning, designates a network architecture with multiple layers (3–8 or more layers "deep"), which likewise consists of a statistical inference method based on pattern identification (LeCun et al., 2015). Quantum machine learning (machine learning applied in a quantum environment) is emerging as one of the first potential general-purpose applications of quantum computation

(Biamonte et al., 2017; Zeguendry et al., 2023). The main architectural approaches in machine learning are neural networks, tensor networks, and kernel learning, each of which has quantum counterparts.

3.4.2.2 Quantum Machine Learning

Classical machine learning is based on probabilistic methods, with quantum machine learning even more so, as it entails being expanded to include quantum probabilistic methods (Swan et al., 2022c). A key advance in this area is the Born machine as the quantum version of the Boltzmann machine (Cheng et al., 2018). Both "machines" are automated energy functions that evaluate probability outputs from machine learning algorithms. Classically, the Boltzmann machine employs an energy-minimizing probability function for sampling (per the Boltzmann distribution in statistical mechanics). The Born machine interprets results using the Born rule (a computable quantum mechanical formulation that evaluates the probability density of finding a particle at a given point as being proportional to the square of the magnitude of the particle's wavefunction at that point) (Swan et al., 2022a; Zurek, 2018).

3.4.2.3 Quantum Machine Learning and Quantum Nanoscience

The complexities of quantum nanoscience suggest the field as an obvious environment for the implementation of machine learning technologies. One study employed artificial neural networks to simulate strongly correlated systems as a means of overcoming the limitations inherent in conventional classical and quantum Monte Carlo approaches (Carrasquilla, 2020). A probabilistic model was used to simulate quantum many-body dynamics (factorized general measurements map quantum states to probability distributions, such that an exact formulation of quantum dynamics can be obtained) with a transformer neural network (contemporary natural language processing model) (Carrasquilla et al., 2021). Another team considered a variational technique (trial-and-error wavefunction guessing based on complex networks of classical spins that were similar to neural networks) to compute the dynamics of quantum many-body systems (Verdel et al., 2021). This technique studied the dynamics of two quantum spin systems: a paradigmatic one-dimensional transverse field Ising model, and an interacting two-dimensional lattice gauge theory (quantum link model). Another study considered quantum matter itself as a computational technology to develop a quantum cellular automata model with symmetry-protected topological order phases (Stephen et al., 2019).

3.4.3 Blockchain

3.4.3.1 Blockchain Definition and Status

A blockchain (distributed ledger technology (DLT)) is a distributed database of asset ownership. Transactions are recorded between parties in a secure and tamper-evident manner to create a single source of trusted data. Blocks of transactions are linked together with each having a hash (cryptographic summary) of the previous block such that a chain is formed ("blockchain"). The chain is secure since altering the transaction data in any block would require rewriting the entire chain and having all peer nodes hosting the database agree. Nested ultra-secure blockchains are likely to be indispensable for the organization, robust and stable operation, quality control, and sustainability of various aspects of massively distributed MM units on a global scale (Guo et al., 2022; Manzoor et al., 2022; Radmanesh et al., 2021).

As of 2022, the earliest blockchain (bitcoin) has been operating for almost fifteen years (transactions began in January 2009). The software architecture is a secure Internet-based money system with a time-stamping feature that checks available balances in real-time whenever transactions are submitted (thus creating a global provisionless cryptocurrency) (Nakamoto, 2008). The bitcoin blockchain has never been hacked (security breaches, when they occur, are typically experienced at the on- and off-ramps of user wallets and exchange websites) (Kumar et al., 2022). Over 12 thousand global nodes were hosting the bitcoin transaction ledger as of February 2022 (https://getaddr.bitnodes.io/).

3.4.3.1.1 Blockchain Mining Nodes

Blockchain mining nodes perform competitive remunerated accountancy functions to validate and record new transaction blocks via an automated software protocol. Added transaction blocks are subsequently confirmed by all other nodes (Qi et al., 2023). Proof-of-work competition between miners ensures the security of the blockchain ledger, albeit this comes under attack by critics for its otherwise wasteful use of electricity resources (Sharma et al., 2023). However, bitcoin produces an automated global network-based financial system, a secure network that has logged over 915,592 transaction blocks to date (BiTBO, 2025). The energy consumption is less than that of the world's data centers (Statista, 2021) and the cost of staffed brick-and-mortar bank branches. One way to understand the cost of mining is in the context of the ongoing transition to digital smart network technologies, which are currently utilized for sensitive banking functions, asset transfer, and legal arrangements (Swan et al., 2020). In ongoing efforts to improve the security, efficiency, and flexibility of bitcoin transactions, a software protocol upgrade (Taproot) was activated in November 2021 (at blockheight 709,632).

3.4.3.1.2 Cryptoeconomics

Whereas cryptocurrencies are employed for immediate payments, smart contracts (programs stored on a blockchain that execute automatically) are deployed for ongoing financial transactions such as mortgage loans. These latter smart contract applications are for more sophisticated blockchain-based lending, insurance, and investment products, constituting software protocols that are collectively known as decentralized finance (DeFi) (Schar, 2021). The decentralized finance industry may be coming under greater regulatory scrutiny as ICOs (Initial Coin Offerings) did in 2018. As with any new "gold rush," the vast majority of early offerings may be flagrant scams and frauds before the industry matures into a regulated business (nearly half of ICOs issued in 2017 failed within four months (Hankin, 2018)).

However, the longitudinal message is the change in mindset that signals the ongoing transition to digital methods encapsulated in the idea that "there's a protocol for that" (digital finance solution), much like "there's an app for that" in the smartphone app store. Cryptoeconomics is an economic system delivered by secure network of cryptographic protocols, which leads to the formation of a global smart network-based economic, financial, and legal infrastructure. The phased realization of cryptoeconomics has been envisioned to consist of blockchain 1.0 (cryptocurrencies), 2.0 (smart contracts), and 3.0 (beyond financial market applications in supply chain, energy, healthcare, and space) (Swan, 2015).

3.4.3.1.3 Status of Blockchain Adoption

Market research firm Gartner estimates that blockchains support the global movement and tracking of $2 trillion of goods and services annually (Gartner, 2021). Further, the business value generated by blockchain will grow rapidly, reaching $176 billion by 2025 and $3.1 trillion by 2030 (Kandaswamy et al., 2021). IDC has estimated that organizations will spend ~$19 billion on blockchain solutions in 2024 (Shirer & Soohoo, 2021). The top industries investing in blockchains include banking (30% of activity; for cross-border payments), manufacturing (20%, for lot lineage/provenance chains), professional services, retail, insurance, and healthcare. Even in an uncertain and rapidly changing business climate, digital transformation spending remained a priority during the COVID-19 pandemic.

Deloitte's 2019 Global Blockchain Survey found that 23% of executives surveyed had begun to deploy blockchain projects (Budman et al., 2019). Another survey of 350 executives found industry ecosystem readiness to be the most important factor in blockchain adoption (Lustenberger et al., 2021). Indeed, highly integrated industries such as international freight and logistics may be early adopters of blockchain solutions. Maersk's TradeLens blockchain platform (in partnership with IBM using the Hyperledger Fabric platform) operates at over 20 worldwide ports such as Hong Kong, Singapore, Halifax, Rotterdam, and Bilbao (Musienko, 2021). However, the platform was discontinued in 2022, citing "the need for full global industry collaboration has not been achieved" (Kjærgaard-Winther, 2022).

3.4.3.1.4 Non-Fungible Token (NFT) Intellectual Property Registries

A trend in intellectual property registration can be seen in blockchain-based asset registries (for both physical and digital assets) in the form of non-fungible tokens (NFTs), which are unique ownership certificates that can be traded. There are digital artwork NFT marketplaces for minting cryptoart, such as OpenSea, Rarible, and Foundation, as well as physical asset-specific platforms such as Candy Digital (baseball cards), VeVe (Marvel comics), and Wax (Hot Wheels cars). (One implication of blockchain-based baseball card registries could be a new era of fantasy football leagues, with real-time updating as games are played, and including smart contract sports betting.) The artworld signaled its acceptance of cryptocurrencies and NFTs in 2021 with auction house Christie's initial sale of an NFT for a digital collage by Beeple for over $69 million (Christie's. Beeple: A Visionary Digital Artist at the Forefront of NFTs, 2021). Linking NFTs with gaming engine Asset Stores, the Akane Unity plug-in allows developers to create NFTs for Unity asset NFTs on the Matic and Ethereum blockchains (Katje, 2021). A further future scenario could see NFT minting as a standard feature in prototyping engine IP-generation ("print to NFT"). NFTs enable digital twins (digital representations of physical assets) to be controlled with blockchain-based smart contracts (Adams & Tomko, 2018). In terms of relevance for MM (beyond its domestic utility for the synthesis of all manner of consumer goods, including nutritious gourmet foods, etc.), it is the establishment of these types of standardized economic models for content creation, merchandizing, licensing, and post-sale follow-up, which could extend into other platforms such as those for MM design.

3.4.3.2 Blockchain Science

An innovative blockchain application may be to conduct science, particularly to address the reproducibility of results, workflow management, and protected data access (Leible et al., 2019), where MM might enact its corpus of scientific activities from the outset in blockchain-documented formats.

3.4.3.2.1 Smart Contract Molecular Dynamics Simulations

In an example of blockchain science, a computational chemistry simulation was run directly on the Ethereum virtual machine using smart contracts to evolve and record progress (Hanson-Heine & Ashmore, 2020). The experiment involved an atomic molecular dynamics simulation that was performed to assess the vibration of a carbon monoxide molecule over a 40-femtosecond (10^{-15}) timescale, using a harmonic potential to model the carbon-oxygen molecular bond. The simulation was executed for 400 timesteps of 0.1 femtoseconds. Ethereum blockchain compatible software was written in the Solidity programming language to run a diatomic molecular dynamics trajectory with a variation of the Verlet velocity algorithm used to integrate Newtonian equations of motion in atomic units. An equivalent simulation was executed on a local machine for comparison (written in the C# programming language). The smart contract code was published with the study.

3.4.3.2.2 BioBlockchain

In another example of blockchain science, researchers observed that cancerous tumors exhibited the same structure as blockchains; namely, a stepwise sequential evolution. A platform-agnostic model for biological blockchains was proposed in work from the Sloan Kettering Cancer Institute (Chin, 2020). Blockchain-based recordkeeping may be used as an automation technology to track cancer evolution and perform lineage tracing (Ray & Majumder, 2023). In the blockchain-based model of cancer evolution, the aim is to obtain a high-fidelity evolutionary data by defining the ledger as the full history of a particular cancer. Each new cancer tumor is examined in a blockchain model to seek its starting configuration, or "genesis block" in blockchain nomenclature. BioBlockchains have also been proposed as a secure operating software for multi-participant (cloudmind) human brain/cloud interfaces (B/CIs) and quantum brain-computer interfaces (BCIs) (Swan, 2025).

3.4.3.3 Applications of Blockchain in Manufacturing Industries

> "With blockchain technology, the entire supply chain can be managed with one piece of software that is shared among authorized stakeholders."
>
> **Challener (2019)**

3.4.3.3.1 Blockchain Manufacturing, Clinical Trials, and Genomics

In manufacturing, blockchain technology is being used for product lifecycle management, supply chain coordination, and provenance chains (cold storage vaccine tracking and food traceability). The time it takes to transfer ownership of goods can be reduced from 10 d to 10 min. In health and biology, an early-adopter application is clinical trials, which employ blockchains for secure data transfer, data privacy, and patient follow-up (Wong et al., 2019). For whole human genome sequencing projects, non-fungible genome tokens (NGTs) from Nebula Genomics allow individuals to control access to their data (Grishin et al., 2019).

In robot swarms (fleets of robotic entities), blockchain consensus algorithms enable a swarm of robots to achieve agreement in the peer-to-peer network even in the presence of non-functioning or malicious nodes (a Byzantine fault tolerant solution) (Strobel et al., 2020). Economic functionality might also be used as a coordination mechanism via robot-to-robot financial transactions in robot swarms. Likewise, blockchain and AI smart network technologies converge in a proposed application to integrate otherwise non-interoperable high-sensitivity cardiology data from multiple sources such as imaging and electrocardiogram (ECG) modalities (Krittanawong et al., 2020, 2022; Spadafora et al., 2023). Blockchains have also been proposed for use in environmental monitoring and climate change, including to address the problem of government corruption (Bublitz et al., 2019; Chapron, 2017). For example, blockchain-based processes in agriculture (Ghorbel et al., 2022; Pranto et al., 2021) might make it easier to collect insurance against crop damage with immediate payments that minimize delay and potential corruption.

3.4.3.3.2 Pharmaceutical Manufacturing and Chemical Industry

Since pharmaceutical manufacturing is one of the most regulated, vertically integrated, and technologically advanced matter-based industries worldwide, it may serve as a precursor model for MM. The chemical industry has similar characteristics of raw materials orchestration, intensive process manufacturing, and deployments of blockchain-AI tools to create the next generation of manufacturing supply chains (e.g., the Wanxiang Blockchain is a dedicated platform for chemical e-commerce) (Elser et al., 2019). Pharmaceutical manufacturing is testing various blockchain solutions such as RegTech provenance chains (KPMG UK, 2022), authorization and access chains, asset registries and inventories, trading chains, and smart contract platforms for product manufacturing, clinical trials, and R&D. A specific industry requirement is serialization regulation, being able to track products at multiple levels of production detail such as raw materials lots, individual units, batch, and shipment, as well as later tied to patients (Omidian & Omidi, 2022; Sim et al., 2022). It is likely the case that MM would also incorporate multiple levels of tracking from feedstock lots to finished products.

PharmaBlockchains manage both manufacturing processes and patient data to provide transparency, traceability, and security to the industry. Examples of industry-wide federated chains include MediLedger (a global pharmaceutical supply chain blockchain with over 20 members, including Pfizer, Amgen, and Gilead) (Sample, 2020) and EcoChain (managing clinical trial patient data). Such blockchains might enable virtual trials, in which there is no centralized trial center and patients receive medications and have samples picked up at home. PharmaBlockchains may also facilitate the 3D printing of personalized drugs in a physician's office or hospital, and eventually even in one's own home (a potential precursor to Factory@Home molecular printing) (see Chapter 2). Aprecia's Spritam is one such FDA-approved 3D-printed drug for the treatment of epilepsy, approved in 2015 and marketed as 3D printed ZipDoseTechnology (Essop, 2020; Kite-Powell, 2016).

3.4.3.4 Molecular Manufacturing Economy

3.4.3.4.1 Participant-Provided Infrastructure and Financing

Blockchain peer-to-peer networks have given way to novel beyond-debt financing models such as participant-owned and contributed infrastructure (Swan, 2019). There are both citizen-provided infrastructure (blockchains, solar panels for power generation, and open-source hosted software) and citizen-provided financing (via regulated decentralized financial instruments in insurance, investments, and capital-providing liquidity pools). Similar peer-provided networks in the support of MM infrastructures might likewise arise. At the global level, coin communities constitute a new form of Kardashev-level (planetary-scale) democracy.

Blockchains offer the type of global-scale automation and orchestration technology platforms that molecular manufacturing may require (Table 3.5). Arbitrarily many items, such as feedstock lots, matter printers, and designs, can be registered to blockchains and tracked for secure and compliant transfer (MM would likely be a highly regulated business). Verification proofs can accompany printed matter outputs as an alongside attestation of the quality and integrity of the manufactured item, confirming atomic precision and intellectual property licensing certification. Programmable smart matter (on-board software functionality) refers to MM objects that have a dual physical-digital nature, meaning that they can be better monitored, controlled, updated, monetized, and retired than "dumb-matter" objects. Once the MM platform is stable, the functionality could be constantly upscaled to deliver new applications (such as personalized cellular therapeutics). Blockchains, in conjunction with other convergent technologies such as AI/AGI/ASI, quantum computation, and IoT sensors, could enable MM to quickly expand into manufacturing and beyond, with a high-impact slate of applications in biology and health, space, and energy.

TABLE 3.5
Molecular Manufacturing Supply Chain Elements

Element	Description
Feedstocks	CHON stream lot serialization, specialty, cell-printing provenance blockchains
Matter Printers	Factory@Home, community, specialty regulated (pharmacy) 3D/4D printers
Designs	Matter design and prototyping with gaming engines (Unreal/Unity VR)
Attestation	Verification proof registers printed objects and attests to atomic precision, regulatory compliance, and intellectual property licensing certification

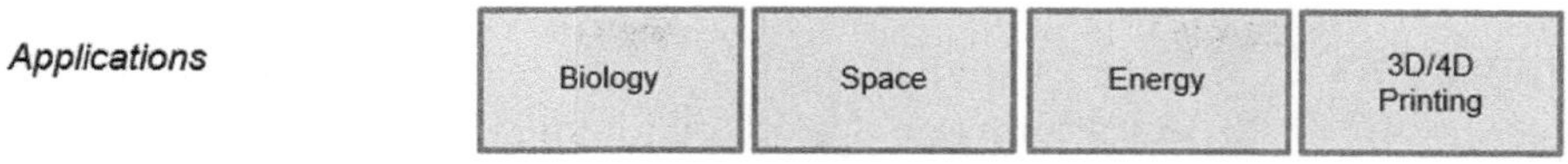

FIGURE 3.5 High-profile MM application areas.

3.5 APPLICATIONS

Aside from manufacturing-related industries in general, the biology, space, energy, as well as 3D and 4D printing domains indicate likely leadership toward MM developments (Figure 3.5). Energy-related applications will have opportunities to exploit smart polymers, coatings, and programmable matter to reduce energy consumption in buildings (regulating solar exposure via dynamic shading systems). In a pilot project, a thermally responsive shape-memory polymer was tested to actuate within the building environment temperature range (Niazy et al., 2021). 3D and 4D printing refer to various on-demand materials printing methods. 3D printing is an additive manufacturing process that builds objects from the bottom up (as opposed to casting them in a mold or machining them from bulk materials). The converse is subtractive manufacturing, such as the photolithography resists that are used in semiconductor manufacturing, renders materials by etching away patterns with stencils, masks, and chemical baths. Further, 4D printing involves the fabrication of programmable smart matter such that objects can respond to environmental stimuli including temperature, light, and other cues (Agarwal et al., 2023; Darmawan et al., 2023).

3.5.1 Biology

"The twenty-first century will be the century of biology."

Dyson, *Our Biotech Future* (2007)

Biology is one of the most important application domains of quantum nanoscience, with obvious potential to significantly benefit human health on a global scale, including the equalization of economic differences in health care standards, with health being the underlying currency that enables all opportunities in life. Quantum nanoscience approaches tend to focus on a variety of problems in genomics, protein folding, neurodegenerative diseases, and neural signaling.

3.5.1.1 The Human Brain

The human brain is considered to be the most complex structure in the known universe, which operates at nine orders-of-magnitude scale tiers, ranging from the central nervous system to the smallest ion channel (Sejnowski, 2020). Although the brain's estimated 86 billion neurons and 242 trillion synapses (Martins et al., 2019) do not constitute large numbers of entities in the big data era, the scope and variability of their interconnections are far more complicated. In a recent experiment, 76% of the total capacity of the Japanese K supercomputer (63,504 of 82,944 computational nodes) was required to simulate just one-third of the human brain's neural spiking activity (Igarashi et al., 2019). Hence, quantum models will be required for the in-depth study of the brain (Swan et al., 2022c).

TABLE 3.6
Neuroscience Physics Interpretations of Foundational Physics Findings

Neuroscience Physics	Description	Reference
AdS/Neuroscience (AdS/Brain, AdS/Memory)	AdS/CFT correspondence theory of bulk-boundary relationships with long/short-range correlations	Swan et al. (2022b), Dvali, (2017)
Neuronal Gauge Theories	Gauge fields reset universal symmetry (free energy)	Sengupta et al. (2016)
Network Neuroscience	Graph theoretical basis for multiscalar brain function	Bassett et al. (2018)
Chern-Simons Neuroscience	Geometrical curvature-based theory of exception	Bajardi et al. (2021)

3.5.1.2 Neuroscience Physics

Neuroscience physics refers to the neuroscience interpretation of foundational physics findings (Table 3.6). As a subfield of neuroscience, and particularly quantum neurobiology, neuroscience physics applies core discoveries in physics to the study of neuroscience. This may include any and all ongoing findings in physics, particularly those with novel elucidations on time, space, dynamics, system criticality, and phase transitions. Multiscalar models, long-range and short-range correlations, data theoretical and graph theoretical formulations, and developments in mathematics, symmetry, and topology might be relevant to characterizing, modeling, and resolving neural behavior and disease. Physics is on the vanguard of the understanding of all manner of physical systems. As information science has become its vernacular, so too information-related approaches to neuroscience are suggested, and developments in entanglement, gravitation, and black holes may likewise inform the study of the brain.

3.5.1.3 Chern-Simons/Biology

An area of neuroscience physics, and physics approaches to biology more generally, that is particularly relevant to quantum nanoscience is the application of the Chern-Simons theory to biological problems. The Chern-Simons theory is a standard mathematical model of topological invariance, where certain aspects of a system do not change when other changes are applied (e.g., a donut can be bent and misformed, but still has a hole in the center). The Chern-Simons theory defines mathematical forms based on secondary characteristics (Chern & Simons, 1974), as is formulated in the quantum setting as a quantum field theory that can be described with a Jones polynomial (knot theory) (Witten, 1989). Chern-Simons is a foundational topological building block that is motivated in the study of biological systems.

The distinguishing aspect of Chern-Simons topological models is a curve-based structure; thus, to the extent that a problem domain can be formulated as a curvature-related problem, the Chern-Simons model might apply. One team was inspired by the Chern-Simons theory (deployed as a gravitational model and measured with the expectation values of Wilson loops) and described biological models as on the basis of curvature problems to investigate various situations. The research examined genomics (DNA mutations in cancer and COVID-19) (Bajardi et al., 2021),

DNA-RNA transcription (errors and drift) (Capozziello et al., 2018), and protein folding (time series knots and links) (Capozziello & Pincak, 2018). For the cancer and COVID-19 investigations, genetic mutations were found to occur at Chern-Simons curvature peaks.

Another team applied the Chern-Simons/Biology approach to protein folding (Melnikov & Neves, 2019). In this case, the curvature-based model was a one-dimensional abelian Higgs model with Chern-Simons interaction as an effective theory of one-dimensional curves embedded in three-dimensional space. The model was able to predict some characteristic relations in the geometries of secondary structure motifs found in proteins. Experimental data was used to set the local geometric parameters (the curvature and torsion of helical motifs) in the model, which left a free parameter that described the abundance and shape of the principal motifs (alpha helices, beta strands, and interconnecting loops).

3.5.2 Space

3.5.2.1 Off-World Manufacturing

The first demonstration of off-world manufacturing was performed by 3D printing in zero gravity on the International Space Station (Prater et al., 2019b). This was the earliest example of employing a fused filament fabrication process for 3D printing in a microgravity environment. The ability to manufacture objects on-demand in space could significantly expand the future of space missions. Flexible technologies are required that have the capability to operate under extreme conditions, including those of zero gravity. Space missions are constrained by payload capacities (it currently costs $10,000 to lift a pound of payload into Earth orbit) (Kramer & Mosher, 2016). It may be that case that MM delivers reliable high performance production at low cost, albeit the technology will also be required to be resistant against radiation damage (Drexler, 1992).

A further point is that given the extreme environment of space, spare materials and parts will inevitably be necessary; however, they mostly (95%) go unused, and deterioration is unpredictable (Prater et al., 2019b). NASA's in-space manufacturing roadmap considers printable electronics such as personal CO_2 monitors and a 3D-printed cortisol biosensor (printed with hardware such as the nScrypt 3D Multi-Material Printer and Voltera electronics printer) (Prater et al., 2019a). Printable electronics (such as the STANLE Flutter Landers for Mars) and even full-scale printed spacecraft and nanosatellites are under consideration (Short, 2012).

A priority for long endurance space missions is the development of capacities for in-space manufacturing. To facilitate such zero-gravity capabilities, one team proposed a handling process for additive manufacturing powders that function independently of the gravitational environment and with no restriction on feedstock powder flowability (D'Angelo et al., 2021). Another project described an atmospheric pressure plasma printing system as an alternative to existing techniques and demonstrated the printing of nanomaterials such as titanium oxide and tin oxide particles on flexible substrates (Ramamurti et al., 2020). The plasma-based printing system offered advantages specific to space, such as a robust print head with integrated fluid delivery that directed the aerosol to a substrate using an electromagnetic field.

3.6 CONCLUSION

3.6.1 Summary

This chapter reviewed the current status of envisaged MM (the atomically or molecularly precise production of physical entities), which integrates quantum methods more intensively; potentially proceeding toward an era of quantum nanoscience that explicitly integrates quantum effects into engineered nanostructures. The key trend is that not only is quantum computation emerging as a post-Moore's law computational infrastructure, but quantum data science is pervasively enabling far more foundational strategies for conceiving, treating, and producing matter, including novel matter phases at low temperature, with topological properties. The traditional vision of the molecular nanotechnology program, currently instantiated as atomically precise manufacturing, is now merging with quantum computation, in a new era of quantum nanoscience. Advances in molecular control, for the positioning, manipulation, and correction of quantum objects in quantum computation, are not only extending to the control of matter but also to the creation of matter.

3.6.2 Risks and Limitations

There will likely be myriad risks and limitations involved with the MM hardware, software, and application proposals described in this work. The discussion leans heavily on the advent of quantum technologies; thus, it may be too early to know if the required technical breakthroughs in quantum error correction will progress from NISQ (noisy intermediate-scale quantum) devices to fully FTQC (fault-tolerant quantum computation) (Preskill, 2021). Even if quantum methods advance as anticipated, the complexities of atomic and molecular matter suggest a slow and costly road to large-scale commercialization. Multidisciplinary knowledge of various domains will be essential. Although the quantum computation industry is achieving a greater mastery of molecular control, upleveling from controlling to creating matter is nontrivial, and may not be a near-term possibility without step-function advances in scientific discovery and engineering practices, particularly toward the realization of significantly beneficial global applications such as personalized nanomedicine.

As the proposed economic automation platform for the MM economy, blockchains themselves are complex, early-stage, and more often deployed for get-rich-quick schemes than the global betterment of humanity. Philosophically, there are substantial risks in the implication of a much deeper move toward technologically controlled human lives. Doctorow (2019) underlines this point in a dystopian science fiction short story in which a completely automated smart apartment refuses to toast "unauthorized bread" in the Boulangism-branded toaster oven. Other science fiction, though, quickly embraces the trope of shipmind-fabbed matter for any circumstance in the interest of focusing on more interesting technical problems such as the Alcubierre-White drive that compacts space in front and stretches it out in back to enable faster-than-light warp bubble travel (Alcubierre, 1994; Bear, 2019). Specifically studying the risks of atomically precise manufacturing, one survey argues that the key challenge will be the potential proliferation of weapons in an era

TABLE 3.7
Shifts to Digital Smart Network Applications

Application	Smart Network Platform	Risk Profile
Digital news	Internet (Yahoo)	Low sensitivity
Digital money	Blockchains	Medium sensitivity
Digital brains	Cloudmind B/CIs	High sensitivity

of low-cost expanded capability manufacturing (Beckstead, 2015). Policy watchdog organizations might be employed to closely watch the progress of MM technologies using similar types of industry self-monitoring mechanisms developed for the emergence of recombinant DNA and bioengineering.

3.6.3 Farther Future

The first class of industries to move to a digital format included news, information, entertainment, stock trading, and mortgage loan shopping; comparatively low sensitivity applications (Table 3.7). The second class of industries currently in the process of digitization includes economic and legal systems related to contractual obligations and the transfer of value; an area of medium sensitivity in this schema. A third class of digitization might include intellectual transfer, individual experience-sharing, and personalized medicine as Internet-connected realities become more intense. In the farther future, smartphones (currently serving as cognitive prosthetics and external memory devices), particularly with the recent emergence of far more powerful AI capabilities, are likely to give way to brain-computer interfaces (BCIs) and brain/cloud interfaces (B/CIs), which will arguably comprise the applications with the highest sensitivity.

Realized through user-directed parameters, MM might be a critical future-class technology that instills trust and equalization in the ongoing trend toward the instantiation of all remaining matter-based industries in a digital format on global smart network platforms. The Internet comprises the first Kardashev-level (planetary-scale) technology, and layers of increasing intelligence continue to be added to it (e.g., blockchain economic automation). Kardashev-level governance too is emerging in the form of blockchain coin communities constituting a new form of digitally based global democracy. The implication is that MM could enable the further transformation of emerging economies in the leapfrog progression of digital methods for efficient infrastructures in the model of cellular telephones, digital banking solutions, and telemedicine diagnostics. Molecular manufacturing may emerge as a potent global scale balancing force that enables more equitable access to opportunities and abundance for all individuals.

REFERENCES

Adams B, Tomko M. A Critical Look at Cryptogovernance of the Real World: Challenges for Spatial Representation and Uncertainty on the Blockchain. *10th International Conference on Geographic Information Science* (GIScience 2018) 2018;18:1–18.

Agarwal N, Solanki VS, Ameta KL, Yadav VK, Gupta P, Wanale SG, et al. 4-dimensional printing: exploring current and future capabilities in biomedical and healthcare systems – a concise review. *Front Bioeng Biotechnol.* 2023 Aug 22;11:1251425.

Albert VV, Covey JP, Preskill J. Robust encoding of a qubit in a molecule. *Phys Rev X.* 2020;10(3):031050.

Alcubierre M. The warp drive: hyper-fast travel within general relativity. *Class Quantum Grav.* 1994;11:L73–L77.

Ashbridge Z, Fielden SDP, Leigh DA, Pirvu L, Schaufelberger F, Zhang L. Knotting matters: orderly molecular entanglements. *Chem Soc Rev.* 2022 Sep 20;51(18):7779–7809.

August DP, Dryfe RAW, Haigh SJ, Kent PRC, Leigh DA, Lemonnier JF, et al. Self-assembly of a layered two-dimensional molecularly woven fabric. *Nature.* 2020;588: 429–435.

Baiardi A. Electron dynamics with the time-dependent density matrix renormalization group. *J Chem Theory Comput.* 2021 Jun 8;17(6):3320–3334.

Bajardi F, Altucci L, Benedetti R, Capozziello S, Sorbo MRD, Franci G, Altucci C. DNA mutations via Chern-Simons currents. *Eur Phys J Plus.* 2021;136(10):1080.

Ball P. Making the world from topological order. *Natl Sci Rev.* 2019 Mar;6(2):227–230.

Bao H, Liu Y, Li H, Qi W, Sun K. Luminescence of carbon quantum dots and their application in biochemistry. *Heliyon.* 2023 Sep 20;9(10):e20317.

Barroso MM. Quantum dots in cell biology. *J Histochem Cytochem.* 2011;59(3):237–51.

Bassett DS, Zurn P, Gold JI. On the nature and use of models in network neuroscience. *Nat Rev Neurosci.* 2018;19(9):566–974.

Bear E. *Ancestral Night.* Saga Press, New York, 2019.

Beckstead N. Risks from Atomically Precise Manufacturing. *Open Philanthropy* June 2015. https://www.openphilanthropy.org/research/cause-reports/atomically-precise-manufacturing (accessed February 28, 2022).

Bernhard S, Tibbitt MW. Supramolecular engineering of hydrogels for drug delivery. *Adv Drug Deliv Rev.* 2021 Apr;171:240–256.

Biamonte J, Wittek P, Pancotti N, Rebentrost P, Wiebe N, Lloyd S. Quantum machine learning. *Nature.* 2017;549(195):195–202.

BiTBO https://bitbo.io/how-many-bitcoin/ (accessed September 20, 2025).

Bouju X, Joachim C, Girard C. Moving gold atoms with an atomic-force-microscope tip: a study of dimer and trimer formation on NaCl(100). *Phys Rev B Condens Matter.* 1994 Sep 15;50(11):7893–7902.

Bracha D, Walls MT, Brangwynne CP. Probing and engineering liquid-phase organelles. *Nat Biotechnol.* 2019;37(12):1435–1445.

Broholm C, Cava RJ, Kivelson SA, Nocera DG, Norman MR, Senthil T. Quantum spin liquids. *Science.* 2020 Jan 17;367(6475):eaay0668.

Bublitz FM, Oetomo A, Sahu KS, Kuang A, Fadrique LX, Velmovitsky PE, Nobrega RM, Morita PP. Disruptive technologies for environment and health research: an overview of artificial intelligence, blockchain, and internet of things. *Int J Environ Res Public Health.* 2019 Oct 11;16(20):3847.

Budman M, Hurley B, Khan A, Gangopadhyay N. Deloitte's 2019 Global Blockchain Survey. *Deloitte Insights.* Deloitte Development LLC, Cambridge, United Kingdom, 2019. https://www.deloitte.com/za/en/Industries/technology/analysis/blockchain-gets-down-to-business.html (accessed September 20, 2025).

Bussmann E, Butera RE, Owen JHG, Randall JN, Rinaldi SM, Baczewski AD, Misra S. Atomic-precision advanced manufacturing for Si quantum computing. *MRS Bulletin.* 2021;46:607–615.

Cao W, Cudney HH, Waser R. Smart materials and structures. *PNAS.* 1999;96:8330–8331.

Capozziello S, Pincak R. The Chern-Simons current in time series of knots and links in proteins. *Ann Phys.* 2018;393:413–446.

Capozziello S, Pincak R, Kanjamapornkul K, Saridakis EN. The Chern-Simons current in systems of DNA-RNA transcriptions. *Annalen der Physik*. 2018;530(4):1700271.

Carrasquilla J. Machine learning for quantum matter. *Adv Phys X*. 2020;5(1):1797528.

Carrasquilla J, Luo D, Perez F, Milsted A, Clark BK, Volkovs M, Aolita L. Probabilistic simulation of quantum circuits with the transformer. *Phys Rev A*. 2021;104:032610.

Chakraborty P, Das SS, Dey A, Chakraborty A, Bhattacharyya C, Kandimalla R et al. Quantum dots: the cutting-edge nanotheranostics in brain cancer management. *J Control Release*. 2022;350:698–715.

Challener CA. Why the Industry Is Moving Toward Blockchain Technology. *Pharma's Almanac*. PAP-Q4-19-NI-007. 6 December 2019. https://www.pharmasalmanac.com/articles/why-the-industry-is-moving-toward-blockchain-technology (accessed February 28, 2022).

Chamorro JR, McQueen TM, Tran TT. Chemistry of quantum spin liquids. *Chem Rev*. 2021 Mar 10;121(5):2898–2934.

Chapron G. The environment needs cryptogovernance. *Nature*. 2017;545:403–405.

Chen F, Tao NJ. Electron transport in single molecules: from benzene to graphene. *Acc Chem Res*. 2009 Mar 17;42(3):429–38.

Chen X, Yan B, Yao G. Towards atom manufacturing with framework nucleic acids. *Nanotechnol*. 2023 Feb 13;34(17).

Cheng S, Chen J, Wang L. Information perspective to probabilistic modeling: Boltzmann machines versus Born machines. *Entropy*. 2018;20:583.

Chern SS, Simons J. Characteristic forms and geometric invariants. *Annals of Mathematics*. 1974;99(1):48–69.

Chin AC. Blockchain biology. *Front Blockchain*. 2020;3:606413.

Choi RY, Coyner AS, Kalpathy-Cramer J, Chiang MF, Campbell JP. Introduction to machine learning, neural networks, and deep learning. *Transl Vis Sci Technol*. 2020 Feb 27;9(2):14.

Choi HS, Liu W, Misra P, Tanaka E, Zimmer JP, Ipe BI, et al. Renal clearance of nanoparticles. *Nat Biotechnol*. 2007;25(10):1165–1170.

Choi DJ, Lorente N, Wiebe J, von Bergmann K, Otte AF, Heinrich AJ. Colloquium: atomic spin chains on surfaces. *Rev Mod Phys*. 2019;91:041001.

Christie's. Beeple: A Visionary Digital Artist at the Forefront of NFTs. 2021. https://www.christies.com/features/Monumental-collage-by-Beeple-is-first-purely-digital-artwork-NFT-to-come-to-auction-11510-7.aspx (accessed February 28, 2022).

Christopoulou E, Xinogalos S. Overview and comparative analysis of game engines for desktop and Mobile devices. *Int J Serious Games*. 2017;4(4):21–36.

Cirac I, Perez-Garcia D, Schuch N, Verstraete F. Matrix product States and projected entangled pair States: concepts, symmetries, and theorems. *Rev Mod Phys*. 2021;93:045003.

Claeys PW, Lamacraft A. Ergodic and nonergodic dual-unitary quantum circuits with arbitrary local Hilbert space dimension. *Phys Rev Lett*. 2021 Mar 12;126(10):100603.

Cong I, Choi S, Lukin MD. Quantum convolutional neural networks. *Nat Phys*. 2019;15(12):1273–1278.

Corke G. Unity for manufacturing. *Develop 3D*. Oct 17 2019. https://develop3d.com/features/unity-visualisation-vr-manufacturing-industrial-design-game-on-simulation (accessed February 28, 2022).

D'Angelo O, Kuthe F, Liu SJ, Wiedey R, Bennett JM, Meisnar M, et al. A gravity-independent powder-based additive manufacturing process tailored for space applications. *Addit Manuf*. 2021;47:102349.

Damioli V, Zorzin E, DeSimone A, Noselli G, Lucantonio A. Transient shape morphing of active gel plates: geometry and physics. *Soft Matter*. 2022 Aug 10;18(31):5867–5876.

Darmawan BA, Park JO, Go G, Choi E. Four-dimensional-printed microrobots and their applications: a review. *Micromachines (Basel)*. 2023 Aug 15;14(8):1607.

De, Roeck W, Imbrie JZ. Many-body localization: stability and instability. *Philos Trans A Math Phys Eng Sci*. 2017 Dec 13;375(2108):20160422.

Del Pino J, Zilberberg O. Dynamical gauge fields with bosonic codes. *Phys Rev Lett*. 2023 Apr 28;130(17):171901.

Ding C, Guo Y, Liu J, Kent GB, Jobson BT, Fu X, Yang X, Zhong WH. A super-breathable "Woven-like" protein nanofabric. *ACS Appl Bio Mater*. 2020 May 18;3(5):2958–2964.

Ding Y, Wang C, Zeng M, Fu L. Atomic manufacturing of advanced nanomaterials. *Adv Mater*. 2023 Sep 18:e2306689.

Ding D, Zhu Q. Recent advances of PLGA micro/nanoparticles for the delivery of biomacromolecular therapeutics. *Mater Sci Eng C Mater Biol Appl*. 2018 Nov 1;92:1041–1060.

Doctorow C. *Radicalized*. Tor Publishing Group, New York, 2019.

Drexler KE. *Engines of Creation: The Coming Era of Nanotechnology*. Doubleday, New York, 1986.

Drexler KE. Molecular manufacturing for space systems – an overview. *J Br Interplanet Soc*. 1992;45:401–405.

Dvali G. Critically excited states with enhanced memory and pattern recognition capacities in quantum brain networks: Lesson from black holes. arXiv:1711.09079v1, 2017.

Dyck O, Ziatdinov M, Lingerfelt D, Unocic RR, Hudak BM, Lupini AR, et al. Atom-by-atom fabrication with electron beams. *Nat Rev Mater*. 2019;4(7):497–507.

Dyson F. Our biotech future. *The New York Review of Books*. July 19, 2007. https://www.nybooks.com/articles/2007/07/19/our-biotech-future/ (accessed September 20, 2025).

Ekimov AI, Onushchenko AA. Quantum size effect in the optical spectra of semiconductor micro-crystals. *Sov Phys Semicond-USSR*. 1982;16(7):775–778.

El-Wajeh YAM, Hatton PV, Lee NJ. Unreal Engine 5 and immersive surgical training: translating advances in gaming technology into extended-reality surgical simulation training programmes. *Br J Surg*. 2022 Apr 19;109(5):470–471.

Else DW, Monroe C, Nayak C, Yao NY. Discrete time crystals. *Annu Rev Condens Matter Phys*. 2020;11:467–499.

Elser B, Ulbrich M, Fasshauer P, Aggarwal V. Artificial Intelligence and Blockchain: Insights and Actions for the Chemical Industry. *Accenture*. 2019. https://www.accenture.com/us-en/insights/chemicals/ai-blockchain-chemical-industry (accessed December 11, 2023).

Engwerda AHJ, Fletcher SP. A molecular assembler that produces polymers. *Nat Commun*. 2020;11:4156.

Erdoğar N, Akkın S, Bilensoy E. Nanocapsules for drug delivery: an updated review of the last decade. *Recent Pat Drug Deliv Formul*. 2018;12(4):252–266.

Erhard A, Poulsen, Nautrup H, Meth M, Postler L, Stricker R, Stadler M, et al. Entangling logical qubits with lattice surgery. *Nature*. 2021 Jan;589(7841):220–224.

Essop A. "Aprecia and Purdue University Partner to Advance 3D Printed Pharmaceuticals" 3D Printing Industry. April 02, 2020. https://3dprintingindustry.com/news/aprecia-and-purdue-university-partner-to-advance-3d-printed-pharmaceuticals-170533/ (accessed December 06, 2023).

Eugene M, Renkin EM. Chapter 42 "Microcirculation and Exchange," in Patton HD, Fuchs AF, Hille B, Scher AM, Steiner R, eds., *Textbook of Physiology*, *21st* Edition, W.B. Saunders, Philadelphia, PA. 1989;860–878.

Ewaschuk R, Turney PD. Self-Replicating Strands that Self-Assemble into User-Specified Meshes. Feb 22, 2005. *arXiv*:cs/0502087v1 [cs.NE].

Ewaschuk R, Turney PD. Self-replication and self-assembly for manufacturing. *Artificial Life*. 2006;12(3):411–433.

Feldman DE, Halperin BI. Fractional charge and fractional statistics in the quantum Hall effects. *Rep Prog Phys*. 2021 Jun 22;84:076501.

Feynman RP. Simulating physics with computers. *Int J Theor Phys*. 1982;21(6):467–488.

Fishlock TW, Oral A, Egdell RG, Pethica JB. Manipulation of atoms across a surface at room temperature. *Nature*. 2000 Apr 13;404(6779):743–5.

Flamini F, Spagnolo N, Sciarrino F. Photonic quantum information processing: a review. *Rep Prog Phys*. 2019;82(1):016001.

Freeney SE, Slot MR, Gardenier TS, Swart I, Vanmaekelbergh D. Electronic quantum materials simulated with artificial model lattices. *ACS Nanosci Au*. 2022 Jun 15;2(3):198–224.

Freitas RA Jr. Welcome to the Future of Medicine. Transhumanism and Personal Identity. In More M, Vita-More N, Eds., *The Transhumanist Reader*. Oxford, UK: Wiley-Blackwell, 2012; 67–72.

Freitas RA Jr, Merkle RC. *Kinematic Self-Replicating Machines*. Landes Bioscience, Georgetown, TX, 2004.

Gaehtgens P. Flow of blood through narrow capillaries: rheological mechanisms determining capillary hematocrit and apparent viscosity. *Biorheology*. 1980;17:183–189.

Gartner. Blockchain Technology: What's Ahead? 2025. https://www.gartner.com/en/information-technology/insights/blockchain (accessed September 20, 2025).

Ghezzi M, Pescina S, Padula C, Santi P, Del Favero E, Cantù L, Nicoli S. Polymeric micelles in drug delivery: an insight of the techniques for their characterization and assessment in biorelevant conditions. *J Control Release*. 2021 Apr 10;332:312–336.

Ghorbel O, Frikha T, Hajji A, Alabdali R, Ayadi R, Abbas Elmasry M. Blockchain-based supply chain system for olive fields using WSNs. *Comput Intell Neurosci*. 2022 Sep 23;2022:9776776.

Gil HM, Price TW, Chelani K, Bouillard JG, Calaminus SDJ, Stasiuk GJ. NIR-Quantum dots in biomedical imaging and their future. *iScience*. 2021 Feb 15;24(3):102189.

Giustino F, Bibes M, Lee JH, Trier F. The 2021 quantum materials roadmap. *J Phys Mater*. 2021;3(4):042006.

Goldstein SC, Campbell J, Mowry TC. Programmable matter. *IEEE Computer*. 2005;38(6):99–101.

Gopalakrishnan S, Vasseur R. Anomalous transport from hot quasiparticles in interacting spin chains. *Rep Prog Phys*. 2023 Feb 2;86:036502.

Gottesman D, Kitaev A, Preskill J. Encoding a qubit in an oscillator. *Phys Rev A*. 2001;64:012310.

Gowtham P, Harini K, Thirumalai A, Pallavi P, Girigoswami K, Girigoswami A. Synthetic routes to theranostic applications of carbon-based quantum dots. *ADMET DMPK*. 2023May 8;11(4):457–485.

Grieves M. "Virtually Intelligent Product Systems: Digital and Physical Twins," in *Complex Systems Engineering: Theory and Practice*, Flumerfelt S, ed., American Institute of Aeronautics and Astronautics, Reston, VA, USA, 2019;175–200.

Grishin D, Obbad K, Church GM. Data privacy in the age of personal genomics. *Nat Biotechnol*. 2019;37:1110–1117.

Gross R, Dorigo M. Self-Assembly at the Macroscopic Scale. *Proceedings of the IEEE*. 2008;96(9):1490–1508.

Guimarães D, Cavaco-Paulo A, Nogueira E. Design of liposomes as drug delivery system for therapeutic applications. *Int J Pharm*. 2021 May 15;601:120571.

Guo X, Zhang G, Zhang Y. A comprehensive review of blockchain technology-enabled smart manufacturing: a framework, challenges and future research directions. *Sensors (Basel)*. 2022 Dec 23;23(1):155.

Hall E. On a new action of the magnet on electric currents. *Am J Math*. 1879;2:287–292.

Han J, Cheng R, Liu L, Ohno H, Fukami S. Coherent antiferromagnetic spintronics. *Nat Mater*. 2023 Jun;22(6):684–695.

Hankin A. Nearly half of all 2017 ICOs have failed. *Fortune*. 26 February 2018. https://www.marketwatch.com/story/nearly-half-of-all-2017-icos-have-failed-2018-02-26 (accessed February 28, 2022).

Hanson-Heine MWD, Ashmore AP. Computational chemistry experiments performed directly on a blockchain virtual computer. *Chem Sci*. 2020 Apr 21;11(18):4644–4647.

Harrison ER. Phenomena, comment & notes. *Smithsonian Magazine*. December 2015.

Hayden P, May A. Localizing and excluding quantum information; Or, how to share a quantum secret in spacetime. *Quantum*. 2019;3:196.

He Z, Shi K, Li J, Chao J. Self-assembly of DNA origami for nanofabrication, biosensing, drug delivery, and computational storage. *iScience*. 2023 Apr 10;26(5):106638.

Herrmann IK, Wood MJA, Fuhrmann G. Extracellular vesicles as a next-generation drug delivery platform. *Nat Nanotechnol*. 2021 Jul;16(7):748–759.

Hess PW, Becker P, Kaplan HB, Kyprianidis A, Lee AC, Neyenhuis B, Pagano G, Richerme P, Senko C, Smith J, Tan WL, Zhang J, Monroe C. Non-thermalization in trapped atomic ion spin chains. *Philos Trans A Math Phys Eng Sci*. 2017 Dec 13;375(2108):20170107.

Hirjibehedin CF, Lutz CP, Heinrich AJ. Spin coupling in engineered atomic structures. *Science*. 2006 May 19;312(5776):1021–4.

Hosur P, Qi XL, Roberts DA, Yoshida B. Chaos in quantum channels. *J High Energy Phys*. 2016;2:4.

Hu G, Xiang B. Recent advances in two-dimensional spintronics. *Nanoscale Res Lett*. 2020 Dec 9;15(1):226.

Huang YY, Xu Y, Wang L, Zhao CC, Tu CP, Ni JM, et al. Heat transport in herbertsmithite: can a quantum spin liquid survive disorder? *Phys Rev Lett*. 2021 Dec 24;127(26):267202.

Hübener H, De, Giovannini U, Sato SA, Rubio A. Floquet band engineering in action. *Sci Bull (Beijing)*. 2023 Apr 30;68(8):751–752.

Hwang D, Ramsey JD, Kabanov AV. Polymeric micelles for the delivery of poorly soluble drugs: from nanoformulation to clinical approval. *Adv Drug Deliv Rev*. 2020;156:80–118.

Igarashi J, Yamaura H, Yamazaki T. Large-scale simulation of a layered cortical sheet of spiking network model using a tile partitioning method. *Front Neuroinform*. 2019;13(71):1–15.

Ippoliti M, Kechedzhi K, Moessner R, Sondhi SL, Khemani V. Many-body physics in the NISQ era: quantum programming a discrete time crystal. *PRX Quantum*. 2021;2:030346.

Jacobs WM. Self-assembly of biomolecular condensates with shared components. *arXiv*:2102. 2021;12547v1.

Ji SC, Schweigler T, Tajik M, Cataldini F, Sabino J, Møller FS, et al. Floquet engineering a bosonic Josephson junction. *Phys Rev Lett*. 2022 Aug 19;129(8):080402.

Jia TZ, Bellini T, Clark N, Fraccia TP. A liquid crystal world for the origins of life. *Emerg Top Life Sci*. 2022 Dec 22;6(6):557–569.

Jiang HC. Superconductivity in the doped quantum spin liquid on the triangular lattice. *Npj Quantum Mater*. 2021;6:71.

Jonay C, Khemani V, Ippoliti M. Tri-unitary quantum circuits. *arXiv*:2106. 2021;07686v2.

Jungck JR, Brittain S, Plante D, Flynn J. Self-assembly, self-folding, and origami: comparative design principles. *Biomimetics (Basel)*. 2022 Dec 27;8(1):12.

Jurková B, Zámečník L. Turing and von Neumann machines: completing the new mechanism. *Biosystems*. 2023 Oct 17;234:105046.

Kaarsberg T. *Innovations in Scanning Tunneling Microscope Control Systems for High-throughput Atomically Precise Manufacturing*. DOE/EE-1842. U.S. Department of Energy, Advanced Manufacturing Office, Washington, DC, USA, 2019.

Kandaswamy R, Furlonger D, Stevens A. Digital Disruption Profile: Blockchain's Radical Promise Spans Business and Society. *Gartner Research*. 2021. https://www.gartner.com/en/doc/3855708-digital-disruption-profile-blockchains-radical-promise-spans-business-and-society (accessed February 28, 2022).

Kaspar C, Ravoo BJ, van der Wiel WG, Wegner SV, Pernice WHP. The rise of intelligent matter. *Nature*. 2021;594:345–355.

Katje C. Is Unity Software Being Overlooked as NFT Play? *Benzinga*. 26 March 2021. https://www.benzinga.com/markets/cryptocurrency/21/03/20329493/is-unity-software-being-overlooked-as-nft-play (accessed February 28, 2022).

Katsonis N, Lancia F, Leigh DA, Pirvu L, Ryabchun A, Schaufelberger F. Knotting a molecular strand can invert macroscopic effects of chirality. *Nat Chem*. 2020;12: 939–944.

Kennes DM, de la Torre A, Ron A, Hsieh D, Millis AJ. Floquet engineering in quantum chains. *Phys Rev Lett*. 2018 Mar 23;120(12):127601.

Khalifa OO, Amirah bt Sharif N, Saeed RA, Abdel-Khalek S, Alharbi AN, Alkathiri AA. Digital system design for quantum error correction codes. *Contrast Media Mol Imaging*. 2021 Dec 15;2021:1101911.

Khalili MH, Williams CJ, Micallef C, Duarte-Martinez F, Afsar A, Zhang R, et al. Nanoindentation response of 3D printed PEGDA hydrogels in a hydrated environment. *ACS Appl Polym Mater*. 2023 Jan 20;5(2):1180–1190.

Kim SK, Beach GSD, Lee KJ, Ono T, Rasing T, Yang H. Ferrimagnetic spintronics. *Nat Mater*. 2022 Jan;21(1):24–34.

Kitaev A. Fault-tolerant quantum computation by anyons. *Ann Phys*. 2003;303:2–30.

Kitaev A, Preskill J. Topological entanglement entropy. *Phys Rev Lett*. 2006;96:110404.

Kite-Powell J. FDA Approved 3D Printed Drug Available in the US. *Forbes*. 22 March 2016. https://www.forbes.com/sites/jenniferhicks/2016/03/22/fda-approved-3d-printed-drug-available-in-the-us/?sh=40bd434a666b (accessed February 28, 2022).

Kjærgaard-Winther C. "A.P. Moller – Maersk and IBM to discontinue TradeLens, a blockchain-enabled global trade platform," A.P. Moller – Maersk, Nov. 29, 2022. https://www.maersk.com/news/articles/2022/11/29/maersk-and-ibm-to-discontinue-tradelens (accessed December 06, 2023).

Ko W, Gai Z, Puretzky AA, Liang L, Berlijn T, Hachtel JA, et al. Understanding heterogeneities in quantum materials. *Adv Mater*. 2023;35(27):e2106909.

KPMG UK. A user's guide to RegTech: Navigating the challenges and what success looks like (2022). https://assets.kpmg.com/content/dam/kpmg/uk/pdf/2022/11/innovate-finance-regtech-industry-and-adoption.pdf (accessed December 07, 2023).

Kramer S, Mosher D. (2016). Here's how much money it actually costs to launch stuff into space. *Tech Insider*. 20 July 2016. https://www.businessinsider.com/spacex-rocket-cargo-price-by-weight-2016-6 (accessed February 28, 2022).

Krittanawong C, Aydar M, Hassan, Virk HU, Kumar A, Kaplin S, Guimaraes L, Wang Z, Halperin JL. Artificial intelligence-powered blockchains for cardiovascular medicine. *Can J Cardiol*. 2022;38(2):185–195.

Krittanawong C, Rogers AJ, Aydar M, Choi E, Johnson KW, Wang Z, Narayan SM. Integrating blockchain technology with artificial intelligence for cardiovascular medicine. *Nat Rev Cardiol*. 2020;17:1–3.

Kubica A, Vasmer M. Single-shot quantum error correction with the three-dimensional subsystem toric code. *Nat Commun*. 2022 Oct 21;13(1):6272.

Kumar A, Singh AK, Ahmad I, Kumar, Singh P, Anushree, Verma PK, et al. A novel decentralized blockchain architecture for the preservation of privacy and data security against cyberattacks in healthcare. *Sensors (Basel)*. 2022;22(15):5921.

Kumar S, Singh P, Sharma S, Pottoo FH, Ali J, Baboota S. Envisioning the future of nanomedicines in management and treatment of neurological disorders. *Comb Chem High Throughput Screen*. 2021;24(10):1544–1556.

Laucht A, Hohls F, Ubbelohde N, Gonzalez-Zalba MF, Reilly DJ, Stobbe S, et al. Roadmap on quantum nanotechnologies. *Nanotechnology*. 2021;32:1–47.

Le N, Zhang M, Kim K. Quantum dots and their interaction with biological systems. *Int J Mol Sci*. 2022 Sep 15;23(18):10763.

LeCun Y, Bengio Y, Hinton G. Deep learning. *Nature.* 2015;521(7553):436–444.
Leible S, Schlager S, Schubotz M, Gipp B. A review on blockchain technology and blockchain projects fostering open science. *Front Blockchain.* 2019;2:16.
Leigh DA, Danon JJ, Fielden SDP, Lemonnier J-F, Whitehead GFS, Woltering SL. A molecular endless (7_4) knot. *Nat Chem.* 2021;13:117–122.
Leinen P, Esders M, Schutt KT, Wagner C, Muller KR, Tautz FS. Autonomous robotic nanofabrication with reinforcement learning. *Sci Adv.* 2020;6:eabb6987.
Leis A, Schleenvoigt M, Cherepanov V, Lupke F, Schuffelgen P, Mussler G, et al. Lifting the spin-momentum locking in ultra-thin topological insulator films. *Adv Quantum Technol.* 2021;4:2100083.
Levine H, Goldman DI. Physics of smart active matter: integrating active matter and control to gain insights into living systems. *Soft Matter.* 2023 Jun 14;19(23):4204–4207.
Lewandowska U, Zajaczkowski W, Corra S, Tanabe J, Borrmann R, Benetti E, et al. A triaxial supramolecular weave. *Nat Chem.* 2017;9:1068–1072.
Li W, Qiu J, Li XL, Aday S, Zhang J, Conley G, et al. BBB pathophysiology–independent delivery of siRNA in traumatic brain injury. *Sci Adv.* 2021;7(1):eabd6889.
Li L, Zou CL, Albert VV, Muralidharan S, Girvin SM, Jiang L. Cat codes with optimal decoherence suppression for a lossy bosonic channel. *Phys Rev Lett.* 2017 Jul 21;119(3):030502.
Lim NCH, Jackson SE. Molecular knots in biology and chemistry. *J Phys: Condens Matter.* 2015;27:354101.
Liu CX, Zhang SC, Qi XL. The quantum anomalous Hall effect: theory and experiment. *Annu Rev Condens Matter Phys.* 2016;7:301–321.
Lo PK, Metera KL, Sleiman HF. Self-assembly of three-dimensional DNA nanostructures and potential biological applications. *Curr Opin Chem Biol.* 2010 Oct;14(5):597–607.
Lodge MS, Yang SA, Mukherjee S, Weber B. Atomically thin quantum spin hall insulators. *Adv Mater.* 2021 Jun;33(22):e2008029.
Loidl A, Lunkenheimer P, Tsurkan V. On the proximate Kitaev quantum-spin liquid α-$RuCl_3$: thermodynamics, excitations and continua. *J Phys Condens Matter.* 2021 Aug 27;33(44).
Lustenberger M, Malesevic S, Spychiger F. Ecosystem readiness: blockchain adoption is driven externally. *Front Blockchain.* 2021;4:720454.
Ma WL, Puri S, Schoelkopf RJ, Devoret MH, Girvin SM, Jiang L. Quantum control of bosonic modes with superconducting circuits. *Sci Bull (Beijing).* 2021 Sep 15;66(17):1789–1805.
Mahmood K, Shamshad S, Saleem MA, Kharel R, Das AK, Shetty S, Rodrigues JJPC. Blockchain and PUF-based secure key establishment protocol for cross-domain digital twins in industrial Internet of Things architecture. *J Adv Res.* 2023 Sep 29:S2090-1232(23)00266-7.
Manzoor R, Sahay BS, Singh SK. Blockchain technology in supply chain management: an organizational theoretic overview and research agenda. *Ann Oper Res.* 2022 Nov 24:1–48.
Martins NRB, Angelica A, Chakravarthy K, Svidinenko Y, Boehm FJ, Opris I, Lebedev MA, Swan M, Garan SA, Rosenfeld JV, Hogg T, Freitas RA Jr. Human Brain/Cloud interface. *Front Neurosci.* 2019;13(112):1–24.
Melnikov D, Neves ABF. Chern-Simons-Higgs model as a theory of protein molecules. *J Appl Phys.* 2019;126(24):244701.
Merkle RC. Self-replicating systems and molecular manufacturing. *J Br Interplanet Soc.* 1992;45:407–413.
Milbradt RM, Scheller L, Aßmus C, Mendl CB. Ternary unitary quantum lattice models and circuits in 2+1 dimensions. *Phys Rev Lett.* 2023 Mar 3;130(9):090601.
Mirrahimi M. Cat-qubits for quantum computation. *Comptes Rendus Physique.* 2016;17(7):778–787.

MIT (Massachusetts Institute of Technology). NASA's TESS hits milestone of 5,000 exoplanet candidates. *Phys.org*. 21 January 2022. https://phys.org/news/2022-01-nasa-tess-milestone-exoplanet-candidates.html (accessed February 28, 2022).

Morningstar A, Hauru M, Beall J, Ganahl M, Lewis AGM, Khemani V, Vidal G. Simulation of quantum many-body dynamics with Tensor Processing Units: Floquet prethermalization. *arXiv*:2111. 2021;08044v1.

Mougkogiannis P, Adamatzky A. Logical gates in ensembles of proteinoid microspheres. *PLoS One*. 2023 Sep 18;18(9):e0289433.

Müller C, Krone M, Huber M, Biener V, Herr D, Koch S, et al. Interactive molecular graphics for augmented reality using HoloLens. *J Integr Bioinform*. 2018 Jun 13;15(2): 20180005.

Musienko Y. Maersk Blockchain Use Case. Merehead. 16 November 2021. https://merehead.com/blog/maersk-blockchain-use-case/ (accessed February 28, 2022).

Nabil M, Megahed F. Quantum dot nanomaterials: preparation, characterization, advanced bio-imaging and therapeutic applications. *J Fluoresc*. 2023 Oct 25;34:2467–2484.

Nakamoto S. Bitcoin: A Peer-to-Peer Electronic Cash System. *Bitcoin Foundation* 1–9, 2008. https://bitcoin.org/bitcoin.pdf (accessed February 28, 2022).

Nance E. The Role of Nanoparticles in Neuroscience. *Scientia Global*. 12 Jul 2019. https://www.scientia.global/dr-elizabeth-nance-the-role-of-nanoparticles-in-neuroscience/ (accessed February 28, 2022).

NASA. Comparison: Webb vs Hubble Telescope – Webb/NASA. 26 August 2021. https://www.jwst.nasa.gov/content/about/comparisonWebbVsHubble.html (accessed February 28, 2022).

Neumann SP, Buchner A, Bulla L, Bohmann M, Ursin R. Continuous entanglement distribution over a transnational 248 km fiber link. *Nat Commun*. 2022 Oct 17;13(1):6134.

Niazy D, Elsabbagh A, Ismail MR. Mono–Material 4D printing of digital shape–memory components. *Polymers*. 2021;13:3767.

Noh K, Girvin SM, Jiang L. Encoding an oscillator into many oscillators. *Phys Rev Lett*. 2020 Aug 21;125(8):080503.

Noshay JM, Walker T, Alexander WG, Klingeman DM, Romero J, Walker AM, et al. Quantum biological insights into CRISPR-Cas9 sgRNA efficiency from explainable-AI driven feature engineering. *Nucleic Acids Res*. 2023 Oct 27;51(19):10147–10161.

Omidian H, Omidi Y. Blockchain in pharmaceutical life cycle management. *Drug Discov Today*. 2022 Apr;27(4):935–938.

Osorno LL, Brandley AN, Maldonado DE, Yiantsos A, Mosley RJ, Byrne ME. Review of contemporary self-assembled systems for the controlled delivery of therapeutics in medicine. *Nanomaterials*. 2021;11:278.

Owen JHG, Campbell Q, Santini R, Ivie JA, Baczewski AD, Schmucker SW, et al. Al-alkyls as acceptor dopant precursors for atomic-scale devices. *J Condens Matter Phys*. 2021;33:46.

Paschall CJ, Rao RPN, Hauptmann J, Ojemann, JG, Herron J. An Immersive Virtual Reality Platform Integrating Human ECOG & sEEG: Implementation & Noise Analysis. *Annu Int Conf IEEE Eng Med Biol Soc*. 2022 Jul;2022:3105–3110.

Peng Y. Topological space-time crystal. *Phys Rev Lett*. 2022;128(18):186802.

Pereira FRF, Mancini S. Entanglement-assisted quantum codes from cyclic codes. *Entropy (Basel)*. 2022 Dec 24;25(1):37.

Pereira FRF, Mancini S, La, Guardia GG. Stabilizer codes for open quantum systems. *Sci Rep*. 2023 Jun 29;13(1):10540.

Pishvar M, Harne RL. Foundations for soft, smart matter by active mechanical metamaterials. *Adv Sci (Weinh)*. 2020 Aug 18;7(18):2001384.

Pleskova S, Mikheeva E, Gornostaeva E. Using of quantum dots in biology and medicine. *Adv Exp Med Biol*. 2018;1048:323–334.

Polini M, Guinea F, Lewenstein M, Manoharan HC, Pellegrini V. Artificial honeycomb lattices for electrons, atoms and photons. *Nat Nanotechnol*. 2013 Sep;8(9):625–33.

Poyhonen K, Varjas D, Wimmer M, Akhmerov AR. Minimal Zeeman field requirement for a topological transition in superconductors. *arXiv*:2011. 2021;08263v3.

Pozo G, de la Presa P, Prato R, Morales I, Marin P, Fransaer J, Dominguez-Benetton X. Spin transition nanoparticles made electrochemically. *Nanoscale*. 2020 Mar 7;12(9):5412–5421.

Pranto TH, Noman AA, Mahmud A, Haque AB. Blockchain and smart contract for IoT enabled smart agriculture. *PeerJ Comput Sci*. 2021 Mar 31;7:e407.

Prater T, Risdon D, Edmundson J, Ledbetter F, Wheeler K, Hafiychuk V, et al. Overview of the In-Space Manufacturing Technology Portfolio. *NASA*. Marshall Space Flight Center. NASA & ISS National Lab Microgravity Materials Science Workshop, 2019a.

Prater T, Werkheiser N, Ledbetter F, Timucin D, Wheeler K, Snyder M. 3D printing in zero G technology demonstration mission: complete experimental results and summary of related material modeling efforts. *Int J Adv Manuf Technol*. 2019b;101:391–417.

Preskill J. Quantum computing 40 years later. *arXiv*:2106. 2021:10522.

Pu F, Ren J, Qu X. Nucleic acids and smart materials: advanced building blocks for logic systems. *Adv Mater*. 2014 Sep 3;26(33):5742–57.

Qi Y, Chen Y, Xia T, Lynch I, Liu S. Extra-pulmonary translocation of exogenous ambient nanoparticles in the human body. *ACS Nano*. 2023 Jan 10;17(1):12–19.

Qi Y, Wu J, Xu H, Guizani M. Blockchain Data Mining With Graph Learning: A Survey. *IEEE Trans Pattern Anal Mach Intell*. 2023 Oct 25;46(2):729–748.

Radmanesh SA, Haji A, Fatahi, Valilai O. Blockchain-based cloud manufacturing platforms: a novel idea for service composition in XaaS paradigm. *PeerJ Comput Sci*. 2021 Dec 8;7:e743.

Ramamurti R, Gandhiraman RP, Lopez A, Doshi P, Nordlund D, Kim B, Meyyappan M. Atmospheric pressure plasma printing of nanomaterials for IoT applications. *IEEE Open J Nanotechnol*. 2020;47–56.

Ramella M, Ribolla LM, de Curtis I. Liquid-liquid phase separation at the plasma membrane-cytosol interface: common players in adhesion, motility, and synaptic function. *J Mol Biol*. 2022 Jan 15;434(1):167228.

Randall JN. ZyVector: STM Control System for Atomically Precise Lithography. Zyvex Labs. 2021. https://www.zyvexlabs.com/apm/products/zyvector (accessed February 28, 2022).

Randall JN, Owen JH, Fuchs E, Lake J, Von Ehr JR, Ballard J, Henriksen E. Digital atomic scale fabrication an inverse Moore's Law: a path to atomically precise manufacturing. *Micro Nano Eng*. 2018;1:1–14.

Rathburn LC. Sizing up a molecule. Nanooze, 2013. https://www.nanooze.org/sizing-up-a-molecule (accessed February 28, 2022).

Ray PP, Majumder P. "Chapter Two – A Review on Cancer Data Management Using Blockchain: Progress and Challenges," in Hurson AR, ed., *Advances in Computers*, Elsevier, Amsterdam, Netherlands, 2023;131:49–78.

Rejali R, Coffey D, Gobeil J, Gonzalez JJW, Delgado F, Otte AF. Complete reversal of the atomic unquenched orbital moment by a single electron. *Npj Quantum Materials*. 2020;5:60.

Ringbauer M, Costa F, Goggin ME, White AG, Fedrizzi A. Multi-time quantum correlations with no spatial analog. *Npj Quantum Inf*. 2018;4:37.

Royer B, Singh S, Girvin SM. Stabilization of Finite-Energy Gottesman-Kitaev-Preskill States. *arXiv*:2009. 2020:07941.

Ru XM, Yang ZY, Ran SY. Lanthanide ions induce DNA compaction with ionic specificity. *Int J Biol Macromol*. 2022 Jun 15;210:292–299.

Safinya CR, Deek J, Beck R, Jones JB, Leal C, Ewert KK, Li Y. Liquid crystal assemblies in biologically inspired systems. *Liq Cryst.* 2013 Jan 1;40(12):1748–1758.

Sample M. MediLedger DSCSA Pilot Project. AmerisourceBergen. 2020. https://www.mediledger.com/dscsa-fda-pilot-project (accessed December 2023).

Sankar S, Sela E, Han C. Measuring topological entanglement entropy using Maxwell relations. *Phys Rev Lett.* 2023 Jul 7;131(1):016601.

Satzinger KJ, Liu Y, Smith A, Knapp C, Newman M, Jones C, et al. Realizing topologically ordered states on a quantum processor. *Science.* 2021;374(6572):1237–1241.

Schar F. Decentralized finance: on blockchain- and smart contract-based financial markets. *Fed Reserve Bank St. Louis Rev.* 2021;103(2):153–174.

Schaufelberger F. Open questions in functional molecular topology. *Commun Chem.* 2020;3:182.

Scheie A, Laurell P, Lake B, Nagler SE, Stone MB, Caux JS, Tennant DA. Quantum wake dynamics in Heisenberg antiferromagnetic chains. *Nat Commun.* 2022 Oct 2;13(1):5796.

Seeman NC. DNA nanotechnology: novel DNA constructions. *Annu Rev Biophys Biomol Struct.* 1998;27:225–48.

Seeman NC. DNA nanotechnology at 40. *Nano Lett.* 2020 Mar 11;20(3):1477–1478.

Sejnowski TJ. The unreasonable effectiveness of deep learning in artificial intelligence. *PNAS.* 2020;117(48):30033–30038.

Semeghini G, Levine H, Keesling A, Ebadi S, Wang TT, Bluvstein D, et al. Probing topological spin liquids on a programmable quantum simulator. *Science.* 2021 Dec 3;374(6572):1242–1247.

Sengupta B, Tozzi A, Cooray GK, Douglas PK, Friston KJ. Towards a neuronal gauge theory. *PLoS Biol.* 2016 Mar 8;14(3):e1002400.

Sessa F, Rahm M. Electronegativity equilibration. *J Phys Chem A.* 2022 Aug 18;126(32):5472–5482.

Shaginyan VR, Msezane AZ, Amusia MY, Clark JW, Japaridze GS, Stephanovich VA, Leevik YS. Thermodynamic, dynamic, and transport properties of quantum spin liquid in herbertsmithite from an experimental and theoretical point of view. *Condens Matter.* 2019;4(3):75.

Shao L, Ma J, Prelesnik JL, Zhou Y, Nguyen M, Zhao M, et al. Hierarchical materials from high information content macromolecular building blocks: construction, dynamic interventions, and prediction. *Chem Rev.* 2022 Dec 28;122(24):17397–17478.

Sharma A, Pu S, Balram AC, Jain JK. Fractional quantum Hall effect with unconventional pairing in monolayer graphene. *Phys Rev Lett.* 2023 Mar 24;130(12):126201.

Sharma A, Sharma P, Bamotra H, Gaur V. An extended approach to appraise electricity distribution and carbon footprint of bitcoin in a smart city. *Front Big Data.* 2023 Feb 2;6:1082113.

Shirer M, Soohoo S. Global Spending on Blockchain Solutions Forecast to be Nearly $19 Billion in 2024, According to New IDC Spending Guide. 19 Apr 2021. https://www.businesswire.com/news/home/20210419005059/en/Global-Spending-on-Blockchain-Solutions-Forecast-to-be-Nearly-$19-Billion-in-2024-According-to-New-IDC-Spending-Guide (accessed February 28, 2022).

Short K, Printable spacecraft: flexible electronic platforms for NASA missions. *NIAC 2012 Phase II Printable Spacecraft Final Report.* https://www.nasa.gov/general/printable-spacecraft-flexible-electronic-platforms-for-nasa-missions/ (accessed September 20, 2025)

Sim C, Zhang H, Chang ML. Improving end-to-end traceability and pharma supply chain resilience using blockchain. *Blockchain Healthc Today.* 2022 Aug 12;5.

Sivak VV, Eickbusch A, Royer B, Singh S, Tsioutsios I, Ganjam S, et al. Real-time quantum error correction beyond break-even. *Nature.* 2023 Apr;616(7955):50–55.

Skalak R, Branemark PI. Deformation of red blood cells in capillaries. *Science*. 1969 May 9;164(3880):717–719.

Smith G. The Scale of the Universe. UCSD Center for Astrophysics and Space Sciences, 1999. https://casswww.ucsd.edu/archive/public/tutorial/scale.html (accessed February 28, 2022).

Spadafora L, Comandini GL, Giordano S, Polimeni A, Perone F, Sabouret P, et al. Blockchain technology in cardiovascular medicine: a glance to the future? Results from a social media survey and future perspectives. *Minerva Cardiol Angiol*. 2024 Feb;72(1):1–10.

Statista. Annual electricity consumption in comparison (in TWh), 2021. https://www.statista.com/chart/18632/estimated-annual-electricity-consumption-of-bitcoin/ (accessed February 28, 2022).

Stephen DT, Nautrup HP, Bermejo-Vega J, Eisert J, Raussendorf R. Subsystem symmetries, quantum cellular automata, and computational phases of quantum matter. *Quantum*. 2019;20(3):142.

Strobel V, Castello FE, Dorigo M. Blockchain technology secures robot swarms: a comparison of consensus protocols and their resilience to Byzantine robots. *Front Robot AI*. 2020;7:54.

Sun Y, Nitiss JL, Pommier Y. SUMO: A Swiss army knife for eukaryotic topoisomerases. *Front Mol Biosci*. 2022 Apr 6;9:871161.

Swan M. "Impact of a B/CI on Humanity – Gateway to the Ten Billion Synapse World Mind," in Boehm F, ed., *Nanomedical Brain/Cloud Interface: Explorations and Implications*. CRC Press, Boca Raton, FL. Nov. 1, 2025.

Swan M. *Blockchain: Blueprint for a New Economy*. O'Reilly Media, Sebastopol, CA, 2015.

Swan M. "Blockchain Economic Theory: Digital Asset Contracting Reduces Debt and Risk (pp. 2–23) and Blockchain Theory of Programmable Risk: Black Swan Smart Contracts," in Swan M, Potts J, Takagi S, Witte F, Tasca P, eds., *Blockchain Economics: Implications of Distributed Ledgers – Markets, Communications Networks, and Algorithmic Reality*. World Scientific, London, 2019;171–194.

Swan M, Witte F, dos Santos RP. Quantum Information Science. *IEEE Internet Computing*. Jan.-Feb 2022a;26(1):7–14.

Swan M. Quantum neurobiology. *Quantum Rep*. 2022b;3:1–30.

Swan M, dos Santos RP, Lebedev MA, Witte F. *Quantum Computing for the Brain*. World Scientific, London, UK 2022c.

Swan M, dos Santos RP, Witte F. *Quantum Computing: Physics, Blockchains, and Deep Learning Smart Networks*. World Scientific, London, 2020.

Tan J, Xu S, Han X, Zhou Y, Li M, Cao W, Zhang Q, Lu P. Resolving and weighing the quantum orbits in strong-field tunneling ionization. *Adv Photonics*. 2021;3(3): 035001.

Tegmark M. *Life 3.0: Being Human in the Age of Artificial Intelligence*. Alfred A. Knopf, New York, 2017.

Thedford RP, Yu F, Tait WRT, Shastri K, Monticone F, Wiesner U. The promise of soft-matter-enabled quantum materials. *Adv Mater*. 2023 Feb;35(5):e2203908.

Thomas M, Schwartz R. Quantitative computational models of molecular self-assembly in systems biology. *Phys Biol*. 2017 May 23;14(3):035003.

Toffoli T, Margolus N. Programmable matter: concepts and realization. *Physica D*. 1991;47(1–2):263–272.

Tully RB, Courtois H, Hoffman Y, Pomarede D. The Laniakea supercluster of galaxies. *Nature*. 2014;513(7516):71.

University of Manchester. Scientists awarded a Guinness World Record for the finest woven fabric. *PhysOrg*. 19 January 2021. https://phys.org/news/2021-01-scientists-awarded-guinness-world-finest.html (accessed February 28, 2022).

Varjas D, Rosdahl TO, Akhmerov AR. Qsymm: algorithmic symmetry finding and symmetric Hamiltonian generation. *New J Phys*. 2018;20:093026.

Veldman LM, Farinacci L, Rejali R, Broekhoven R, Gobeil J, Coffey D, Ternes M, Otte AF. Free coherent evolution of a coupled atomic spin system initialized by electron scattering. *Science*. 2021;372:964.

Verdel R, Schmitt M, Huang YP, Karpov P, Heyl M. Variational classical networks for dynamics in interacting quantum matter. *Phys Rev B*. 2021;103:165103.

Vilbrandt T, Malone E, Lipson H, Pasko A. "Universal desktop fabrication," in Pasko A, Adzhiev V, Comninos P, eds., *Heterogeneous Objects Modeling and Applications: Collection of Papers on Foundations and Practice*, Springer, Berlin, 2008;259–284.

von Neumann J *Theory of Self-Reproducing Automata*. Burks AW, ed. University of Illinois Press, Urbana, IL, 1966.

Wallauer R, Raths M, Stallberg K, Munster L, Brandstetter D, Yang X, et al. Tracing orbital images on ultrafast time scales. *Science*. 2021;371(6533):1056–1059.

Wang L, Urbas AM, Li Q. Nature-inspired emerging chiral liquid crystal nanostructures: from molecular self-assembly to DNA mesophase and nanocolloids. *Adv Mater*. 2020 Oct;32(41):e1801335.

Ward DR, Schmucker SW, Anderson EM, Bussmann E, Tracy L, Lu TM, et al. Atomic precision advanced manufacturing for Si quantum computing. *Electron Device Failure Anal (EDFAAO)*. 2020;22(1):4–10.

Webber MJ, Pashuck ET. (Macro)molecular self-assembly for hydrogel drug delivery. *Adv Drug Deliv Rev*. 2021;172:275–295.

Whitesides GM, Grzybowski B. Self-assembly at all scales. *Science*. 2002;295:2418–2421.

Wiesing M, Fink GR, Weidner R. Accuracy and precision of stimulus timing and reaction times with Unreal Engine and SteamVR. *PLoS ONE*. 2020;15(4):e0231152.

Wilson DN, Doudna, Cate JH. The structure and function of the eukaryotic ribosome. *Cold Spring Harb Perspect Biol*. 2012 May 1;4(5):a011536.

Winogradoff D, Li PY, Joshi H, Quednau L, Maffeo C, Aksimentiev A. Chiral systems made from DNA. *Adv Sci (Weinh)*. 2021 Jan 21;8(5):2003113.

Witten E. Quantum field theory and the Jones polynomial. *Commun Math Phys*. 1989; 121(3):351–399.

Wölfle P. Quasiparticles in condensed matter systems. *Rep Prog Phys*. 2018 Mar;81(3): 032501.

Wong DR, Bhattacharya S, Butte AJ. Prototype of running clinical trials in an untrustworthy environment using blockchain. *Nat Commun*. 2019;10:917.

Xia K, Shen J, Li Q, Fan C, Gu H. Near-atomic fabrication with nucleic acids. *ACS Nano*. 2020 Feb 25;14(2):1319–1337.

Xu Q, Gao J, Wang S, Wang Y, Liu D, Wang J. Quantum dots in cell imaging and their safety issues. *J Mater Chem B*. 2021 Jul 28;9(29):5765–5779.

Yang Y, Zhong L, Li S, Yu A. Research on the perceived quality of virtual reality headsets in human-computer interaction. *Sensors (Basel)*. 2023 Jul 31;23(15):6824.

Yazdani A, von Oppen F, Halperin BI, Yacoby A. Hunting for Majoranas. *Science*. 2023 Jun 23;380(6651):eade0850.

Yin Y, Gu Q, Liu X, Liu F, McClements DJ. Double network hydrogels: design, fabrication, and application in biomedicines and foods. *Adv Colloid Interface Sci*. 2023;320:102999.

Yin J, Li YH, Liao SK, Yang M, Cao Y, Zhang L, et al. Entanglement-based secure quantum cryptography over 1,120 kilometres. *Nature*. 2020 Jun;582(7813):501–505.

Zaguri D, Shaham-Niv S, Chakraborty P, Arnon Z, Makam P, Bera S, et al. Nanomechanical properties and phase behavior of phenylalanine amyloid ribbon assemblies and amorphous self-healing hydrogels. *ACS Appl Mater Interfaces*. 2020 May 13;12(19): 21992–22001.

Zeguendry A, Jarir Z, Quafafou M. Quantum machine learning: a review and case studies. *Entropy (Basel)*. 2023 Feb 3;25(2):287.

Zhan P, Peil A, Jiang Q, Wang D, Mousavi S, Xiong Q, et al. Recent advances in DNA origami-engineered nanomaterials and applications. *Chem Rev*. 2023 Apr 12;123(7): 3976–4050.

Zhang ZH, Andreassen BJ, August DP, Leigh DA, Zhang L. Molecular weaving. *Nat Mater*. 2022:1–9.

Zhang J, Hess PW, Kyprianidis A, Becker P, Lee A, Smith J, et al. Observation of a discrete time crystal. *Nature*. 2017 Mar 8;543(7644):217–220.

Zhao J, Gulan U, Horie T, Ohmura N, Han J, Yang C, et al. Advances in biological liquid crystals. *Small*. 2019 May;15(18):e1900019.

Zielińska A, Eder P, Rannier L, Cardoso JC, Severino P, Silva AM, Souto EB. Hydrogels for modified-release drug delivery systems. *Curr Pharm Des*. 2022;28(8):609–618.

Zurek WH. Quantum theory of the classical: quantum jumps, Born's rule and objective classical reality via quantum Darwinism. *Philos Trans A Math Phys Eng Sci*. 2018 Jul 13;376(2123):20180107.

GLOSSARY

3D printing (additive manufacturing) building objects with subsequent layers of deposition.

4D bioprinting biofabrication technology enabling stimulus-responsive materials or objects.

4D printing transformation of 3D printed objects per external stimulus (e.g., temperature, light).

Abelian commuting (order of terms does not matter) (v. non-abelian, non-commuting).

Additive manufacturing adding elements in the manufacturing process (as 3D printed layers).

Alexander polynomial knot classification algorithm; (4_1) the first number indicating the crossing number (minimum number of crossings in a projection of the chain onto a plane), the second subscripted index denoting the knot's order among all knots with that crossing number.

Amphiphile ("both"-loving) chemical compound with both hydrophilic (water-loving) and lipophilic (fat-loving) properties (soaps, detergents, phospholipid bilayers in cell membranes).

Angular momentum measure of a body's tendency to roll, spin, or orbit; measured by the amount (magnitude), direction (projection), and intrinsic angular momentum (spin).

Atomic precision advanced manufacturing (APAM) ultradoping of surface chemistry layer.

Atomically precise manufacturing (APM) single atom precision fabrication of objects.

Artificial lattice microscopy-applied atomic-scale grid designed to confine electrons or spins.

Autologous therapies personalized therapies using an individual's own cells or DNA.

Band gap gap in electron energy levels (bands) between the "lower" stable bands in which most electrons vibrate, and the "upper" conductance band in which fewer electrons vibrate.

Biological liquid crystals liquid-crystal matter phases with properties of liquids and crystals, abundant in biology: membrane phospholipid bilayer, DNA, RNA, proteins, spider silk (protein).

Bio-MEMS/Bio-NEMS millimeter-scale/nanometer-scale devices used in biology applications.

Biomolecular condensate non-membrane-bounded cellular organelle for intracellular processes.

Biopolymer natural polymer (covalently bonded) produced by cells to make DNA, proteins.

Bloch theorem solution to the Schrödinger equation for periodic systems.

Block copolymers copolymers of blocks of monomers linked by covalent bonds (AAAA-BBBB) (vs. alternating copolymers of regular alternating A and B units (A-B-A-B).

Blockchain (distributed ledger technology) distributed database of asset ownership; transactions are recorded between parties in a tamper-evident manner to create a single source of trusted data.

Blockchain science blockchains employed for the conduct of science (automation, replicability, protected data access, time-stamped records of intellectual property development).

Boson force particle (photon, gluon) with integer spin (compared with fermion: matter particle).

Bosonic code photon-based system in which physical and protective logical qubits can be encoded in a self-contained continuous value system in a single bosonic mode (state).

Bosonic mode photon state controllable with standard Gaussian operations such as squeezing, displacement, phase rotation, and beam splitting.

Cat state (Schrödinger's cat) superposition (quantum state) of two or more states.

Charge electrical valence of a particle (particle properties: spin, charge, angular momentum).

Chern-Simons theory mathematical model of topological invariance.

Code (error correction) allowed values (or structure of values) for data in a system, may include parameters re: how many ancilla (extra) bits protect one logical bit over what distance.

Codespace error correction domain; possibly denoted by lattice grid states or graph states.

Coherent state oscillatory quantum state (quantum state of the harmonic oscillator).

Copolymer polymer derived from more than one species of monomer (linear, branched, graft).

Covalent bond (strongest) chemical bond between atoms by sharing electrons in a stable balance of attractive and repulsive forces, each atom maybe with equivalent of a full valence shell.

Cryptoeconomics economic system instantiated by secure network cryptographic protocols.

Crystal atoms organized in a regular array (lattice).

Decentralized finance (DeFi) protocols for blockchain-based lending, insurance, investments.

Deep learning advanced machine learning architecture that is 3–8 or more layers "deep."

Digital twin virtual representation serving as a real-time digital counterpart of a physical object or process, possibly in constant communication and updating via Internet of Things sensors.

Dimer a molecule or molecular complex consisting of two identical molecules linked together.

Discrete time crystal non-equilibrium state of matter that breaks time translation symmetry (repeating structure in time vs. spatial crystals with repeating structure in space).

DNA nanotechnology nucleic acids used as building materials not information carriers.

Doping introduction of impurities to modulate properties (electrical, optical, structural).

Eigenvalues values at allowable scale tiers in a system, levels; characteristic system values.

Entanglement quantum property of correlated physical attributes between particles.

Factory@home proposal for 3D/4D molecular printing units at the home or community scale.

Fault-tolerant quantum computing large-scale universal error-corrected quantum computing.

Fermion matter particle (electron, quark) with half-odd integer (1/2, 3/2, etc.) spin (vs. boson).

Fractional quantum Hall effect topological matter phase obtained by applying a strong magnetic field perpendicular to a two-dimensional electron system at low temperature such that electrons come together to create quasiparticles (collective states) that have a fraction of the charge of a single electron and obey anyonic statistics (use case:fault-tolerant computing).

Floquet engineering management of periodically driven time cycles in quantum matter systems.

Floquet theorem transforms periodic system (Bloch) into solvable linear differential equations.

Frontier orbitals the highest occupied and lowest unoccupied orbitals of a molecule.

Gapless no band gap (no tiered spacing between energy levels in a system).

GKP codes (Gottesman, Kitaev, Preskill) error-corrected qubit encoding in an oscillator with superpositions of squeezed states, protected against shifts in position and amplitude damping.

Graph code error correction method with the codes being the graph states (codespace = graph).

Graph state multi-qubit state that can be represented with a graph; vertices are qubits, edges are qubit interactions (the graph is an efficient way to represent (certain types of) entangled states).

Grid states structure for the ideal basis states (corrections) to bodies (qubits) in the codespace, as periodically spaced points in the position or momentum space of an oscillator.

Hall effect (Hall conductance) production of a voltage difference (the Hall voltage) across an electrical conductor that is transverse (perpendicular) to an electric current in the conductor, and to an applied magnetic field perpendicular to the current (discovered by Edwin Hall, 1879).

Hamiltonian an operator used to calculate the energy levels of a quantum system.

Herbertsmithite mineral with quantum spin liquid magnetic properties (neither ferromagnet nor antiferromagnet), magnetic particles have constantly fluctuating scattered kagome orientations.

Hilbert space infinite-dimensional space of quantum mechanics (vs 3D Euclidean space).

Homeomorphism an instance of topological equivalence to another space or figure.

Hydrogel crosslinked polymer that can absorb significant amounts of water and not dissolve.

Interference quantum property of waves reinforcing or canceling each other out (cohering or decohering); the observed pattern of phase difference between two wavefunction amplitudes.

Ionic bond chemical bond involving electrostatic attraction between oppositely charged ions or between two atoms with sharply different electronegativities.

Ionization process of removing an electron (to ionize); photoionization microscopy.

Jones and HOMFLY polynomials advanced algorithms to discriminate complex knot types.

Josephson junction quantum tunneling superconducting device used in quantum computing.

Josephson junction-based superconducting circuits superconducting qubits controlled with microwave photons (quantized electromagnetic fields stored in the superconducting circuits).

Kagome lattice (trihexagonal tiling) uniform tiling pattern by equilateral triangles and hexagons.

Kitaev honeycomb exactly solvable spin model in two dimensions; spins are on the vertices of a honeycomb lattice and have (link-orientation dependent) nearest neighbor interactions.

Lattice regular geometric arrangement of matter in a space, grid, scaffolding (e.g., crystal).

Lattice surgery switching between error correcting codes on the fly.

Levitated nanoparticle laser-levitated nanoscale object cooled to near ground state temperature with sensitivity to read superposition and entanglement (standard quantum device building block).

Liquid crystal matter state between liquid-crystal, flows like a liquid, crystal-based molecules.

Machine learning artificial intelligence method in which a computer algorithm writes an equation (function) to describe some input data, iteratively improving, and identifying new data.

Majorana (bound) states (zero modes) non-abelian anyons bound to a defect at zero energy.

Many-body localization (MBL) many-body interactions causing quantum particles to be localized and maintained in an out-of-equilibrium state.

Many-body problem physical problem in systems with many interacting particles (3 or more).

Matter current amount of mass passing through an area in a given amount of time.

Micelle sphere with hydrophilic heads on the outside and hydrophobic tails in the center.

Microelectromechanical systems (MEMS) millimeter scale machines with moving parts: devices 20–1000 μm (components 1–100 μm); silicon, polymers, metals.

Mode state, characteristic state, normal frequency, allowable value (e.g., count the photons in a certain mode (state) characteristic expression sequences (e.g., single-photon wave sequences)).

Molecular code GKP codes extended to asymmetric bodies (molecules) in free space.

Molecular dressing constrained, configured, non-free floating; dressed states (vs. bare states).

Molecular knot mechanically interlocked molecular architecture analogous to a macroscopic knot; naturally found in organic molecules (DNA, RNA, proteins) and can be synthesized.

Molecular manufacturing atomically or molecularly precise production of physical objects.

Molecular rotor molecular scale rotational assembly; assembly with rotary movement.

Molecular self-assembly molecules adopting a defined arrangement without external guidance.

Momentum *see* Angular momentum.

Nanoelectromechanical systems (NEMS) devices with electrical and mechanical functionality on the nanoscale (transistor-like nanoelectronics with mechanical actuators, pumps, motors); physical, biological, chemical sensors (possible quantum harmonic oscillation effects).

Nanotechnology study of materials and devices with features on a scale below 100 nm.

NISQ (Noisy Intermediate-Scale Quantum) Devices currently available quantum computers, generally with 50–100 qubits, that can perform tasks that do not require error correction.

Non-abelian non-commuting (order of terms matters) (v. abelian, commuting).

Non-fungible token (NFT) tradable blockchain-based ownership record of intellectual property.

Noncovalent interaction (weaker chemical bond) via electromagnetic interactions between molecules per van der Waals forces, hydrogen bonding, electrostatic interactions (ionic bonding).

Particle properties (encode qubits) spin, charge, angular momentum, polarization, spatial mode.

Piezoelectricity electricity from pressure and latent heat; electric charge that accumulates in certain solid materials (ceramics, bone, DNA, proteins) in response to applied mechanical stress.

Plasmon a collective excitation ("quasiparticle" of bosons), which is the quantum of plasma oscillations (wherein all electrons simultaneously oscillate with respect to all ions).

Plasmonics computation via optical energy concentration below the diffraction of light.

Polymer (artificial or natural) repeating chain of many repeating polymeric subunits.

Printed electronics printing methods used to create electrical devices on various substrates.

Protein folding physical process of protein chain "folding" into a final useful conformation.

Quantum dot semiconductor nanoparticle (a few nm in size) with quantum-mechanical optical and electrical properties, standard quantum nanoscience building block.

Quantum error correcting code logical codespace corresponds to subspace of a lattice model.

Quantum Hall effect quantum version of the Hall effect, observed in two-dimensional electron systems in which low temperatures and strong magnetic fields are applied for Hall conductance phase transition to take on quantized values; quantized system plateaus seen at values of charge.

Quantum machine learning machine learning applied in a quantum environment.

Quantum microscopy various techniques to image, measure, and control quantum particles.

Quantum nanotechnology nanostructured systems that incorporate quantum effects.

Quantum phase transition phase transition between different quantum phases (phases of matter at zero temperature) via physical parameter changes such as magnetic field or pressure.

Quantum spin Hall states quantum Hall effect per flow of spin currents (vs. charge currents).

Quantum spin liquid magnetic system that does not settle into a long-range ordered configuration, even at zero temperature, residing in a nontrivial quasi-disordered ground state.

Quantum topology novel bend-but-not-break properties of zero-temperature matter phases.

Quasiparticles (fermions) and collective excitations (bosons) emergence of collective interactions/particle excitations in a many-body system; a low energy excited state.

Rényi entropy composite of Shannon, Hartley, collision, and minimum entropy.

"Schrödinger cat" states superpositioned states.

Self-assembly non-covalent individual-unit grouping into arranged patterns (organic/inorganic).

Serialization product identification at various levels (raw materials lots, individual units, batch).

Shape memory polymers material returning to original shape after deformed by stimulus.

Smart contract computer program or transaction protocol intended to automatically execute according to the terms of an agreement (thus reducing the need for trusted intermediaries).

Smart network technologies (blockchain, deep learning) intelligent self-orchestrating software.

Spectroscopy light-based analysis of a sample, identify matter based on absorbed wavelengths.

Spin intrinsic form of angular momentum carried by elementary particles; particles with spin can possess a magnetic dipole moment (exploited in electronic devices).

Spin chain collection of magnetic moments, arranged linearly, interacting w spin-spin coupling.

Spin engineering manipulation of quantum spin systems (magnetic atoms) in devices/materials.

Spinors more complicated version of vectors or tensors required to describe particle spin.

Spintronics (spin electronics) devices with electron spin (magnetic moment) + electronic charge.

Squeezed (coherent) state oscillatory quantum state that is squeezed (quantum noise-reduced).

Stabilizer codes quantum error correction code (quantum version of linear codes), based on X, Y, Z axis Pauli operators to measure entangled states and correct (bit flip, spin flip) a corrupt quantum state to its original state (examples: toric code, surface code).

Subtractive manufacturing removing (etching away) elements during the manufacturing process.

Supercluster extended region with a high concentration of galaxies (e.g., Laniakea supercluster).

Superposition quantum property of an unobserved particle existing in all possible states simultaneously, until collapsing to only one when measured.

Supramolecular (many molecules) structures composed of several or many molecules.

Surface code stabilizer code, topology-based, defined on a 2D spin lattice, taking various shapes (not necessarily toroidal as in the toric code).

Surfactant compounds that lower the surface tension between two domains (liquid, gas, solid).

Symmetry features of spacetime and particles which are unchanged under some transformation; the property of a system looking the same from different points of view (whether a face, a cube, or the laws of physics) and its partner, symmetry breaking (phase transition).

Symmetry-breaking phase transition, breaking a system's symmetry, time-translation, chirality.

Symmetry-protected topological (SPT) order quantum matter phases with trivial topological order (short-range entanglement), symmetry, and a finite energy gap (e.g., topological insulator).

Time-reversal symmetry breaking system property in which the dynamics of a process remain well-defined when the sequence of time-states is reversed.

Topological entanglement entropy topology-based measure of entanglement entropy specific to quantum matter and phase transition (tripartite information: 2-space/1-time dimension).

Topological insulator material with a conducting surface and an insulating interior; surface states are symmetry-protected (e.g., by time-reversal, particle-hole, chiral symmetry).

Topological quantum field theory quantum field theory which focuses on topological invariants, used to explain physical systems such as quantum matter phases, low-energy effective theories of topologically ordered states (fractional quantum Hall states, quantum liquid states).

Topological qubits computational qubits made with quantum spin liquid matter phases (by putting the quantum spin liquids in a geometrical array).

Topological strings strings that connect atoms entangled in a quantum spin liquid.

Topology property of geometric form being preserved under deformation (such as bending, stretching, twisting, and crumpling, but not cutting or gluing).

Toric code stabilizer code defined on a 2D lattice with periodic boundary conditions (torus-shaped) with stabilizer operators on the spins around each vertex and plaquette (face).

Trotterization many-step operation to simulate the evolution of a Hamiltonian.

Two-dimensional electron gas physics model in which an electron gas is free to move in two dimensions but tightly confined in the third, resulting in quantized energy levels for motion in the third direction; the electrons appear to be a 2D sheet embedded in a 3D world.

Vesicle hollow sphere that can deliver hydrophilic (core) and hydrophobic (bilayer) compounds.

Von Neumann entropy taking the minimum overall measurement bases of Shannon entropy.

Voxel 3D datapoint; unit of graphic information defining a point in three-dimensional space.

Wavefunction quantum system state description (positions or speeds (momenta) of entire system configurations); calculated with the generally intractable Schrödinger equation (complex-valued probability amplitudes with real and imaginary wave-shaped components).

4 DNA-Based Manufacturing
Potential Strategies and Prospects

Frank J Boehm

(Derived from "An Investigation of Nucleic Acid/DNA-Based Manufacturing" by Frank Boehm ©Aug/2004. Research contribution to the Center for Responsible Nanotechnology: Thirty Essential Nanotechnology Studies – #10 (What will be required to develop nucleic acid manufacturing and products?) https://www.researchgate.net/publication/215836586_An_Investigation_of_Nucleic_AcidDNA-Based_Manufacturing)

CONTENTS

4.1 DNA Structure and Function ... 101
4.1.1 DNA Data Storage ... 102
4.2 Survey of Nanodevices Based on DNA ... 103
4.3 A Toolbox of Nucleic Acid Devices ... 106
4.3.1 B-Z DNA Torsion Device ... 106
4.3.2 DNA Actuation Devices ... 107
4.3.3 DNA Tweezers ... 108
4.3.4 DNA Scissors ... 109
4.3.5 Deoxyribozymes ... 110
4.3.6 DNA – Three State Device ... 110
4.3.7 DNA Crossover Junction-Based Device ... 111
4.3.8 DNA G-Quartet Device ... 111
4.3.9 DNA Motors – Free-Running and Clocked ... 111
4.3.10 Precise Manipulation of Proteins Using DNA ... 113
4.3.11 DNA Walkers ... 113
4.4 DNA Manufacturing Complex: Infrastructure Elements ... 114
4.4.1 Discussion ... 117
4.4.2 Replication of DNA Structural Elements – Octahedron ... 118
4.4.3 DNA Origami ... 119
4.5 DNA Adhesion To/Dissociation from Nano-Entities and Substrates ... 120
4.5.1 DNA Binding to Nanotubes ... 120
4.5.2 Thiolated DNA ... 121
4.5.3 DNA Adhesion to Dendrimers ... 121

 DOI: 10.1201/9781003724896-4

4.5.4 DNA "Velcro" 122
4.5.5 DNA Adhesion to Surfaces 122
4.6 Potential Computational Requirements for a DNA-Based Manufacturing Complex 123
4.6.1 Fractal Sierpinski Triangle Self-Assembly Using DNA 125
4.7 DNA Nanoelectronics 126
4.7.1 DNA Nanowires 126
4.7.2 DNA Conductivity 126
4.8 Prospective DNA Manufacturing Strategy 128
4.9 Conclusion 129
References 130

4.1 DNA STRUCTURE AND FUNCTION

The most ubiquitous structural type of DNA is right-handed (Watson-Crick Model) B-DNA (Watson & Crick, 1953), which is configured as a Ø ~2 nm double-helix (duplex) with a 34 Å helical pitch, 10 base pairs (bp) per turn, and an adjacent base pair distance of 3.4 Å. The stability of B-DNA is attributed to the "spine of hydration" that structurally supports it along the length of its major groove and backbone (Phan et al., 1999; Ventra & Zwolak, 2004; Zhang et al., 2019). Other major species of DNA include A-DNA (right-handed with 11 bp per double helical turn with a 2.6 Å/bp rise) (Franklin & Gosling, 1953; Harvey, 2015), Z-DNA (left-handed with 8 bp per double helical turn), which requires a high salt concentration (Leng, 1985), and a rarer H-DNA (triple helix) (Travers, 1993).

Distinct nucleic acid binding facilitates the synthesis of myriad novel and complex structures. Since B and Z DNA possess various π-stacking modes via base pair π electron overlap, it is presumed that they have distinct electronic properties. Thus, they may serve as one-dimensional conduits for the transport of electrons through their central helical axes (Yang et al., 2003). Single-stranded oligonucleotides are linear polymers comprised of a linkage of 2′-deoxyribose nucleotides attached to one of four different bases (adenine (A), cytosine (C), thymine (T), and guanine (G)). They are also asymmetric and bound via a phosphodiester link that spans the 5′ carbon atom in the sugar of one nucleotide and the 3′ carbon atom of its neighboring counterpart. Thus, the chemical structure of one end of the molecule is distinct from its opposite end.

Complementary DNA base pairs (A-T) (C-G) are tethered by weak hydrogen bonds (Figure 4.1), where when a hydrogen atom binds to an electronegative atom, a percentage of its charge density is contributed to the bond. This creates a positive charge domain on the outward-facing side of the hydrogen atom relative to the bond, resulting in weak interactions with the lone pair electrons of an adjacent atom (Seeman, 2004). The force required to separate two strands of DNA is on the order of ~70–75 pN/bond, whereas the linear tension threshold of the DNA backbone is ~10 nN (Lee et al., 1994). The creation of a stable duplex structure at room temperature requires the association of approximately twelve base pairs (Alivisatos, 2000).

The exterior geometry of DNA is comprised of sequentially occurring sugar-phosphate backbones, which are segregated by two grooves (major and minor) whose profiles are specified by their base pair orientations. The exposed N and C atoms

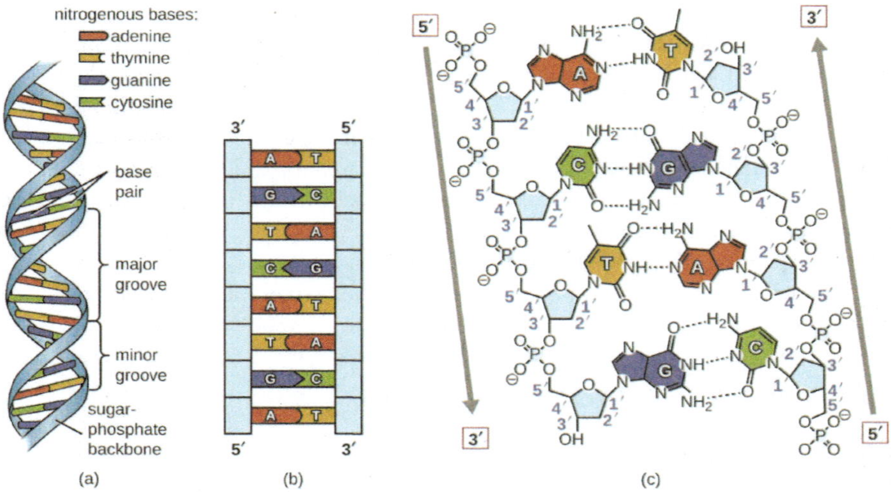

FIGURE 4.1 DNA double helix AT and CG nucleobase pairing. (Parker et al., 2016, Creative Commons License https://creativecommons.org/licenses/by/4.0/ (accessed April 01, 2024)).

that face into the major groove might be employed as anchors for the synthesis of pre-designed constructs (Travers, 1993). Investigations have shown that DNA can be linearly extended up to ~2.1 times its initial length before rupturing (Lebrun & Lavery, 1996). Further, its tensile strength was estimated to be ~270 pN (lower value) (Harrington & Zimm, 1965), while from ~476 +/– 84 pN (Bensimon et al., 1995) to ~500 pN (Lebrun & Lavery, 1996) was estimated to be the force required to break a strand of DNA. In contrast, the tensile strength of the thinnest spider silk (~Ø20 nm) is ~1.3 Gpa (Shao & Vollrath, 2002), whereas that of polysilicon is 1.45 ± 0.19 Gpa (Sharpe, 1999), and up to ~130 Gpa (Yakobson & Smalley, 1997) for single and multi-walled carbon nanotubes.

4.1.1 DNA Data Storage

In terms of the amount of data that can be stored by DNA (Travers & Muskhelishvili, 2015) state, "The information encoded by DNA is both digital – the precise base specifying, for example, amino acid sequences – and analogue. The latter determines the sequence dependent physicochemical properties of DNA, for example, its stiffness and susceptibility to strand separation." It is estimated that under ideal conditions the data storage capacity of DNA might attain two bits per nucleotide. Yin et al. (2022) posit that "If a DNA device were used to store all information on the Internet, a device smaller than one cubic inch is all that would be needed …" (Castillo, 2014). In terms of physical volume, it is thought that DNA can store ~60 petabytes/cm^3 of data (Ceze et al., 2019; Erlich & Zielinski, 2017; Goldman et al., 2013; Nguyen et al., 2021), while in terms of weight, data storage estimates are ~455 exabytes/gram (Church et al., 2012). Beyond its benefits of extreme density and relative simplicity of replication via the polymerase chain reaction (PCR) is its longevity, which can endure for thousands of years (Grass et al., 2015; Kohll et al., 2020).

4.2 SURVEY OF NANODEVICES BASED ON DNA

In all living organisms, an abundance of infinitesimal machines oversees the conversion of chemical energy to mechanical actuation (Bray, 2001; Howard, 2001). It is conceivable that the functionalities of these entities might be extrapolated to guide the design and development of synthetic nanoscale analog devices that are able to perform similar operations. In the biological domain, the primary purveyors of motion are molecular motors, which include dynein (activates cilia and flagella), actin, myosin (muscle protein filaments), and kinesin (transfers cellular elements).

The molecular replication of DNA may be considered an intricate assembly process through which DNA generates almost perfect copies of itself, while engaging in error correction and proofreading (Alberts et al., 1994). Synthetic nanodevices may be designed with the capacity for actuation through the integration of molecular motors. Chemical energy sources such as ATP (adenosine triphosphate) and GTP (guanosine triphosphate) might be employed to activate these entities, as may glucose oxidation/reduction or hydrolysis. The powering of bacterial flagella motors involves the extraction of energy from protons that transit through an electrically charged cell wall (Bray, 2001).

In terms of synthetic nanoscale motors, it is conceivable that they may operate using sources of energy that are quite distinct from those observed in nature. Thus far, several molecular actuators have been developed, including a light-powered rotary motor (Balzani et al., 2000; Koumura et al., 1999). Furthermore, catalytic electrodes could be utilized to induce redox reactions (Katz & Bollella, 2021). Giant unilamellar vesicles (GUVs) comprised of liposomal membranes were employed to encapsulate kinesin and myosin molecular motors (Shoji & Kawano, 2020). Additionally, encapsulated actin was induced to alter its morphology via increased temperature, which enabled the capacity to deform the GUV membrane (Honda et al., 1999; Hotani et al., 1999). Reversible GUV membrane deformation (from a spherical to spindle shape) was driven by fluorescent-labeled actin that was actuated by modifying the osmotic pressure or through light exposure (Hayashi et al., 2016; Tanaka et al., 2018).

Several mechanical DNA-based actuation mechanisms have been created by exploiting the intrinsic motive forces involved in base pair binding, which results in the well-known double-helix structure. Although currently achievable, these entities represent very preliminary steps toward the ultimate realization of mature nucleic-acid-based fabrication technologies. In the coming decades, similar devices may comprise the requisite infrastructures that enable molecular manufacturing (MM) systems. As a leading visionary and tireless contributor toward this goal, the late Dr. Nadrian Seeman (New York University) designed and synthesized myriad extraordinary DNA structures and feasible actuation mechanisms. Seeman has led the way toward the emergence of genuine and robust DNA-based MM (Chen & Seeman, 1991; Lu et al., 2021; Seeman, 2004, 2005, 2010; Shen et al., 2004; Wang et al., 2021; Zhu et al., 2021).

"Due to its powerful molecular recognition savvy, exquisite versatility, and relative stiffness (in its double-helix form), DNA is an extraordinary molecule with which to construct a myriad of novel nanometric tools" (Boehm, 2004). Biotechnology, as manifested in the constituents of DNA, encompasses the possibility to design predictable structures by virtue of the specific pairing and sequences of its base pairs. Interactions between oligonucleotides may be engineered to create elemental structures that can self-assemble in a programmable highly orchestrated manner. With current technologies, customized oligonucleotides (typically a maximum of ~200 nucleotide residues long) can be generated, inclusive of molecular accessory nodes and myriad additional modifications. For instance, this might enable the tethering of dye molecules at specific sites along an oligonucleotide, as well as catalytic domains to facilitate substrate binding (Simmel & Yurke, 2004).

Certain enzymes have the capacity to precisely modify DNA (Keller & Marx, 2011) (e.g., polymerase enables amplification, ligase supports covalent bonding, and endonucleases very accurately cleave DNA strands), as instructed by base pair sequences. These capabilities have facilitated the synthesis (by nature) of the most complex nanometric mechanisms imaginable (Simmel & Yurke, 2004). Complimentary base-pairing (e.g., A-T, C-G) drives the anti-parallel orientation of two single DNA strands. Unstable DNA molecules are the consequence of base pair mismatches, in which case the adhesive forces between them are far more tenuous in contrast to when they bind to their true complements. Thus, when the number of mismatches is too great, the duplex will not form; when immersed in solution, the two strands will separate.

This formidable DNA recognition attribute will be essential for the design and manufacture of DNA-based structures, as complementary strands will join very precisely; gravitating to one another and engaging in Brownian motion-mediated self-assembly. This automated procedure of matching strands is referred to as "hybridization" (Simmel & Yurke, 2004). Complementary base pair hybridization is a second-order chemical reaction, where its rate is proportional to the concentration of the square of one reactant, or the product of the concentration of two reactants (Helmenstine, 2019). In contrast, a first-order reaction entails the "unzipping" of complementary DNA strands into two distinct components. The main factors that induce duplex dissociation into single strands include temperature and its length; however, they have less influence on their hybridization (Morrison & Stols, 1993).

At low temperatures, there are negligible detachment rates and the hybridization levels of strands of sufficient length are acceptable. Brackish aqueous media is essential for DNA hybridization, as immediate dehybridization will proceed with the removal of salt (Atkinson, 2003). The time required for the assembly of nanostructures, or the actuation of nanoscale motors driven by the programmable hybridization of nucleic acids is primary guided by the inherent hybridization kinetics of DNA. However, it is conceivable that the hybridization of synthetic structural DNA segments might be expedited by various means (Kang et al., 2014; Wang et al., 2005; Wang et al., 2012).

The potentially vast numbers of base pair sequences translate to an immensity of possible complex self-assembled DNA structures, including wireframe polyhedra (Jun et al., 2019), octahedrons (Shih et al., 2004), tetrahedral DNA nanostructures (Li et al., 2019), DNA nanocages (Duanghathaipornsuk et al., 2020), DNA origami structures (Endo et al., 2013), DNA tile-lattices (Ke et al., 2006), DNA bricks/Legos (Ke et al., 2012), tessellated 3D DNA meshes (Zhang et al., 2016), DNA crystals (Winfree et al., 1998), and hybrid RNA/DNA nanostructures, etc. (Ko et al., 2010)

(Krissanaprasit et al., 2021). Molecular computational elements based on DNA may also be synthesized (Adleman, 1994, 1998; Mao et al., 2000). Self-assembled DNA nanostructures can be decorated with multivalent aptamers to construct logic gate-based mechanisms. These entities have the potential for myriad applications, ranging from the performance of computations for neural networks (via strand displacement reactions) to the controlled programmable transport of molecular cargo (e.g., therapeutics) in vivo (Pei et al., 2012; Qian et al., 2011; You et al., 2014).

The comparative stiffness of molecular DNA in contrast to other biopolymers, makes it an excellent candidate as a nanometric structural element in multi-strand arrangements. Single-strand polymers are continuously jostled about in solution via Brownian motion (e.g., short segments will be incessantly agitated and twisted, whereas longer sections will respond through bending). The persistence length of a polymer is determined through the decorrelation of two points along its chain, which indicates its bending stiffness and tendency to remain stable. In the case of single-stranded DNA, its persistence length was observed to range from ~1.5 nm (~4 bp) in 2 M NaCl to 3 nm (~9 bp) in 25 mM NaCl (Tinland et al., 1997; Murphy et al., 2004), whereas in duplex form it was ~50 nm (150 bp) (Bloomfield et al., 2000; Rittman et al., 2009; Zaichuk & Marko, 2021). Although the DNA duplex has increased stiffness it remains quite malleable; a 5 nm long segment can still bend by 25° and a 1 nm long segment by 11° (Smith et al., 1996; Round et al., 2002).

These variations in stiffness may conceivably be exploited to enable flexible joints within DNA structures, whereas the free energy available via base pair hybridization might be utilized to spatially manipulate molecular-scale entities. This aspect could be specified through the strategic design and arrangement of specific base pairs relative to their adjoining counterparts. Molecular dynamics simulations were employed to calculate the Watson-Crick pairing free energies of adenine-thymine and guanine-cytosine base pairs in water, which were revealed to be –4.3 and –5.8 kcal/mol, respectively (Stofer et al., 1999). Single-stranded DNA may be employed to displace a load at an increment of twice the 0.34 nm nucleotide spacing through hybridization (Figure 4.2).

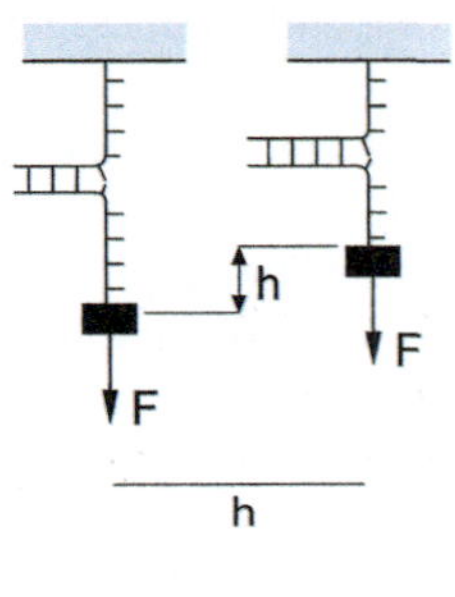

FIGURE 4.2 Theoretical illustration of DNA-driven displacement of a load being drawn against the gravitational force F. The free energy of hybridization is converted to work through base pair binding, whereupon the load is relocated by double the gap Δx between bases in ssDNA, $h = 2\Delta x$. (Reproduced from Simmel FC, Yurke B. "DNA-based nanodevices," in Nalwa HS, ed., *Encyclopedia of Nanoscience and Nanotechnology*. Vol. 2. American Scientific Publishers, Stevenson Ranch, CA, 2004;495–504. With permission.)

Various instrumentation, including Atomic Force Microscopy (AFM) and optical traps, has been employed to quantify the force required to stall a DNA motor (~15 pN), which correlated with the forces required to decouple DNA duplex strands and those produced by nanoscale biomotors (Essevaz-Roulet et al., 1997; Liphardt et al., 2001; Rief et al., 1999). The stall force of the *Bacillus subtilis* phage (Ø29) ranges from 40 to 70 pN (~57 pN average), making it one of the most robust molecular motors ever observed (Smith et al., 2001; Chemla et al., 2005). This is far greater than that required to stall a walking kinesin on a microtubule (~5–6 pN) (Coppin et al., 1997; Svoboda & Block, 1994), a myosin motor (2.5 pN) (Finer et al., 1994; Mehta et al., 1999; Uemura et al., 2004), or RNA polymerase (25 pN) (Wang et al., 1998).

The kinetics behind DNA hybridization power the synthesis of nucleic acid constructs, and drive DNA-based molecular motors (Simmel & Yurke, 2004; Michaelis et al., 2009; Blanchard et al., 2019). Myriad nanometric components and tools are being devised that may facilitate the emergence of robust and precise DNA-based manufacturing capacities (Dai et al., 2020; Fritzsche, 2002; Jiang et al., 2019; Li et al., 2019; Liu et al., 2021; Ma et al., 2020; Sameiyan et al., 2019; Stoltenberg & Woolley, 2004; Zhang et al., 2015). What is referred to as "branch migration" has the ability to reposition specific DNA strands within nanostructures, which arises when two single strands that possess the same base sequences attempt to hybridize concurrently with one or two corresponding strands. Subtle deviations in temperature can cleave base pairs at sites where they compete to unite with their counterparts. The subsequent creation of base pairs can shift the branch point by one base pair in either direction from the original location, and can also be utilized for three-stranded branch migration. This process can be constrained or stopped under high concentrations of magnesium (Panyutin & Hsieh, 1994; Simmel & Yurke, 2004).

Branch point migration is an intrinsic aspect in the repair and recombination of DNA across biological domains. A mobile four-way connection known as a Holliday junction is generated via strand exchange between two equivalent sequences (Heo et al., 2022; Lilley, 2000; Mao et al., 1999a). The Holliday junctions of non-equivalent DNA duplexes can result in static structures, as the branch points are immobilized. These entities may serve as basic structural units for the manufacture of nanoscale machines as they possess both double and triple multiple crossover nodes (Winfree et al., 1998; LaBean et al., 2000b).

A recently developed strategy referred to as cooperative branch migration (CBM) has the capacity to control the displacement of DNA through the regulation of branch migration domain complementarity. This approach can facilitate the synthesis of more complex DNA mechanisms for applications in biomedicine and nucleic acid research (Weng et al., 2022), which may also conceivably serve to advance the possibility of a high-precision DNA-based nanofactory.

4.3 A TOOLBOX OF NUCLEIC ACID DEVICES

4.3.1 B-Z DNA Torsion Device

The nanomechanical transition from the right-handed B-form of DNA to the left-handed Z-form is referred to as the B-Z transition, which may be exploited for movement of nanoscale entities (Bhanjadeo et al., 2019; Nayak et al., 2016). A B-Z transition device may be activated through chemical changes, where an "active" DNA region interacts within stiff and level duplex elements. "A DNA strand containing methylated cytosine residues (which are more prone to B-Z transition) are connected to the two stiff DX structures. By altering the concentration of specific chemical groups, the B-Z device can be actuated (the two-level members rotate one-half turn with respect to each other's original orientation) repeatedly" (Boehm, 2004; Mao et al., 1999b).

A general biophysical technique referred to as fluorescence resonance energy transfer (FRET) can be employed to indirectly verify B-Z transition cycling. FRET is typically used to measure and confirm nanoscale movement, and is based on differences in the efficacy of energy transfer when the distance (Förster distance) is varied between two fluorescent dye molecules, which typically ranges from ~2 to 6 nm (Arai & Nagai, 2013; Long et al., 2021; Simmel & Yurke, 2004). One disadvantage of this otherwise robust apparatus is that when multiple such devices are combined, only the two (B or Z) states would be available for the entire construct. Thus, to enable the distinct actuation of each singular unit would necessitate the integration of separate activation strands to trigger them (Seeman, 2004).

4.3.2 DNA Actuation Devices

DNA-based actuation devices utilize external engineered "fuel strands" to create motion via the hybridization of single strands of DNA, which merge with flexible complementary oligonucleotides, integrating stiff elements that are tethered to a substrate (Simmel & Yurke, 2004). Anchored DNA can provide a stable work platform (akin to securing an industrial robot to a factory floor), which is critical when planning to orchestrate a range of precise movements along multiple axes to achieve highly accurate component placement. The repeatable actuation of DNA devices is contingent on the capacity to return them to their original positions and orientations via fuel strand displacement, which may be achieved through DNA branch migration. "For example, a strand X (the fuel strand) can be peeled off its complement Y (displaced) by another strand Z (the removal strand) if X and Z are complementary, and X and Y are only partially complementary so that a 'toehold' of X remains dangling, unattached to Y. Z binds to the toehold and proceeds to unzip X from Y. At the end of the process, X and Z are bound together, and Y is free" (Boehm, 2004; Phoenix, 2004).

Achieving high levels of precision for nanoscale fabrication becomes exponentially more challenging when DNA fabrication units are free-floating in aqueous media. Nevertheless, it has been demonstrated that precise nanostructures can be assembled by DNA within microfluidic compartments that are the size of human cells. "These encapsulated DNA nanostructures not only exhibit structural reconfigurability owing to their pH-sensitive molecular switches upon external stimuli but also possess optical feedback enabled by the integrated plasmonic probes" (Göpfrich et al., 2020). In biological milieu, DNA can be tethered to lipid membranes/bilayers using single or multiple anchors comprised of cholesterol, tocopherol, or various combinations of the two (Rubio-Sánchez et al., 2021a). Further, membrane-bound DNA nanostructures can be programmatically laterally distributed (Rubio-Sánchez et al., 2021b).

There are numerous other DNA-based actuation devices that may facilitate the precise fabrication of nanostructures, including (Freitas & Merkle, 2004):

- silicon cantilevers activated by DNA hybridization (Fritz et al., 2000; Wu et al., 2001; Majumdar, 2002; Liu et al., 2003; Stachowiak et al., 2006; Zhang et al., 2006).
- DNA pistons that exhibit linear actuation (Alberti & Mergny, 2003; Bleha & Cifra, 2020; Bayoumi et al., 2021; Lim & Montague-Cardoso, 2021).
- molecular DNA "inchworm" motors (Li & Tan, 2002; Draper & Rao, 2007; Nichols & Corces, 2018; Castillo et al., 2021).
- DNA nano-springs (Saran et al., 2020; Wang et al., 2010; Wang et al., 2014).
- DNA-based gears (Tian & Mao, 2004).
- DNA-based sensors (Freeland et al., 2020; Liu et al., 2016; Lymperopoulos et al., 2010).
- hybrid RNA/DNA rotary motors (Shu & Guo, 2003; Valero et al., 2018).
- continually running DNA motors, and "sliding strut" actuators (Merkle, 1999).
- "molecular sleds" running along DNA, which may be useful for transferring feedstock materials to assemblies (Xiong et al., 2016).
- combined X-shaped DNA tiles may be induced to form grids that can be expanded and retracted via chemical activation (Feng et al., 2003; Yan et al., 2003).
- a RNA polymerase positioning motor (15–20 pN) was induced to "walk" along DNA with nanoscale precision (Pomerantz et al., 2003).
- autonomous nanomechanical DNA devices (Bath et al., 2007; Chen et al., 2004; Feng et al., 2003; Nummelin et al., 2020; Surana et al., 2011).

Several DNA-based mechanisms are expanded on below.

4.3.3 DNA Tweezers

DNA tweezers comprised of three single strands (e.g., A, B, and C) were the first such nanodevices to integrate branch migration as essential elements of their operation (Yurke et al., 2000; Landon et al., 2012; He et al., 2021; Wu et al., 2021) (Figure 4.3). In one model, the B and C strands were engineered to hybridize with strand A for an 18 bp length each. The final device consisted of two 18 bp duplex sections spanned by a 4 bp long single-stranded section. Projecting from the two double-helical arms were 24 bp-long unbonded single-stranded segments of B and C. As the DNA duplex persistence length was substantial, the 18 bp long elements were considered as stiff, while the single-stranded segments served as flexible joints (Simmel & Yurke, 2004).

The introduction of a (F1) fuel strand (48 bases comprising two 20-base long segments) to the open position induced the closure of the two-armed device, which corresponded to the unbound B and C segments. The integration of an overhanging eight-base long "toehold" segment facilitated their extraction by a "removal" strand (F2), which decoupled the F1 strand (via branch migration) to return the tweezers to the open position. Thus, the mechanism could be repeatedly cycled between the

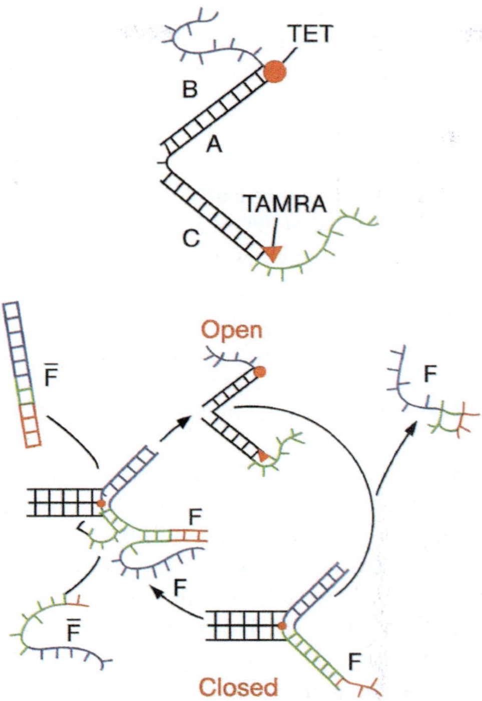

FIGURE 4.3 Structure and operation of molecular tweezer via oligonucleotide (A, B, C) strand hybridization. Closing strand F connects with floppy ends of B (blue) and C (green) strands initiating closed tweezer position. Overhanging section of F (red) connection enables removal of F, inducing open tweezer position. (Reproduced from Yurke et al., *Nature.* 2000 Aug 10;406(6796):605–608. With permission.)

two configurations. The tweezer opened or closed states could be verified using the FRET procedure (Simmel & Yurke, 2004).

It is worth noting that there were two functional drawbacks for this apparatus, including inadequate flexibility when the tweezers were in their open state, and the (per cycle) accretion of waste products that could stall the device when the F2 and F1 strands dissociated. The latter issue might be resolved through the intermittent removal of waste strands. Also of note is that disequilibrium can be sustained in some biomechanisms. A case in point is the inorganic phosphate that is dissociation during ATP to ADP conversion, which is reassociated due to the phosphate of these organic compounds being linked to metabolic processes (McLeish & Kenyon, 2005; Simmel & Yurke, 2004).

4.3.4 DNA Scissors

Hybridization was also exploited to drive the scissor-like closing action of one side of a DNA-based device and transfer it to the opposite side (13 nm distant). This motion was achieved through the addition of short segments, which served as linked rigid hinges that facilitated free rotation. One drawback in terms of the force transfer was that the scissor "handle" side closed more efficiently than the "jaw" side (Mitchell & Yurke, 2002).

4.3.5 Deoxyribozymes

DNA-based deoxyribozymes are catalytically active entities with excellent selectivity and very high product yields (10^{10} over background). In contrast to numerous natural counterparts, deoxyribozymes are single-turnover enzymes, albeit with multiple-turnover capacities. Although there remains much to learn in terms of their exact structures and kinetics, several perceived catalytic reactions include DNA phosphorylation (Li & Breaker, 1999) DNA adenylation (capping) (Li et al., 2000), DNA deglycosylation (Sheppard et al., 2000), porphyrin metalation (Li & Sen, 1996), thymine dimer photoreversion (Chinnapen & Sen, 2004), and DNA cleavage (Carmi et al., 1996).

These deoxyribozymes had the capacity to fuse two DNA substrates through a reaction that contained an activated 3′-phosphorimidazolide 5′-hydroxyl group, which required either Zn^{2+} or Cu^{2+} for activation (Silverman, 2004). Numerous challenges remain to be addressed before these catalysts have the practical utility of linking DNA substrates. These include that the activated termini cannot be easily attained (Cuenoud & Szostak, 1995) and the required sequences can greatly limit the types of substrates that may be used (Sreedhara et al., 2004). However, catalytic improvements can be realized through the chemically mediated addition of functional groups (Bittker et al., 2002) including metal chelators (e.g., pyridine, imidazole), which may be incorporated by employing altered DNA polymerases (Silverman, 2004).

Catalytic reactions can occur on various substrates when nucleic acids are tethered to them with the sole purpose of binding with DNA enzymes via typical Watson-Crick base pairing. This arrangement may include linked cleavable segments to facilitate nucleic acid removal following the catalytic reactions. Alternate strategies for securing DNA to substrates without the use of base pairs may involve the identification of certain sequences that attach to substrates and simultaneously perform catalysis, or through the creation of DNA aptamers that are employed to facilitate binding (Famulok et al., 2000; Silverman, 2004). Hypothetically, it is conceivable that modular catalytic hybrids might be developed through the combination of DNA aptamers and enzymes, where the aptamer components bind with a substrate while the enzyme elements catalyze/generate the intended products (Silverman, 2004).

4.3.6 DNA – Three State Device

A modified scissor apparatus was created that exhibited two mechanically robust configurations (when the concentration of salt was appropriate) and a transitional state. In its default mode, two single-strand extensions met to create a roughly circular geometry. The removal of F1 and F2 fuel strands toggled the device between stretched and closed configurations. By introducing the proper removal strands, the three positions of this apparatus could be cycled, each of which (as described above) was verified using the FRET technique (Simmel & Yurke, 2004). It is possible that the force transduction between these two mechanical states might be translated to other components (Simmel & Yurke, 2004).

4.3.7 DNA Crossover Junction-Based Device

A DNA crossover junction-based device utilizes branch migration induced by strand displacement to transition between configurations. This entity involves more complex double crossover molecules and is comprised of multiple coupled four-way junctions. The dual DNA duplex actuating portion of the mechanism is linked by multiple crossovers, such that when the lower portion is rotated by 180° it can be employed as a molecular switch (Yan et al., 2002). Due to the rigidity of the crossover molecule, it is quite robust with the capacity to enhance the transduction of its generated force. Copper ions have been integrated into DNA double-crossover tiles, which were shown to impart greater thermal stability (Rousina-Webb et al., 2020).

Biotin can be used to activate the fuel strands for this apparatus as it rapidly connects with streptavidin, which is the bacterial counterpart of biotin that is typically employed as a molecular linker (Hess & Vogel, 2001; Niemeyer et al., 1994). The non-covalent bond between biotin and streptavidin can withstand 5 pN of force for 1 min; however, under 200 pN, the bond will break within 1 ms (Grubmüller et al., 1996; Merkel et al., 1999). This intense affinity may enable a potential strategy for the sustainable removal of waste products from a DNA manufacturing complex using magnetic beads that are coated with streptavidin (Chodosh, 2001; Niemeyer et al., 1994; Paul et al., 2009; Yamamoto et al., 2015).

4.3.8 DNA G-Quartet Device

A DNA G-quartet device is based on the unique attributes of guanine that allow it to not only bind with the three other bases, but also to itself. When domains rich in guanine sequences are in the presence of monovalent cations, G-quartets can form, including within non-coding telomeres (Simmel & Yurke, 2004). Hydrogen bonds serve to support these structures comprised of four in-plane-aligned guanine bases, whereas potassium ions assist in their assembly (Bloomfield et al., 2000; Chan et al., 2020; Jin et al., 1992; Yatsunyk et al., 2013). A G-quartet consists of a single (17 base sequence) DNA strand that creates a tetraplex geometry containing two stacked G-quartets constrained by one lower and two upper loops. The device may be activated through the addition or removal of an external complement strand, which transforms the structure to a duplex (Pu et al., 2015; Simmel & Yurke, 2004; Tang et al., 2013). A conductive matrix containing G-quartet wires was deposited onto nanofibers, which facilitated the construction of electrochemical sensors (Zhong et al., 2018).

4.3.9 DNA Motors – Free-Running and Clocked

The nanodevices described above are rudimentary DNA-based molecular actuators that convert the free energy associated with the activities of fuel and removal strands to mechanical actuation. Molecular-scale free-running motors (Bath et al., 2007) are triggered via integrated switches, while the activation sequences and cycles of clocked motors are stepped through using alternating fuel inputs (Phoenix, 2004). The practical utility of clocked motors is clear when the precise placement of nanoscale components must be synchronized with their defined movements. A bio-analog of a clocked motor is a GTP hydrolysis-modified regulatory protein (Bourne et al., 1991).

The concurrent input of fuel and removal strands to the molecular motors described above negates their state-cycling, which is due to the direct binding of these strands with each other prior to engaging with the motor. For the most part, biomotors are free-running and will cycle incessantly through their full repertoire of conformational states on a single source of fuel (e.g., ATP or GTP). As fuel molecules are metastable, they do not degrade and will release their stored energy only when they bind with molecular motors.

DNA-based molecular motors have been synthesized that employ accessory strands to inhibit the explicit binding of fuel and removal strands (Turberfield et al., 2000; Turberfield et al., 2003). This was achieved by "shielding" the fuel strand (F1) with an accessory strand segment (P), which connects with it and creates an intermediary structure that has two duplex arms and a little single-strand loop. Kinetics investigations revealed that this preventative strategy delayed binding with the removal strand (F2) by a factor of 100, in contrast to one that was exposed (Simmel & Yurke, 2004).

A catalytic strand segment (C) was then introduced, which was engineered to more easily bind with the F1/P complex than the F2 removal strand. When this segment connected with the F1/P via branch migration the loop was opened, which enhanced its affinity for F2. Subsequently, F2 connected with the "triplet" (F1/P/C) fuel strand to supplant both P and C to generate F1/F2 and P waste products. The C strand (a very simple molecular machine) was recovered following one cycle. When coupled with the F1/P/C complex this basic molecular motor toggled between relaxed single-stranded and extended duplex states, which might be considered as a new type of deoxyribozyme [39] (Breaker, 2000; Turberfield et al., 2003).

Conceivably, multiple molecular motor structures might be designed wherein each motor is actuated by distinct F1 and F2 strands. These systems could facilitate the creation of nanostructures that integrate arrays of complex assembly sequences, whereby actuators are programmatically orchestrated through the introduction of a cascading series of F1 and F1 strands (Simmel and Yurke, 2004). While the nucleic acid devices described above were utilized to manipulate only dye molecules in experiments, they inspire prospects for the emergence of a feasibly controllable and broadly versatile set of nanometric tools (Saper & Hess, 2020).

These tools, working cooperatively in a massively parallel fashion might be employed for the precisely organized assembly of nano- and microstructures that could incorporate a wide variety of molecules (Diez et al., 2003; Hiyama et al., 2008; Jia et al., 2015; Kumar et al., 2012; Kumar et al., 2016; Ramachandran et al., 2006), nanoparticles and microparticles (Månsson et al., 2004; Bachand et al., 2005; Persson et al., 2013), cellular elements, organelles, and even living cells (Takatsuki et al., 2011). These sub-assemblies might then be combined and integrated to ultimately culminate in macroscale (albeit nanoscopically precise) cost effective but high-value products (e.g., static/functionalized materials with unique properties, hybrid inorganic/organic materials, complex systems replete with nanoelectronic and quantum computational elements and infrastructures, enhanced and novel clinically relevant biomaterials, etc.). Artificial intelligence (AI/AGI/ASI), quantum computing, and advanced modeling/simulation capacities will be critical facilitators that significantly expedite the design and validation of DNA-based mechanisms, and how they might be optimally orchestrated to realize a viable strategy toward nanoscale, and ultimately, MM (Khare et al., 2018; Kotnala & Zheng, 2019; Seeman et al., 2002; Tosello, 2017; Zadpoor, 2019).

4.3.10 Precise Manipulation of Proteins Using DNA

The intentional control of DNA hybridization can influence the self-assembly of protein arrays (Niemeyer et al., 1994; Smith et al., 1997). The DNA Guided Assembly of Proteins (DGAP) proposes that DNA may be utilized as a fixture to facilitate the nanoscale fabrication of proteins in aqueous media (Lewis et al., 1995). For this process, multiple DNA sequences are attached to precise surface sites of target proteins. Various protein components (e.g., fibers, enzymes, antibodies, biomolecular motors) may then be arranged to generate predesigned structures through complementary sequence pairing. Subsequently, the stabilization of these constructs may be achieved by covalent inter-protein binding. This precise assembly strategy might also be utilized to fabricate myriad other entities, which are multi-functionalized using explicit DNA sequences (Freitas & Merkle, 2004).

4.3.11 DNA Walkers

In terms of facilitating subnanometer resolution assembly, kinesin (a natural cellular walker protein powered by ATP hydrolysis) inspired DNA walkers may exhibit utility for the precise acquisition, transfer, and placement of nanoscale components to fabricate myriad products using various materials. Single kinesin, myosin, and actin walkers have the ability to traverse their tracks (e.g., cytoskeletal polymeric filaments) at velocities of up to ~1000 nm/s with remarkable processivity (Coy et al., 1999). Processivity is the capacity of an enzyme to complete multiple catalytic reactions without disengaging from its substrate. In the case of kinesin, it can travel >100 steps over a 1.5 μm distance on a track before extricating itself (Andreasson et al., 2015). Investigations into artificial DNA walkers focus on potential power sources and strategies for the optimization of walking speeds and processivity. In terms of speed, the fastest DNA walker to date (~5 nm/s) is a "cartwheeling DNA acrobat," which still lags far behind kinesin (Li et al., 2018; Valero & Škugor, 2020).

A bipedal and bidirectional DNA walker was devised that traveled using inchworm actuation (Sherman & Seeman, 2004), while a bipedal DNA nanomotor was developed that proceeded one foot in front of the other, which could be monitored in real time via "multiplexed fluorescence quenching" (Shin & Pierce, 2004) (Figure 4.4). Non-autonomous DNA walkers depend on a continuous supply of chemical fuel strands, whereas autonomous DNA walkers rely on the enzyme-mediated cleavage of either DNA or RNA strands (Bath et al., 2005; Tian et al., 2005; Yin et al., 2004). Further drivers of these systems include photochemical reactions (You et al., 2012) and specific chemical stimuli (Wang et al., 2011).

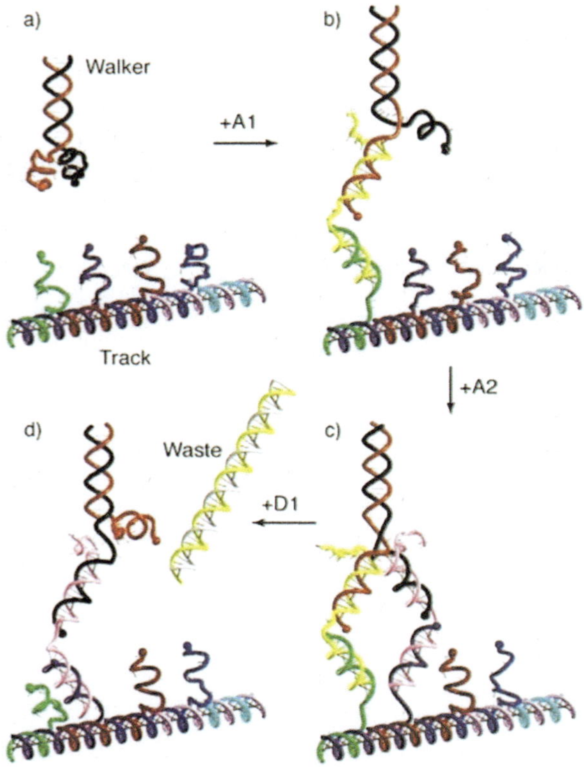

FIGURE 4.4 Representation of DNA walker locomotion. Variously colored strands signify different dyes used to detect walker travel. (a) disengaged walker, (b) walker tethered to strand 1, (c) walker tethered to strands 1 and 2, and (d) walker freed from strand 1 yielding waste duplex. (Reproduced from Shin JS, Pierce NA. *J Am Chem Soc.* 2004 Sep 8;126(35):10834–10835. With permission.)

In contrast to step-passing or inchworm-type gaits, autonomous DNA walkers proceed along their tracks as would a bucket being passed along a fire brigade (Kelly, 2005; von et al., 2011).

4.4 DNA MANUFACTURING COMPLEX: INFRASTRUCTURE ELEMENTS

Stable infrastructures with suitable scaffolds/substrates may enable the potential for nucleic acid-based MM using positional cycling systems akin to those described above. The fabrication of nanometrically precise macroscale products might be more likely to meet with success when DNA-based manufacturing complexes are as structurally stable/mechanically constrained to a solid substrate as possible to minimize the omnipresent effects of Brownian motion (Phoenix, 2004). This approach may enable the more accurate placement of components of certain sizes, in contrast to their being continuously randomly jostled in aqueous media. However, it may be of benefit for the self-assembly of some types of more diminutive and simple elements to free float in solution.

Existing microfabrication strategies facilitate the integration of biomolecules with non-organic substrates, which are employed when interfacing DNA microassays with glass or plastic wafers. The photolithographic procedures used to generate computer chips can coat these surfaces with high populations of DNA strands. Further, photosensitive compounds assist with this process as do multiple thin layers doped with specific chemicals (Henke, 1998). Once the challenges with transitioning from 2D to 3D fabrication are resolved, a new and extensive range of DNA manufacturing possibilities will emerge (Seeman, 2004). Scalable DNA scaffolds may play critical roles in the manufacture of progressively more complex nanoscale components and eventually, entire functional systems.

One may surmise that assemblages of autocatalytic polynucleotides could be integrated with elaborate three-dimensional functional DNA-based components (e.g., actuators, motors, switches) to produce sophisticated molecular mechanisms with the ability to self-replicate (von Kiedrowski, 1986, 2000; von Kiedrowski et al., 1989,1991, 2003; Achilles & von Kiedrowski, 1993; Luther et al., 1998; Scheffler et al., 1999; Sievers et al., 1994; Freitas & Merkle, 2004). Three-dimensional non-covalent nanostructures were formed by exploiting the self-assembly of trisoligonucleotide covalent junctions, wherein the 3′-ends of three oligonucleotides were linked by three-armed synthetic junctions. The reproduction of certain entities may then proceed by utilizing an electrophoretic procedure referred to as surface-promoted replication and exponential amplification of DNA analogs (eSPREAD), which is also known as cloning and amplification technology (CAT). This process can be employed for nanodevice cloning on electrophoresis chip surface (Scheffler et al., 1999; von et al., 2003).

Two strong candidates as rigid substrate-bound support structures for a DNA-based manufacturing platform include trihelical DNA and a novel four-stranded form of DNA ("G wire") (Hu et al., 2017; Luhrs et al., 2002; Marsh & Henderson, 1994; Marsh et al., 1995; Uddin et al., 1996; Vesenka et al., 1996). These components might serve as ultra-stiff construction elements, providing a reasonably solid foundation to which prefabricated/self-assembled parts are added to create a facilitative support structure to accommodate fabrication tools. A number of simple DNA-based structural elements may include cubes with corners consisting of triple-and six-armed junctions and two-dimensional crystals, which can be combined to construct scaffolding, and octahedrons (Seeman, 2004). A four-way junction might also serve as a useful stabilizing interconnect. The specific rigidity of these elements in their basic form would be quantified to determine whether they are too floppy and in need of reinforcement.

These structural components may be robustly joined using the sticky ends of DNA; the single strands that project beyond the primary duplex, which can directly bind to mirror image strands. Two important tools for assisting in the construction of a DNA manufacturing complex framework include restriction enzymes (to precisely cleave DNA) and ligases (that covalently bond DNA duplexes). Further, the "decoration" of DNA duplexes/oligonucleotides with diverse molecules, functional proteins, dendrimers, and fullerenes, etc., may be achieved by attaching them to the externally available atoms of phosphate backbones (Cassell et al., 1998; Seeman, 2004; Shchepinov et al., 1999; Smith et al., 1997).

Polymerases can facilitate the synthesis of unique molecular devices (e.g., single strands of DNA), which may be used to control the reasonably complex mechanical actuation of DNA loops arranged by octahedron scaffolds. This directed molecular evolution process (Harris et al., 1999) might facilitate the creation of molecular assemblers/disassemblers, akin to chaperones or polymerase clamp loaders (Freitas & Merkle, 2004). Polymerases are distinct enzymatic catalysts that employ DNA as a template in chemical reactions to direct the structural arrangement of other DNA-based products. Clamp loaders serve as molecular coordinators that extract energy from ATP binding and hydrolysis to connect with and "load" a sliding (closed protein ring) clamp onto primed DNA, which topologically links the two (Hingorani & O'Donnell, 2000; Sancar & Hearst, 1993).

The edges of discrete DNA structures (molecular tiles) can be configured to connect in tessellated patterns (Rothemund et al., 2000; Zhang et al., 2016). Increasingly complex patterns have been devised, which may have potential for extensive utility in myriad real-world applications (Parker, 2003; Tang et al., 2023). Programmable tiles might be induced to self-assemble into anchored substrates or jigs onto which suitable DNA assembly mechanisms may be installed. One strategy for a DNA-based fabrication complex may involve the mounting of substrates (populated with various assembly mechanisms and workpiece conveyance systems) to a submerged surface in a suitable aqueous medium in shallow segmented trays with open channels between the different workspaces. This could facilitate the unencumbered movement of the required discrete construction elements and fuel strands to minimize the possibility of aggregation or erosion. The substrates could be engineered with large surface areas to optimize production efficiency.

A rudimentary DNA decamer (10-unit monomer chain) was designed and synthesized to form a staggered/extended quadruplex array, which attained lengths of ~1000 nm (Luhrs et al., 2002; Marsh & Henderson, 1994; Marsh et al., 1995), as well as large DNA arrays with diverse connectivities and erratic interstices (von et al., 2003). As described above, more diminutive elements (comprised of carbon, metals, etc.) might be prefabricated in thousands of tiny fluid-filled cavities in preassembly trays (e.g., Seeman et al. integrated a microscopic fragment of nylon on a nucleic acid backbone (Seeman, 2004; Zhu et al., 2003). These components may feature unique guide strands, which ensure their correct placement by interlocking with complementary docking strands that are tethered to the substrate.

Further confirmation of the correct location for a given component might be achieved through the use of DNA "barcodes." These molecular identifiers can be synthesized by connecting double-crossover (DX) units (representing 0) and double-crossover and junction (DX+J) units (representing 1) in specifically designed arrangements, and attached as parallel constructs (Seeman, 2004). These barcodes might be laser scanned and verified against original designs. Once all inorganic components are confirmed to be in their correct positions and orientations, a nuclease could be added to induce the dissolution of the original DNA infrastructure, leaving behind the desired products (e.g., nanoelectronic, photonic, quantum computation elements, various nanostructured materials, etc.). This procedure could be duplicated as multiple DNA substrate (assembly seed) layers are added and stacked in place, one on top of the other, after which the DNA dissolution solvent could be added to reveal the resulting functional macroscale product. A useful preparatory step to facilitate larger-scale production may be to build custom DNA substrates beforehand, and then flash freeze and store them (for years if required as they would still be viable) pending assembly. DNA can be quickly reconstituted, with its structure relying primarily on the solution composition and conditions (Seeman, 2004). DNA concentrates may be re-formed through precipitation with isopropanol or ethanol (Green & Sambrook, 2016, 2017). Characterizing the electron dynamics of enzymatic catalytic reactions may be an important use case for DNA-based scaffolding (e.g., using DX and DX+J linkages), which might be employed to physically secure reactive enzymes in place for use in electron transfer systems (Seeman, 2004). These enzymes may catalyze certain molecules and transfer electrons to nanowire probes (for example) that test nanoelectronic systems and components.

4.4.1 Discussion

Investigations into/analyses of diamondoid and sapphire have determined that they are optimal materials for the fabrication of advanced nanomedical devices (Drexler, 2004; Freitas, 2002, 2003, 2013). It remains to be seen whether a DNA-based fabrication complex might have the capacity to build with these particular materials at the nanoscale (Luu et al., 2024). One challenge involves disparities in scale, as DNA dimensions range from 20 to 50 Å, while diamondoid subunits (adamantane) are at 1.5 Å (Fort, 1976), with sapphire crystal structures at 4.7 Å × 12.9 Å (Wen et al., 2020). Ultimately, dedicated engineered DNA accoutrements may have the ability to manipulate these diminutive structures; however, energy issues are more prominent. Most DNA modifications engage low free energy hydrogen bonds, which are several kcal/mol (Dong et al., 2007). However, hydrocarbon carbon-carbon bonds are ~80 kcal/mol (Yale University Department of Chemistry, 2024); thus, it is unlikely that the slight forces produced by DNA will have the capacity to precisely maneuver carbon atoms to create diamond lattices (Seeman, 2004). Further, integrated molecular fragments might have adverse impacts on DNA (Drexler, 2004).

Although DNA is exceptionally compliant and efficient for the fabrication of soft molecules, it is unlikely that it will have the capacity to directly assemble diamondoid entities (Freitas, 2004). Various fullerenes might be structurally modified by enzymes that have been spatially configured by DNA (Yang et al., 2014). Customized graphene edges might be grown through the precise arrangement of enzymes at specific sites (Phoenix, 2004). Graphene and buckyballs have been synthesized underwater (Sygula et al., 2001). The acceleration of currently sluggish material deposition processes may inspire the improved fabrication efficacies of engineered constructs (Drexler, 2004). Further, the prospective creation of complex silica structures via DNA choreographed fabrication would be an alluring study domain (Numata et al., 2004) (e.g., silica-based diatoms with complex geometries are assembled at low energies) (Phoenix, 2004; Scala & Bowler, 2001; Smith, 2025).

To counteract the incessant damage to DNA in biological systems, robust repair mechanisms evolved to maintain its functional integrity. Consequently, a derivative DNA system may be more efficacious for dedicated nanomanufacturing (Seeman, 2004); potentially, one that is less prone to damage. Further issues to explore include the precisely controlled introduction of feedstocks into the complex and efficient removal of waste products. Myriad useful products could be created using DNA, which may be utilized as both a structural material and programmable jig to facilitate the assembly (aligning, joining, etc.) of organic or inorganic structures (Freitas, 2004).

4.4.2 Replication of DNA Structural Elements – Octahedron

The self-assembly of DNA duplexes and oligonucleotide fragments, as well as some degree of self-replication, will be important to consider when aspiring to fabricate products intended for real-world applications. Further to having the capacity to assemble prefabricated microscale and nanoscale structural elements and modular components, DNA can facilitate molecular, atomic aggregates, and ultimately "atom by atom" resolution manufacturing (Cai et al., 2023). Nucleic acids may exhibit greater proficiencies for preprogrammed self-assembly than directed component-by-component spatial placement and might be constrained to feedstock inputs that can endure aqueous environments (Freitas, 2004). However, it may be possible to encase superhydrophobic materials (Sun et al., 2020a) in superhydrophilic packages, decorated with oligonucleotides that are targeted to attach to specific sites (via complementary binding) on DNA manufacturing complexes, for integration into assemblies in progress.

Multiple millions or even billions of duplicate units interacting with equivalent populations of other dedicated devices will be required to realize the bottom-up DNA-based fabrication of useful microscale and macroscale products. Molecular biology cloning strategies might inspire and guide progress in this area (Andreou & Nakayama, 2020; Zhang, 2021). A single oligonucleotide was designed by Shih et al. (2004) to induce the self-assembly of a nanometric octahedron (Ø22 nm). Due to its single-stranded starting configuration, the resulting structure was easily amplified, cloned, and evolved for a multitude of applications. The self-assembly procedure required heating and consecutive cooling steps, where certain segments were designed to fold over on themselves to stabilize the entire structure. The twelve external edges of the octahedron contained specific addressable sequences to facilitate the creation of higher-order structures with greater complexity and the encapsulation of different molecular species, while providing the ordered scaffolding necessary for photonics and nanoelectronics (Bardi, 2004).

The primary structural DNA strands could be copiously amplified using polymerases (albeit this technique can result in errors) or synthesized in bacteria and purified (Phoenix, 2004). Despite there being no intrinsic covalent bonds between the various DNA segments, the complete tetrahedral geometry was robustly rigid. Greater structural stiffness can be imparted to the entire assembly by adding smaller fragments of looped DNA (Phoenix, 2004). The internal void of the octahedron accommodated a ~Ø14 nm sphere, while ~Ø8 nm spheres would traverse the triangular openings.

These structures utilized molecular detection to convert the DNA sequence encoded data to 3D positional data (Shih et al., 2004).

4.4.3 DNA Origami

Nascent branched DNA-based structural elements (DNA tiles) featured sticky ends, which enabled the assembly of complex constructs in three dimensions (Yang et al., 2021). DNA origami technology was initially advanced by Rothemund (2006) (Figure 4.5), who formed predesigned structures using hundreds of small single-stranded DNA (ssDNA) staples to manipulate long ssDNA scaffolds. This enhanced the range of possibilities in the development of polymorphic DNA structures to

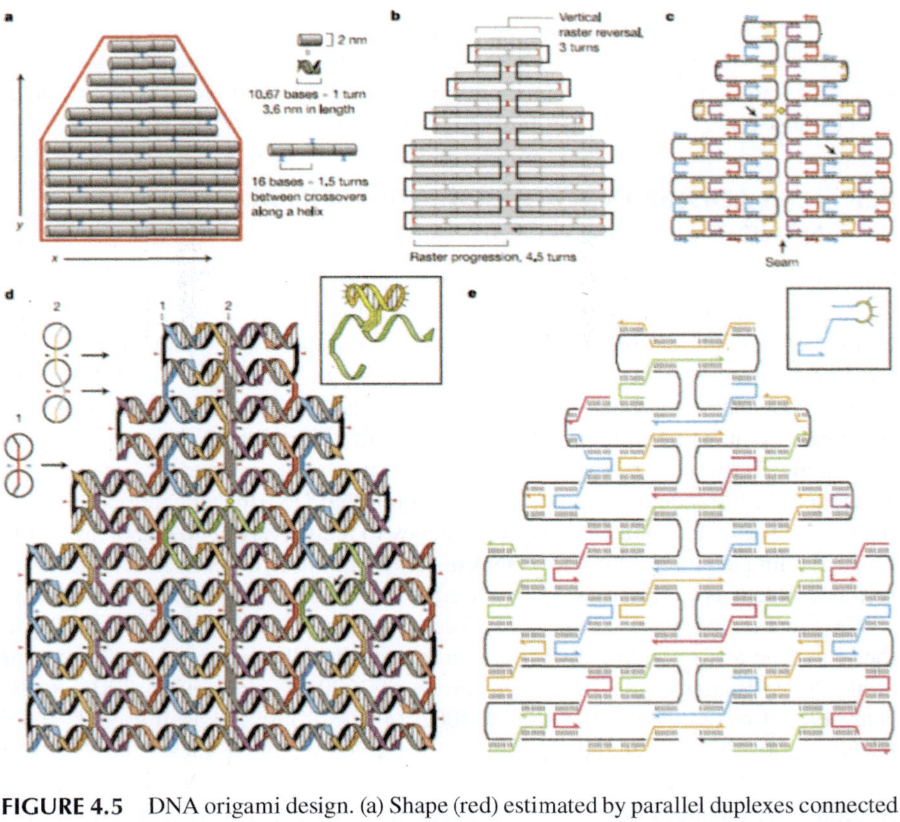

FIGURE 4.5 DNA origami design. (a) Shape (red) estimated by parallel duplexes connected by regular crossovers (blue). (b) Scaffold (black) traverses all helices to create additional crossovers (red). (c) Initial design where 16-mer staples typically bind two helices. (d) Strands shown as helices. Red → (scaffold crossovers), black → (recurring crossovers, minor grooves top face), blue → (recurring crossovers, minor grooves bottom face). Crossover sections (1, 2) showing backbones (colored lines), and major/minor grooves. (e) Complete design following reordering along seam, with most staples being 32-mers that span three duplexes. Insets (dumbbell hairpins (d) and 4-T loops (e) may be employed for larger constructs. (Reproduced from Rothemund, *Nature*. 2006 Mar 16;440(7082):297–302. With permission.)

facilitate the accurate placement of functional elements. Knudsen et al. (2015) utilized a DNA origami platform to fabricate potentially conductive DNA-conjugated polymer wires of various shapes, while Schill et al. (2021) assembled supramolecular polymers. Further, polymer nanopatterns with a negative charge were generated in situ on DNA origami. This may serve as a precursor for the development of elegant charge-separated substructures to facilitate a deeper elucidation of the distribution of surface charges as they relate to interactions between nanomaterials and biosystems (Shi et al., 2022). DNA origami has also been employed for the assembly and investigation of liposomes (Loffler et al., 2017; Yang et al., 2016; Yang et al., 2021; Zhang et al., 2017).

A series of DNA origami structures was designed that guided the fabrication of positively changed assemblies of gold nanoparticles having nanodonut and nanocuboid geometries, as well as micron-long polymerized nanorods (Schreiber et al., 2011). Furthermore, single-walled carbon nanotube structures were self-assembled using DNA origami templates (Maune et al., 2010). A 3D grooved DNA origami structure was also created to confine carbon nanotubes in the construction of field-effect transistors (Sun et al., 2020b; Zhao et al., 2020; Cai et al., 2023).

4.5 DNA ADHESION TO/DISSOCIATION FROM NANO-ENTITIES AND SUBSTRATES

The real potency of DNA-based MM systems may reside in their ability to enable the repeatable manipulation and placement of myriad inorganic elements to ultimately build useful functional (nano/micro/macroscale) products with practically flawless nanometric accuracy. The elegant organizing strategies employed in DNA hybridization may enable an extensive range of precisely designed nanometric elements to be closely interfaced and bonded, which might be facilitated through mechanical, chemical, thermal, nanoelectronic, magnetic, or photonic means. For example, MacDonaill (2002) demonstrated "remote" controlled DNA hybridization by applying the magnetic field of a GHz radio frequency to inductively couple and covalently link a 1.4 nm gold nanocrystal to a DNA duplex. Nanotubes comprised of DNA were self-assembled, as were several DNA-based dendrimers (Li et al., 2004; Nilsen et al., 1997). Further, numerous studies have been conducted involving natural and synthetic nucleic acids combined with inorganic elements (Benner et al., 1998; Berger et al., 2000; Berger et al., 2002; Fujiwara et al., 2001; Garner et al., 2001; Kanai et al., 1969; Loakes, 2001; MacDonaill, 2002; Mirkin, 2000; Nakashima et al., 1997; Nielsen, 2001; Nielsen & Egholm, 1999; Switzer et al., 1989; Tackett et al., 2001; Obika et al., 2002; Service, 2002; Wu et al., 2002; Ball, 2003; Chaput et al., 2003). Additionally, quantum dots can be tethered to DNA using a transitional organic layer, and nanowires may be generated through the fusion of metallic nanoparticles (Seeman, 2004).

4.5.1 DNA Binding to Nanotubes

It was revealed by Williams et al. (2002) that DNA could be bound to the carboxyl groups that were added to the tips of oxidized nanotubes using an acid treatment

(Liu et al., 1998). This technique was also used to add carboxyl groups to defects in the sidewalls of nanotubes (Chen et al., 1998), to which functional linkers were employed to attach DNA molecules (Baker et al., 2002; Cai et al., 2003; Dwyer et al., 2002; Hazani et al., 2003; Nguyen et al., 2002). This system may facilitate the precise orientation and integration of nanotubes with nanodevices.

The surface coating of nanotubes with DNA results in a negative charge that induces the stable dispersion of nanotubes in aqueous media, which is required for directed DNA-based assembly. Conversely, if only the tips of the nanotubes are coated, a surfactant needs to be added to prevent their aggregation, which is caused by hydrophobic interactions between their uncovered surfaces. The adequate coverage of nanotube sidewalls with DNA enables more flexible allowances in nanotube lengths for their positioning within an assembly (e.g., linking the space between two electrodes with complementary oligonucleotides). An alternative method for the adhesion of DNA to nanotube tips and sidewalls involves a solid-state azide-photochemical reaction. This enables DNA oligonucleotides to be synthesized from the reactive groups of each photo-adduct (Moghaddam et al., 2004).

4.5.2 Thiolated DNA

Gold nanoparticles have an attraction for and readily bind with thiolated DNA strands (a thiol is the sulfur analogue of alcohol) (Jin et al., 2003; Mirkin et al., 1996; Mucic et al., 1998; Carey, 2004). The modification of one terminal of the DNA sequence with thiol (also viable for RNA and PNA (peptide nucleic acid) is necessary to secure the gold nanoparticles, after which they can self-assemble due to the intrinsic gold-thiol affinity. Further assembly options include the possibility of decorating DNA duplexes with proteins that contain exposed histidine chain imidazole rings, which also have affinities for several transition metal ions (e.g., Co^{2+}, Cu^{2+}, Fe^{3+}, Mn^{2+}, and Zn^{2+}) (Rau & Chiang, 2004). In addition, a synthesized trithiol-covered oligodeoxyribonucleotide was used to stabilize ~Ø30 nm gold nanoparticles.

Various 2D and 3D DNA/nanoparticle combinations have been achieved due to their unique catalytic, electrical, and optical attributes (Loweth et al., 1999; Mirkin et al., 1996; Mucic et al., 1998; Mitchell et al., 1999; Mbindyo et al., 2001; Taton et al., 2000; Willner et al., 2001). Strategies for the creation of hybrid DNA/nanoparticle assemblies include the direct adsorption of oligonucleotides terminated with alkylthiol or disulfide to the surfaces of metal nanoparticles (Mirkin et al., 1996); covalent attachment of oligonucleotides to preactivated surfaces of nanoparticles (Li et al., 2002); and adhesion of biotinylated oligonucleotides on avidin-coated nanoparticles (Alivisatos et al., 1996; Niemeyer et al., 1998; Li et al., 2002).

4.5.3 DNA Adhesion to Dendrimers

Dendrimers are typically synthesized in increments, layer by layer, where every layer redefines their molecular dimensions. Dendrimer molecules have average diameters of several nanometers, albeit some have been formed up to Ø30 nm. The peripheral layer of dendrimers consists of densely layered molecular group arrays, which serve as anchoring sites for the attachment of other useful molecules (e.g., DNA) in their peripheral branches (Bielinska et al., 1997; Bielinska et al., 1999; Freitas, 2003; Freitas & Merkle, 2004).

4.5.4 DNA "Velcro"

It was initially demonstrated by Hazarika et al. (2004) that gold nanoparticles could be both combined and dissociated using ssDNA, the technologies behind which have become increasing adept (Dey et al., 2024). The binding/separation capacities of DNA strands enabled the editing of prefabricated structures (e.g., component swapping, substrate modification/expansion, enhancement/general repair). The ~Ø15 nm gold nanoparticles were bound to the ssDNA backbone via sulfur and joined through the addition of a third ssDNA. This third strand hybridized with half of each gold-tethered strand and connected to both; thus, creating a link, one end of which was longer than the other and hanging free. To dissociate the nanoparticles a fourth ssDNA segment was introduced to bind with the overhanging end, which peeled it away akin to the separation of Velcro strips.

4.5.5 DNA Adhesion to Surfaces

The modification of surfaces with diverse DNA composites has been demonstrated with DNA-carbon nanotubes (Wang et al., 2002) and additive layer-by-layer oligocationic polymer-DNA films (Luo et al., 2001) (Pei et al., 2001). To assist in the fabrication of microarrays, DNA was anchored to glass (both phosphate and borosilicate). The glass was initially cleaned with an acid and base etch and then coated with a 3-aminopropyltrieethoxysilane layer. Viable DNA adhesion to silicate glass was confirmed as its features were well distributed and defined (Eyster, 2019).

Certain surface proteins may be employed as linkers for the attachment of DNA to molecular substrates and/or templates, and the combinatorial biology methods of biomimetic studies can identify entities with robust adhesion properties (Brown, 1997; Giver & Arnold, 1998). Gu et al. (2019) investigated the binding kinetics of DNA duplexes to two types of palladium surfaces (Pd(111) and Pd(100)) toward the design of prospective metal nanostructures for biomedical applications. Simulations revealed that DNA binding modes were quite distinct in terms of strength and orientation, with DNA/PD(111) binding being much stronger than that of DNA/Pd(100), to the point of partially denaturing the DNA. Further, the duplexes were horizontally oriented on the PD(111), while they were vertically oriented on the Pd(100) (Figure 4.6). A Large Random Peptide Library can be used to identify peptides having the same number of amino acids with varied sequences, which robustly bind to specific inorganic surfaces (Brown, 1997; Brown et al., 2000; Giver & Arnold, 1998; Naik et al., 2002; Schembri et al., 1999; Whaley et al., 2000).

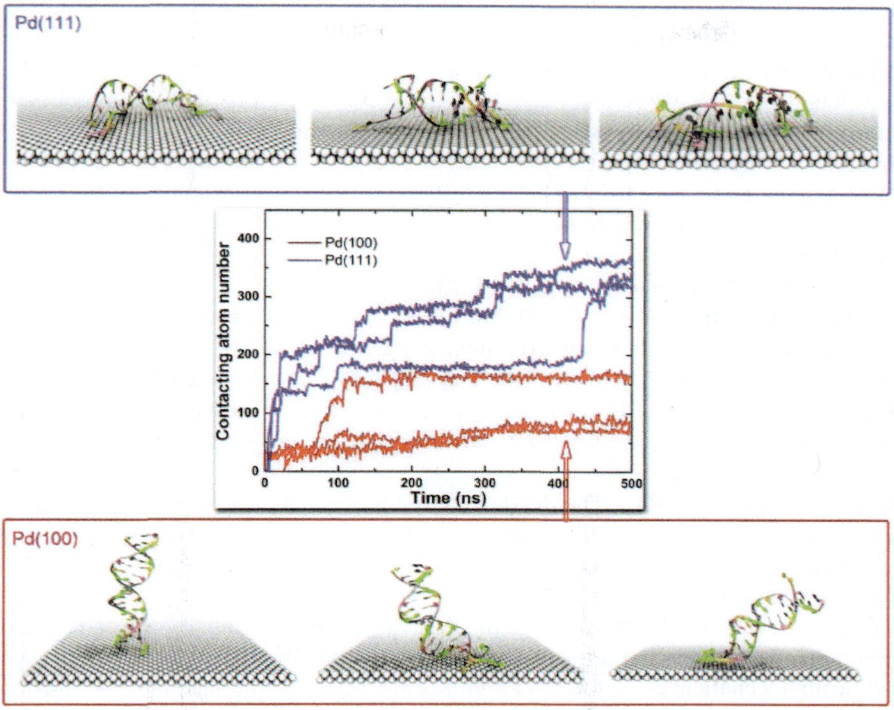

FIGURE 4.6 Distinct binding modes and orientation of DNA duplexes to two types of palladium surfaces (Pd(111) and Pd(100)). (Reproduced from Gu et al., *Nanoscale.* 2019 Jan 23;11(4):1827–1836. With permission.)

4.6 POTENTIAL COMPUTATIONAL REQUIREMENTS FOR A DNA-BASED MANUFACTURING COMPLEX

It may be anticipated that a nucleic acid-based manufacturing system will require varying degrees of computational support for specific operations. It might likely be the case that DNA-based computers will facilitate molecular identification and self-assembly processes for the efficient fabrication of products. However, the diffusing molecule strategy they currently utilize would be incompatible with the level of sensing and control required for nanomanufacturing (Drexler, 2004). Computation will likely be required for the coordinated placement of components (e.g., onsite allocation, reading, and analysis of DNA bar code data) and release of actuation strands to the complementary receptors of mechanisms affixed to substrates. Further, DNA-based computers may be employed to verify the correct placement of components and the management of stringent testing protocols.

One DNA computation strategy involves the manipulation of three or more interlocking Borromean rings, where the removal of one would result in the collapse of the entire structure. Logical statement proofs could be embodied by the linkage integrity (Freitas, 1999; Liang & Mislow, 1994; Mao et al., 1997). The nucleic-acid-mediated self-assembly of magnetic nanoparticles can be arranged to form nanoscale memory units (Ventra & Zwolak, 2004). Adleman (1994, 1998) developed the first DNA computer with the capacity to solve complex problems at computational speeds estimated to be ~10^{13} operations/s (100 teraflops (100 trillion floating-point operations/s)). Currin et al. (2017) estimated that a desktop DNA computer might feasibly perform ~10^{20} operations/s (~2×10^{19} operations per joule). The fastest supercomputer to date (Frontier) can perform a ~quintillion (10^{18}) calculations/s (ORNL, 2022).

Parker (2003) stated that a tangle of 1,018 DNA strands might conceivably achieve a computational pace that is 10,000 times faster than its most nimble silicon equivalent. However, unique DNA-based computers will require algorithms to complete useful computations (Phoenix, 2004). The transfer of data to generate molecules in vivo in the human body is quick and very efficient for the most part. For instance, hemoglobin molecules, comprised of 500 amino acids, are synthesized 500 trillion times a second, which translates to (15×10^{18}) ribosome "read" operations every 60 seconds (Kurzweil, 1990). The distinction in computational strategies between silicon and DNA is that, to arrive at solutions, silicon chips must separately test every possibility until the correct answer is found, while DNA has the capacity to simultaneously test every possibility.

The area required for data storage in DNA for one bit is ~1 nm^3; however, its computational data output is very sluggish. Further, due to DNA pair mismatching there are accuracy challenges, as there is a significant reliance on the specificity of enzyme activities. These issues will increase exponentially with more complex calculations. Additionally, versatile system designs may be at the expense of instability, as integrated fault-tolerant redundancy may significantly constrain DNA reactivity; thus, fabrication processes (Gacs, 2004). While it may be the case that DNA computational efficacy will never equal that of silicon-based quantum computers (Mądzik et al., 2022), the true dominion might reside in the realm of on-site/in vivo computation; to establish precise control over cellular/molecular/atomic domains (Parker, 2003).

An intriguing possibility for the long-term protection and storage of DNA data may involve its insertion within extremophile bacterium such as *Deinococcus radiodurans*. This bacterium, which is thought to be practically immortal (albeit all feasible bacteria are theoretically immortal, yet all may be destroyed) (Drexler, 2004), can endure UV radiation, partial vacuum, dehydration, extreme temperatures, and immense levels of ionizing radiation. It possesses the uncanny capacity to repair and reconstitute even severely damaged DNA (White et al., 1999). An additional extremophile bacterium (*Halomonas bluephagenesis*) can accommodate dual-mode storage, dynamic data conservation, rapid readout, and robust recovery (Sun et al., 2023), with ~1 billion of these bacteria residing in 1 mL of water.

A dramatic surge in data storage capacities might be realized using this strategy when different DNA sequences are stored in separate bacteria. Specific DNA sequences can be stored via their insertion between dual "marker" sequences, integrated as a complete strand with a cloning vector (a circular self-replicating DNA molecule), and then retained for long-term storage within the extremophile bacteria. To access the data, the DNA would be copied by the PCR to detect the marker sequences. Subsequently, the sequence strands would be replicated for analysis and reading (Sun et al., 2023).

4.6.1 Fractal Sierpinski Triangle Self-Assembly Using DNA

Self-assembled two-dimensional DNA tiles were employed to develop cellular automata (Raza et al., 2021), where the XOR (exclusive or) binary function was calculated through its update regime, which then created a fractal triangular Sierpinski pattern as it developed. This was achieved through the translation of abstract tiles to DNA counterparts using double crossovers. Algorithmic crystals were created by guiding the growth of nucleated DNA tiles using long single-stranded DNA molecules, with error levels of ~1%–10% in the resulting pattern (Rothemund et al., 2004). This work confirmed that the programmed self-assembly of DNA could be regarded as a biomolecular universal Turing machine with potential to be applied to any algorithm necessary for computation, or to gain insights into self-assembly tasks. Adaptive algorithms that guide molecular-scale fabrication may integrate fundamental energy and entropy models to elucidate how order emerges from physical dynamics (Rothemund et al., 2004). Numerous classes of DNA tiles were created to generate tessellated two-dimensional micron-scale crystals (LaBean et al., 2000a; Wang, 1961; Wang, 1962; Wang et al., 2024; Winfree, 1998). The DNA tile interfaces were encoded using sequence-specified hybridization, where single-stranded patterns functioned as inputs to facilitate self-assembly (LaBean et al., 2000b; Mao et al., 2000; Winfree, 1998; Yan et al., 2003).

A novel class of irregular crystal was assembled using Sierpinski tiling, wherein each molecular (Wang) tile consisted of a DNA-double crossover molecule that contained four "sticky ends" (five base single-stranded overhangs) which functioned as automated binding sites (Fu & Seeman, 1993; Rothemund et al., 2004). Primary computational self-assembly outputs are required using a suitable nucleating substrate. Input tiles and rows thereof are assembled when guided by four long ssDNA scaffolds. The undersides of the tiles are supplied by nucleating strands, while additional capping strands are employed to create inter-tile duplexes on the nucleating strands (Rothemund et al., 2004). All four prerequisites for universal Turing computation via crystallization are fulfilled with self-assembled DNA-based Sierpinski triangles. (1) Creation of expanded crystals. (2) Programmed exchanges between tiles mediated by (sticky end) DNA sequences. (3) Selective correlations between tiles facilitated by collective binding of multiple sticky ends. (4) Organized nucleation for growth induced by template input data (Rothemund et al., 2004). When applied to molecular-scale fabrication, DNA self-assembly-based computation may guide the directional control and spatial parameters of workpieces to enable precision geometries (Soloveichik & Winfree, 2005). However, several challenges still need to be addressed toward the viability of this type of system. Various errors might include displaced lattice fragments, errantly placed crystals, mismatched tiles, and misguided nucleation; thus, stringent assembly conditions will reduce these types of errors. An alternate, and potentially more dependable pathway may entail the prefabrication of tile sets that are fault-tolerant, which execute similar logic (Chen & Goel, 2005; Winfree & Bekbolatov, 2004; Reif et al., 2003; Schulman & Winfree, 2010).

4.7 DNA NANOELECTRONICS

One class of MM systems based on nucleic acids may, for example, be dedicated to the generation of pristine nanomaterials to incorporate into larger functional products, utilizing complementary DNA nucleotide binding to handle and connect molecular entities. Other fabrication complexes might be devoted to fabrication of integrated nanoelectronics components, as their interconnectivity, and that with computational elements, will be essential for the operation of nanoscale transistors, circuits, molecular/quantum computers, etc. (Ben-Jacob et al., 1999; Hermon et al., 1998; Ventra & Zwolak, 2004). Thus, proficiency in the creation of nanoelectronics will be a powerful feature of DNA-mediated assembly (Ventra & Zwolak, 2004; Dai et al., 2020). Concurrent with these advances, it is not difficult to envision that nanoelectronics and the DNA-based MM complexes employed for their assembly would soon be ubiquitously employed to produce the majority of IOT devices and systems (Sivasankari & Varalakshmi, 2022; You et al., 2019).

4.7.1 DNA Nanowires

DNA templates can be utilized to guide nanoscale metal deposition to generate nanowires comprised of palladium (Richter et al., 2000), silver (Braun et al., 1998; Vittala et al., 2022), gold (Bayrak et al., 2018a), and various other metals (Bayrak et al., 2018b). Further, hybrid organic/inorganic nanowires of various materials are possible (Vittala et al., 2017; Zhou et al., 2015), as well as nanowire multiplexes (Hessari et al., 2014). DNA templated gold nanowires (Ø2 nm) may be synthesized by the linear positioning of Au quantum dots. The two free ends of the DNA-guided nanowires can be designed with sequences that spontaneously hybridize in solution with their counterparts, which are anchored to connector pads. This approach might also be modified using ssDNA for the self-assembly of complex 3D nanowires, for which gold quantum dots could catalyze their additive deposition for continuous constructs (Alivisatos et al., 1996; Hainfeld et al., 2001). DNA serves as an excellent template for nanowires due to its nanoscale dimensions, pliability, and precise end joining capacity. Gold clusters were induced to adhere to a DNA duplex and spaced at ~2 nm, which is close to the electron tunnelling distance between plasmonic nanoparticles (Scholl et al., 2013). Binding options for the gold clusters included photoreactions, intercalation, and covalent attachment, while those for nanowire insulation encompassed surface oxidation, the creation of glass coatings via silane reactions, or alkane thiol coatings (Hainfeld et al., 2001). It is also possible to synthesize DNA-based molecular devices with integrated engineered variances in conductivity (Aich et al., 1999; Lee et al., 1994; Rakitin et al., 2001; Starikov et al., 2008).

4.7.2 DNA Conductivity

It remains uncertain whether DNA is electrically conductive, semiconducting, or insulating. Fink (2001) suggests the possibility that it may have the capacity to accommodate all three states, contingent on how molecules or metals are arranged upon its structure or integrated within it. Charge transfer within DNA can span up to tens of nanometers, albeit the transport rate is strongly impacted by the arrangement of its base sequences. Attached ions can alter its electronic structure, as can phosphate groups that function as tunnel junctions. A further aspect to consider is that when DNA is in contact with a surface its structure becomes compressed, which can modify its electronic attributes (Muir et al., 1998). When thin metallic films are applied to completely cover DNA to synthesize nanowires, its ability to selectively bind with other molecules can be diminished or negated.

DNA can adsorb cobalt, nickel, and zinc into its central helical axis when immersed in high pH (8.5) media. This configuration is referred to as M-DNA (formed from B-DNA), which is electrically conductive but retains its molecular binding capacity and can engage in DNA hybridization (Wettig et al., 2003; Brancolini & Di Felice, 2011). These M-DNA class elements might be applied to self-assemble the ultrafine nanoelectronic infrastructure required for the circuitry produced by a nucleic acid-based fabrication complex. Quantum dynamics may assist with elucidating and monitoring the transport of electrons in M-DNA, which may "hop" between base pairs as suggested might be the case for B-DNA (Shimazaki et al., 2005). The electron transport rate determined for M-DNA, however, is several orders of magnitude higher than for B-DNA (Bixon et al., 1999; Gómez-Navarro et al., 2002).

Separate DNA strands were used to create a triple junction to which various functional groups could be connected (Rhee et al., 1999; Stühmeier et al., 1997). For example, the binding of a thiol functional group will facilitate the attachment to a gold electrode (Gómez-Navarro et al., 2002). Open-ended single-walled carbon nanotubes were functionalized by DNA when amino-terminated DNA strands reacted with terminal carboxylic acid groups, which covalently bound the DNA to the nanotubes. This composite could then be secured (via complementary DNA base pairing) to surfaces, or integrated within structured networks (Dwyer et al., 2002; McLaughlin & Stewart, 2003; Yan et al., 2003). To streamline the assembly of nanoelectronic circuits, molecular lithography might be employed to deposit DNA onto suitable surfaces (Keren et al., 2002). By exploiting DNA self-assembly, metallic and semiconducting nanotubes may initially be sorted, used to populate various substrates, and configured as structural networks (Keren et al., 2003; Zheng et al., 2003).

Since the DNA-based assembly of complex nanoelectronic circuitry may require precise DNA scission, hybrid DNA-fullerenes (for example) can accurately cleave DNA at guanine residues when exposed to light (Boutorine et al., 1995). DNA cleavage may also be achieved using fullerene derivatives with efficacy to generate singlet oxygen (Cook et al., 1997; Donnet & Bansel, 1998; Kadish & Ruoff, 2000). Further, it is possible to directly "write" DNA onto a substrate via dip-pen nanolithography (Blättler et al., 2006; Li et al., 2021). This technique is flexible, as patterns can be deposited using diverse oligonucleotide elements, which may serve as foundations for the creation of increasing complex (conceivably, layered and/or 3D) circuits (Demers et al., 2002). The use of DNA to modify semiconducting substrates suggests potential for the creation of sophisticated nanoelectronics (Pike et al., 2002; Watson et al., 2014).

Thin films of nanocrystalline diamond (NCD) grown on silicon may be functionalized with DNA oligonucleotides, resulting in a platform with excellent utility for customized dynamically selective hybridization processes. This system may be produced through the chemical modification of pristine H-terminated NCD surfaces to generate a consistent layer of amine groups with which DNA can bind, providing a stable interface linking organic elements and synthetic nanoelectronic circuits (Yang et al., 2002, 2004).

4.8 PROSPECTIVE DNA MANUFACTURING STRATEGY

The initial thoughts of the author from "An Investigation of Nucleic Acid/DNA-Based Manufacturing" (2004) are preserved here in their original form.

> There are several criteria that may facilitate and define the emergence and practical application of a nucleic acid-based manufacturing complex. A distinction should be made as to what caliber of precision the products fabricated by this manufacturing capability might possess. Would they be nanometrically precise, or atomically precise? Dr. Drexler points out that as a matter of terminology, "molecular manufacturing" is defined in terms of productive molecular machine systems building complex structures with atomic precision (Drexler, 2004). The following preliminary thoughts reflect only this authors' present perception, which offers but one of many potential organizational strategies, involving many additional unexplored elements. One ideal worth striving for would be toward the development of a completely programmable, and totally automated system to keep the fabrication process free of any contaminants and debilitating effects of cumulative waste products. A cascading sequence of events would most likely entail numerous coordinated functions to occur at very specific times. A perceived composition might be:
>
> **Initially, the self-assembly of smaller modular entities in massively parallel arrays comprised of "sub-component fusing wells."** These wells would be fed by exquisitely directed flows of prepared highly pure feedstock concentrations of the required elemental materials (perhaps extrapolating fluidic array designs and technologies). Nanoparticles comprised of specific elements would be bonded to single-stranded DNA. Nanoparticles might be drawn together in certain orientation, contingent on single-strand hybridization modes. A potential technique for positioning atoms or clusters of atoms in this way may be discovered in the future. There may be a requirement for many sub-components to be built up from metallic constituents as parts of nanoelectronic interconnects, transistors, capacitors, etc. Affinity studies between metallic and non-metallic nanoparticles and DNA will require investigation. Capabilities for directing localized electronic charges may play an important role toward enhancing and assisting the bonding of these infinitesimally minute entities in conjunction with DNA. DNA barcodes, or single-strand "identifiers," might also be added at this stage for part identification and tracking, as well as perhaps several encoded single-strand "guiding handles." The fusing wells might contain an exit valve for purging the contents when this assembly is complete. Conversely, they could be removed by arrays of end-of-arm, robotic mounted pipettes.
>
> **Sub-components would now be transported to an intermediate assembly chamber, and then conceivably to many subsequent intermediate assembly chambers (dependent on the complexity of the desired end-product) to draw these sub-components together for precise mounting on a substrate**. This might be accomplished by incorporating dedicated "guide strands" residing in the aqueous mixture

contained within the chamber. For example, assuming a guide strand sequence of 20 bases, 10 of these bases would hybridize with their complements attached to the sub-component, and the remaining 10 would hybridize with their complements tethered to the substrate at a specific location. Subsequent single strands dangling from the sub-component could find use as fine-positioning tools as they would bond with their complements also located in precise locations on the substrate thereby ensuring correct orientation of the part. Final placement verification might occur by introducing very short single DNA strands combined with addressable gold nanoshells (varying wall thickness), or tuned quantum dots (dimensional variations) that would hybridize with identification strands, tagging the parts. Once in place, the patterns formed by nanoshells or quantum dots might be "read" by external scanners/computers, and digitally superimposed on preconfigured patterns conveying correct positions for all entities. This process may be repeated along and up the assembly sequence, involving assembly of larger components. Corrective mechanisms employed in cases of misplaced entities would be an area to explore. It may turn out to be easier to discard faulty assemblages, rather than institute correction protocols in these cases. If all "identifiers and verifiers" are stringently checked prior to assembly, there should be very low off-pattern assembly incidences.

Verified substrates, studded with finished components would now be transferred to a final "assembly tray." These substrates might still be microns to millimeters in size, so again they might be moved by fluidic means, by pipettes, (or if larger) by grasp-force sensitive robotic placement fingers and arms. The tray would house a well-organized and stratified infrastructure composed of a stable physical framework, immersed and securely anchored to the bottom surface, containing again, a finely regulated (chemically, thermally, controlled turbulence, etc.) aqueous environment. The framework might be comprised of "basement" scaffolding, interspersed at various levels and orientations (dictated by the design of desired end product) with supportive substrates that have been prefabricated and functionalized prior to installation into the assembly mainframe. These substrates may be stockpiled and stored (by freezing, etc.), selected by their inherent actuation and assembly capabilities, mounted to the framework (via hybridization) by pylons, and reactivated by the chemical mix of the buffer solution. The final phase of assembly would proceed in the same manner as described above, giving the nanoscopically precise final product. The completed product would now be transferred by robotic means to final testing and packaging stations.

Boehm, 2004

4.9 CONCLUSION

The potential viability of nanometric resolution fabrication through the development of nucleic acid manufacturing systems may derive from/rely on an initial focus on the exceptional DNA functionalities. Its capacities for molecular detection and precise hybridization may best facilitate the organization, positioning, and verification of the orientation and placement of microscale elements (perhaps atomically preassembled using other strategies) into larger constructs. These DNA fabrication units may serve as critical standardized modules for composite nanomanufacturing complexes. The powerful capacities of DNA self-assembly and adaptable programming may emerge as essential aspects that drive this type of nanofabrication. These, in terms of the development of structural scaffolding, immobilized enzymatic lattices, functional substrates incorporating actuators, positioning devices, etc. Furthermore, DNA computation may enable data transmission, verification, and storage, as well as product identification and tracking. The above are but a few exemplars of likely myriad potential capabilities that are yet to be revealed. In any case, a thrilling and fascinating pathway has yet to unfold with the investigation, development, and eventual implementation of nucleic acid nanomanufacturing. "A most important and essential aspect of the use of DNA in nanometric/molecular fabrication is the idea that here exists a very elegant and potent means for interfacing our macro world designs and dreams with that of the nanometric domains, directly and intelligently" (Boehm, 2004).

REFERENCES

Achilles T, von Kiedrowski G. A self-replicating system from three starting materials. *Angew Chem Int Ed Engl.* 1993;32:1198–1201.

Adleman LM. Molecular computation of solutions to combinatorial problems. *Science.* 1994 Nov 11;266(5187):1021–1024.

Adleman LM. Computing with DNA. *Sci Am.* 1998;279(2):54.

Aich P, Labiuk SL, Tari LW, Delbaere LJ, Roesler WJ, Falk KJ, Steer RP, Lee JS. M-DNA: a complex between divalent metal ions and DNA which behaves as a molecular wire. *J Mol Biol.* 1999 Nov 26;294(2):477–485.

Alberti P, Mergny JL. DNA duplex-quadruplex exchange as the basis for a nanomolecular machine. *Proc Natl Acad Sci U S A.* 2003 Feb 18;100(4):1569–1573.

Alberts B, Bray D, Lewis J, Raff M, Roberts K, Watson JD. *Molecular Biology of the Cell, 3rd ed.*, Garland Publishing, New York, 1994.

Alivisatos P. Colloidal quantum dots. From scaling laws to biological applications. *Pure and Applied Chemistry.* 2000;72(1-2):3–9.

Alivisatos P, Johnsson KP, Peng X, Wilson TE, Loweth CJ, Bruchez MP Jr, Schultz PG. Organization of "nanocrystal molecules" using DNA. *Nature.* 1996 Aug 15;382(6592): 609–611.

Andreasson JO, Milic B, Chen GY, Guydosh NR, Hancock WO, Block SM. Examining kinesin processivity within a general gating framework. *Elife.* 2015 Apr 22;4:e07403.

Andreou AI, Nakayama N. Mobius Assembly. *Methods Mol Biol.* 2020;2205:201–218.

Arai Y, Nagai T. Extensive use of FRET in biological imaging. *Microscopy (Oxf).* 2013 Aug;62(4):419–428.

Atkinson WI. *Nanocosm.* AMACOM Books, Kaduna, Nigeria, 2003.

Bachand M, Trent AM, Bunker BC, Bachand GD. Physical factors affecting kinesin-based transport of synthetic nanoparticle cargo. *J Nanosci Nanotechnol.* 2005 May;5(5):718–722.

Baker SE, Cai W, Lasseter TL, Weidkamp KP, Hamers RJ. Covalently bonded adducts of deoxyribonucleic acid (DNA) oligonucleotides with single-wall carbon nanotubes: synthesis and hybridization. *Nano Letters.* 2002;2(12):1413–1417.

Balzani VV, Credi A, Raymo FM, Stoddart JF. Artificial molecular machines. *Angew Chem Int Ed Engl.* 2000 Oct 2;39(19):3348–3391.

Bardi JS. Nano-origami: Scientists create single, clonable strand of DNA that folds into an octahedron. *The Scripps Research Institute – News and Views.* 2004 Feb 23;4(6). https://www.scripps.edu/newsandviews/e_20040223/nano.html (accessed January 16, 2024).

Bath J, Green SJ, Turberfield AJ. A free-running DNA motor powered by a nicking enzyme. *Angew Chem Int Ed Engl.* 2005 Jul 11;44(28):4358–4361.

Bath S, Dirks RM, Rothemund PW, Winfree E, Pierce NA. An autonomous polymerization motor powered by DNA hybridization. *Nat Nanotechnol.* 2007 Aug;2(8):490–494.

Bayoumi M, Nomidis SK, Willems K, Carlon E, Maglia G. Autonomous and active transport operated by an entropic DNA piston. *Nano Lett.* 2021 Jan 13;21(1):762–768.

Bayrak T, Helmi S, Ye J, Kauert D, Kelling J, Schönherr T, Weichelt R, Erbe A, Seidel R. DNA-mold templated assembly of conductive gold nanowires. *Nano Lett.* 2018a Mar 14;18(3):2116–2123.

Bayrak T, Jagtap NS, Erbe A. Review of the electrical characterization of metallic nanowires on DNA templates. *Int J Mol Sci.* 2018b Oct 3;19(10):3019.

Ben-Jacob E, Hermon Z, Caspi S. DNA transistor and quantum bit element: realization of nano-biomolecular logical devices. *Physics Letters A.* 1999;263:199–202.

Benner SA, Battersby TR, Eschgfaller B, Hutter D, Kodra JT, Lutz S et al. Redesigning nucleic acids. *Pure Appl Chem.* 1998 Feb;70(2):263–266.

Bensimon D, Simon AJ, Croquette VV, Bensimon A. Stretching DNA with a receding meniscus: experiments and models. *Phys Rev Lett.* 1995 Jun 5;74(23):4754–4757.

Berger M, Luzzi SD, Henry AA, Romesberg FE. Stability and selectivity of unnatural DNA with five-membered-ring nucleobase analogues. *J Am Chem Soc.* 2002 Feb 20;124(7):1222–1226.

Berger M, Wu Y, Ogawa AK, McMinn DL, Schultz PG, Romesberg FE. Universal bases for hybridization, replication and chain termination. *Nucleic Acids Res.* 2000 Aug 1;28(15):2911–2914.

Bhanjadeo MM, Baral B, Subudhi U. Sequence-specific b-to-z transition in self-assembled DNA: a biophysical and thermodynamic study. *Int J Biol Macromol.* 2019 Sep 15;137:337–345.

Bielinska AU, Chen C, Johnson J, Baker JR. DNA complexing with polyamidoamine dendrimers: implications for transfection. *Bioconjug Chem.* 1999 Sep-Oct;10(5): 843–850.

Bielinska AU, Kukowska-Latallo JF, Baker JR. The interaction of plasmid DNA with polyamidoamine dendrimers: mechanism of complex formation and analysis of alterations induced in nuclease sensitivity and transcriptional activity of the complexed DNA. *Biochim Biophys Acta.* 1997 Aug 7;1353(2):180–190.

Bittker JA, Phillips KJ, Liu DR. Recent advances in the in vitro evolution of nucleic acids. *Curr Opin Chem Biol.* 2002 Jun;6(3):367–374.

Bixon M, Giese B, Wessely S, Langenbacher T, Michel-Beyerle ME, Jortner J. Long-range charge hopping in DNA. *Proc Natl Acad Sci U S A.* 1999 Oct 12;96(21):11713–11716.

Blanchard AT, Bazrafshan AS, Yi J, Eisman JT, Yehl KM, Bian T, Mugler A, Salaita K. Highly polyvalent DNA motors generate 100+ pN of force via autochemophoresis. *Nano Lett.* 2019 Oct 9;19(10):6977–6986.

Blättler T, Huwiler C, Ochsner M, Städler B, Solak H, Vörös J, Grandin HM. Nanopatterns with biological functions. *J Nanosci Nanotechnol.* 2006 Aug;6(8):2237–2264.

Bleha T, Cifra P. Compression and stretching of single DNA molecules under channel confinement. *J Phys Chem B.* 2020 Mar 5;124(9):1691–1702.

Bloomfield VA, Crothers DM, Tinoco I Jr. *Nucleic Acids: Structures, Properties, and Function.* University Science Books, Sausalito, CA, 2000.

Boehm FJ. An Investigation of Nucleic Acid/DNA-Based Manufacturing. Original research contribution to the Center for Responsible Nanotechnology: *Thirty Essential Nanotechnology Studies* (#10 – What will be required to develop nucleic acid manufacturing and products) Aug. 2004. https://www.scribd.com/document/183051565/DNA-MANUFACTURING-pdf (accessed April 01, 2024).

Bourne HR, Sanders DA, McCormick F. The GTPase superfamily: conserved structure and molecular mechanism. *Nature.* 1991 Jan 10;349(6305):117–127.

Boutorine AS, Takasugi M, Hélène C, Tokuyama H, Isobe H, Nakamura E. Fullerene–Oligonucleotide conjugates: photoinduced sequence-specific DNA cleavage. *Angewandte Chemie – International Edition.* 1995 Jan 3;33(23-24):2462–2465.

Brancolini G, Di Felice R. Combined effects of metal complexation and size expansion in the electronic structure of DNA base pairs. *J Chem Phys*. 2011 May 28;134(20):205102.

Braun E, Eichen Y, Sivan U, Ben-Yoseph G. DNA-templated assembly and electrode attachment of a conducting silver wire. *Nature*. 1998 Feb 19;391(6669):775–778.

Bray D. *Cell Movements, 2nd ed.*, Garland Publishing, New York, 2001.

Breaker RR. Tech.Sight. Molecular biology. Making catalytic DNAs. *Science*. 2000 Dec 15;290(5499):2095–2096.

Brown S. Metal-recognition by repeating polypeptides. *Nat Biotechnol*. 1997 Mar;15(3):269–272.

Brown S, Sarikaya M, Johnson E. A genetic analysis of crystal growth. *J Mol Biol*. 2000 Jun 9;299(3):725–735.

Cai H, Cao X, Jiang Y, He P, Fang Y. Carbon nanotube-enhanced electrochemical DNA biosensor for DNA hybridization detection. *Anal Bioanal Chem*. 2003 Jan;375(2):287.

Cai Z, Yang B, Gao R, Liu H. Opportunities and challenges for DNA in atomic and close-to-atomic scale manufacturing. *Nanotechnology*. 2023 Jul 24;34:412501.

Carmi N, Shultz LA, Breaker RR. In vitro selection of self-cleaving DNAs. *Chem Biol*. 1996 Dec;3(12):1039–1046.

Cassell AM, Scrivens WA, Tour JM. Assembly of DNA/Fullerene hybrid materials. *Angew Chem Int Ed Engl*. 1998 Jun 19;37(11):1528–1531.

Castillo JP, B Tong A, Tafoya S, Jardine PJ, Bustamante C. A DNA packaging motor inchworms along one strand allowing it to adapt to alternative double-helical structures. *Nat Commun*. 2021 Jun 8;12(1):3439.

Castillo M. From hard drives to flash drives to DNA drives. *AJNR Am J Neuroradiol*. 2014 Jan;35(1):1–2.

Ceze L, Nivala J, Strauss K. Molecular digital data storage using DNA. *Nat Rev Genet*. 2019 Aug;20(8):456–466.

Chan MS, Leung HM, Wong SW, Lin Z, Gao Q, Chang TJH, Lai KWC, Lo PK. Reversible reconfiguration of high-order DNA nanostructures by employing g-quartet toeholds as adhesive units. *Nanoscale*. 2020 Jan 28;12(4):2464–2471.

Chaput JC, Ichida JK, Szostak JW. DNA polymerase-mediated DNA synthesis on a TNA template. *J Am Chem Soc*. 2003 Jan 29;125(4):856–857.

Chemla YR, Aathavan K, Michaelis J, Grimes S, Jardine PJ, Anderson DL, et al. Mechanism of force generation of a viral DNA packaging motor. *Cell*. 2005;122(5):683–692

Chen HL, Goel A. "Error Free Self-Assembly Using Error Prone Tiles," in Ferretti, C., Mauri, G., Zandron, C. eds., *DNA Computing. DNA 2004. Lecture Notes in Computer Science*, Springer, Berlin, Heidelberg. 2005;3384.

Chen J, Hamon MA, Hu H, Chen Y, Rao AM, Eklund PC, Haddon RC. Solution properties of single-walled carbon nanotubes. *Science*. 1998 Oct 2;282(5386):95–98.

Chen JH, Seeman NC. Synthesis from DNA of a molecule with the connectivity of a cube. *Nature*. 1991 Apr 18;350(6319):631–633.

Chen Y, Wang M, Mao C. An autonomous DNA nanomotor powered by a DNA enzyme. *Angew Chem Int Ed Engl*. 2004 Jul 5;43(27):3554–3557.

Chinnapen DJ, Sen D. A deoxyribozyme that harnesses light to repair thymine dimers in DNA. *Proc Natl Acad Sci U S A*. 2004 Jan 6;101(1):65–69.

Chodosh LA. Purification of DNA-binding proteins using biotin/streptavidin affinity systems. *Curr Protoc Mol Biol*. 2001 May; Chapter 12:Unit 12.6.

Church GM, Gao Y, Kosuri S. Next-generation digital information storage in DNA. *Science*. 2012 Sep 28;337(6102):1628.

Cook J, Sloan J, Green MLH. Opening and filling carbon nanotubes. *Fullerene Science and Technology*. 1997;5(4):695–704.

Coppin CM, Pierce DW, Hsu L, Vale RD. The load dependence of kinesin's mechanical cycle. *Proc Natl Acad Sci U S A*. 1997 Aug 5;94(16):8539–44.

Coy DL, Wagenbach M, Howard J. Kinesin takes one 8-nm step for each ATP that it hydrolyzes. *J Biol Chem*. 1999 Feb 5;274(6):3667–3671.

Cuenoud B, Szostak JW. A DNA metalloenzyme with DNA ligase activity. *Nature*. 1995 Jun 15;375(6532):611–614.

Currin A, Korovin K, Ababi M, Roper K, Kell DB, Day PJ, King RD. Computing exponentially faster: implementing a non-deterministic universal Turing machine using DNA. *J R Soc Interface*. 2017 Mar;14(128):20160990.

Dai X, Li Q, Aldalbahi A, Wang L, Fan C, Liu X. DNA-based fabrication for nanoelectronics. *Nano Lett*. 2020 Aug 12;20(8):5604–5615.

Demers LM, Ginger DS, Park SJ, Li Z, Chung SW, Mirkin CA. Direct patterning of modified oligonucleotides on metals and insulators by dip-pen nanolithography. *Science*. 2002 Jun 7;296(5574):1836–1838.

Dey S, Rivas-Barbosa R, Sciortino F, Zaccarelli E, Zijlstra P. Biomolecular interactions on densely coated nanoparticles: a single-molecule perspective. *Nanoscale*. 2024 Feb 29;16(9):4872–4879.

Diez S, Reuther C, Dinu C, Seidel R, Mertig M, Pompe W, Howard J. Stretching and transporting DNA molecules using motor proteins. *Nano Lett*. 2003;3:1251–1254.

Dong H, Hua W, Li S. Estimation on the individual hydrogen-bond strength in molecules with multiple hydrogen bonds. *J Phys Chem A*. 2007 Apr 19;111(15):2941–2945.

Donnet JB, Bansal RC. *Carbon Fibers, 3rd* ed., CRC Press, Boca Raton, FL. 1998..

Draper B, Rao VB. An ATP hydrolysis sensor in the DNA packaging motor from bacteriophage T4 suggests an inchworm-type translocation mechanism. *J Mol Biol*. 2007 May 25;369(1):79–94.

Drexler KE. Private communication. 2004.

Duanghathaipornsuk S, Shen B, Cameron BD, Ijäs H, Linko V, Kostiainen MA, Kim D. Aptamer-embedded DNA origami cage for detecting (glycated) hemoglobin with a surface plasmon resonance sensor. *Mater Lett*. 2020 Sept 15;275:128141.

Dwyer C, Guthold M, Falvo M, Washburn S, Superfine R, Erie D. DNA-functionalized single-walled carbon nanotubes. *Nanotechnology*. September 6 2002;13(5):601–604.

Endo M, Yamamoto S, Tatsumi K, Emura T, Hidaka K, Sugiyama H. RNA-templated DNA origami structures. *Chem Commun (Camb)*. 2013 Apr 11;49(28):2879–2881.

Erlich Y, Zielinski D. DNA Fountain enables a robust and efficient storage architecture. *Science*. 2017 Mar 3;355(6328):950–954.

Essevaz-Roulet B, Bockelmann U, Heslot F. Mechanical separation of the complementary strands of DNA. *Proc Natl Acad Sci U S A*. 1997 Oct 28;94(22):11935–11940.

Eyster KM. Protocol for DNA microarrays on glass slides. *Methods Mol Biol*. 2019;1986:17–33.

Famulok M, Mayer G, Blind M. Nucleic acid aptamers-from selection in vitro to applications in vivo. *Acc Chem Res*. 2000 Sep;33(9):591–599.

Feng L, Park SH, Reif JH, Yan H. A two-state DNA lattice switched by DNA nanoactuator. *Angew Chem Int Ed Engl*. 2003 Sep 22;42(36):4342–4346.

Finer JT, Simmons RM, Spudich JA. Single myosin molecule mechanics: piconewton forces and nanometre steps. *Nature*. 1994 Mar 10;368(6467):113–119.

Fink HW. DNA and conducting electrons. *Cell Mol Life Sci*. 2001 Jan;58(1):1–3.

Fort RC. *Adamantane: The Chemistry of Diamond Molecules*. Vol. 5 of Studies in Organic Chemistry. M. Dekker. 1976.

Franklin RE, Gosling RG. Molecular configuration in sodium thymonucleate. *Nature*. 1953 Apr 25;171(4356):740–741.

Freeland J, Zhang L, Wang ST, Ruiz M, Wang Y. Bent DNA bows as sensing amplifiers for detecting DNA-interacting salts and molecules. *Sensors (Basel)*. 2020 May 31;20(11):3112.

Freitas RA Jr., *Nanomedicine, Volume I: Basic Capabilities*, Landes Bioscience, Georgetown, TX, 1999. http://www.nanomedicine.com/NMI.htm

Freitas RA Jr., Is Sapphire Biocompatible With Living Cells? *Foresight Update*. 2002 Nov;50:15–17.

Freitas RA Jr., *Nanomedicine, Volume IIA: Biocompatibility*. Landes Bioscience, Georgetown, TX, 2003. https://www.nanomedicine.com/NMIIA.htm

Freitas RA Jr. Private communication. 2004.

Freitas RA Jr. "Chapter 6. Diamondoid Nanorobotics," in Constantinos Mavroidis, Antoine Ferreira, eds., *NanoRobotics: Current Approaches and Techniques*, Springer, New York. 2013.

Freitas RA Jr., Merkle RC. *Kinematic Self-Replicating Machines*, Landes Bioscience, Georgetown, TX, 2004. http://www.MolecularAssembler.com/KSRM.htm

Fritz J, Baller MK, Lang HP, Rothuizen H, Vettiger P, Meyer E, Güntherodt H, Gerber C, Gimzewski JK. Translating biomolecular recognition into nanomechanics. *Science*. 2000 Apr 14;288(5464):316–318.

Fritzsche W. DNA-Based Molecular Construction. *International Workshop*. 23–25, May 2002, Jena, Germany, Institute for Physical High Technology.

Fu TJ, Seeman NC. DNA double-crossover molecules. *Biochemistry*. 1993 Apr 6;32(13):3211–3220.

Fujiwara T, Kimoto M, Sugiyama H, Hirao I, Yokoyama S. Synthesis of 6-(2-thienyl)purine nucleoside derivatives that form unnatural base pairs with pyridin-2-one nucleosides. *Bioorg Med Chem Lett*. 2001 Aug 20;11(16):2221–2223.

Gacs P. Private communication. 2004.

Garner P, Sherry B, Moilanen S, Huang Y. In vitro stability of alpha-helical peptide nucleic acids (alphaPNAs). *Bioorg Med Chem Lett*. 2001 Sep 3;11(17):2315–2317.

Giver L, Arnold FH. Combinatorial protein design by in vitro recombination. *Curr Opin Chem Biol*. 1998 Jun;2(3):335–338.

Goldman N, Bertone P, Chen S, Dessimoz C, LeProust EM, Sipos B, Birney E. Towards practical, high-capacity, low-maintenance information storage in synthesized DNA. *Nature*. 2013 Feb 7;494(7435):77–80.

Gómez-Navarro C, Moreno-Herrero F, de, Pablo PJ, Colchero J, Gómez-Herrero J, Baró AM. Contactless experiments on individual DNA molecules show no evidence for molecular wire behavior. *Proc Natl Acad Sci U S A*. 2002 Jun 25;99(13):8484–8487.

Göpfrich K, Urban MJ, Frey C, Platzman I, Spatz JP, Liu N. Dynamic actuation of DNA-assembled plasmonic nanostructures in microfluidic cell-sized compartments. *Nano Lett*. 2020 Mar 11;20(3):1571–1577.

Grass RN, Heckel R, Puddu M, Paunescu D, Stark WJ. Robust chemical preservation of digital information on DNA in silica with error-correcting codes. *Angew Chem Int Ed Engl*. 2015 Feb 16;54(8):2552–2555.

Green MR, Sambrook J. Precipitation of DNA with Ethanol. *Cold Spring Harb Protoc*. 2016 Dec 1;2016(12):pdb.prot093377

Green MR, Sambrook J. Precipitation of DNA with Isopropanol. *Cold Spring Harb Protoc*. 2017 Aug 1;2017(8):pdb.prot093385.

Grubmüller H, Heymann B, Tavan P. Ligand binding: molecular mechanics calculation of the streptavidin-biotin rupture force. *Science*. 1996 Feb 16;271(5251):997–999.

Gu Z, Zhao L, Ge C, Liu S, Fang G, Chen SS, Yang Z, Zhou R. Facet-regulated adhesion of double-stranded DNA on palladium surfaces. *Nanoscale*. 2019 Jan 23;11(4): 1827–1836.

Hainfeld JF, Furuya FR, Powell RD, Liu W. DNA Nanowires. *Microscopy and Microanalysis*. 2001;7(S2):1034–1035.

Harrington RE, Zimm BH. Degradation of polymers by controlled hydrodynamic shear. *The Journal of Physical Chemistry*. 1965;69:161–175.

Harris LF, Sullivan MR, Hatfield DL. Directed molecular evolution. *Orig Life Evol Biosph.* 1999 Aug;29(4):425–435.

Harvey SC. The scrunchworm hypothesis: transitions between A-DNA and B-DNA provide the driving force for genome packaging in double-stranded DNA bacteriophages. *J Struct Biol.* 2015 Jan;189(1):1–8.

Hayashi M, Nishiyama M, Kazayama Y, Toyota T, Harada Y, Takiguchi K. Reversible morphological control of tubulin-encapsulating giant liposomes by hydrostatic pressure. *Langmuir.* 2016 Apr 19;32(15):3794–802.

Hazani M, Naaman R, Hennrich F, Kappes MM. Confocal fluorescence imaging of DNA-functionalized carbon nanotubes. *Nano Lett.* 2003;3(2):153–155.

Hazarika P, Ceyhan B, Niemeyer CM. Reversible switching of DNA-gold nanoparticle aggregation. *Angew Chem Int Ed Engl.* 2004 Dec 3;43(47):6469–6471.

He JY, Shang X, Yang CL, Zuo SY, Yuan R, Xu WJ. Antibody-responsive ratiometric fluorescence biosensing of biemissive silver nanoclusters wrapped in switchable DNA tweezers. *Anal Chem.* 2021 Aug 24;93(33):11634–11640.

Helmenstine AM. How to Classify Chemical Reaction Orders Using Kinetics. ThoughtCo. 2019. https://www.thoughtco.com/chemical-reaction-orders-608182 (accessed January 23, 2024).

Henke C. DNA-chip technologies (Parts 1–3). *IVD Technology Magazine.* 1998.

Heo W, Hasegawa K, Okamoto K, Sako Y, Ishii K, Tahara T. Scanning two-dimensional fluorescence lifetime correlation spectroscopy: conformational dynamics of DNA Holliday junction from microsecond to subsecond. *J Phys Chem Lett.* 2022 Feb 10;13(5):1249–1257.

Hermon Z, Caspi S, Ben-Jacob E. Prediction of charge and dipole solitons in DNA molecules based on the behaviour of phosphate bridges as tunnel elements. *Europhys. Lett.* 1998;43:482–487.

Hess H, Vogel V. Molecular shuttles based on motor proteins: active transport in synthetic environments. *J Biotechnol.* 2001 Nov;82(1):67–85.

Hessari NM, Spindler L, Troha T, Lam WC, Drevenšek-Olenik I, da Silva MW. Programmed self-assembly of a quadruplex DNA nanowire. *Chemistry.* 2014 Mar 24;20(13):3626–3630.

Hingorani MM, O'Donnell M. DNA polymerase structure and mechanisms of action. *Curr Org Chem.* 2000;4(9):887–913.

Hiyama S, Inoue T, Shima T, Moritani Y, Suda T, Sutoh K. Autonomous loading, transport, and unloading of specified cargoes by using DNA hybridization and biological motor-based motility. *Small.* 2008 Apr;4(4):410–415.

Honda M, Takiguchi K, Ishikawa S, Hotani H. Morphogenesis of liposomes encapsulating actin depends on the type of actin-crosslinking. *J Mol Biol.* 1999 Mar 26;287(2): 293–300.

Hotani H, Nomura F, Suzuki Y. Giant liposomes: from membrane dynamics to cell morphogenesis. *Current Opinion in Colloid and Interface Science.* 1999;4:358–368.

Howard J. *Mechanics of Motor Proteins and the Cytoskeleton*, Sinauer Associates, Sunderland, MA. 2001.

Hu Y, Cecconello A, Idili A, Ricci F, Willner I. Triplex DNA nanostructures: from basic properties to applications. *Angew Chem Int Ed Engl.* 2017 Nov 27;56(48):15210–15233.

Jia Y, Dong W, Feng X, Li J, Li J. A self-powered kinesin-microtubule system for smart cargo delivery. *Nanoscale.* 2015 Jan 7;7(1):82–85.

Jiang Q, Zhao S, Liu J, Song L, Wang ZG, Ding B. Rationally designed DNA-based nanocarriers. *Adv Drug Deliv Rev.* 2019 Jul;147:2–21.

Jin R, Gaffney BL, Wang C, Jones RA, Breslauer KJ. Thermodynamics and structure of a DNA tetraplex: a spectroscopic and calorimetric study of the tetramolecular complexes of d(TG3T) and d(TG3T2G3T). *Proc Natl Acad Sci U S A.* 1992 Sep 15;89(18): 8832–8836.

Jin R, Wu G, Li Z, Mirkin CA, Schatz GC. What controls the melting properties of DNA-linked gold nanoparticle assemblies? *J Am Chem Soc*. 2003 Feb 12;125(6): 1643–1654.

Jun H, Shepherd TR, Zhang K, Bricker WP, Li S, Chiu W, Bathe M. Automated sequence design of 3D polyhedral wireframe DNA origami with honeycomb edges. *ACS Nano*. 2019 Feb 26;13(2):2083–2093.

Kadish KM, Ruoff RS, editors. *Fullerenes: Chemistry, Physics, and Technology*. Wiley-Interscience, New York, 2000.

Kanai T, Ichino M, Kojima T. Pyrimidine nucleosides. II. The synthesis of unnatural pyrimidine nucleosides saturated at 5,6-double bond. *Chem Pharm Bull (Tokyo)*. 1969 Apr;17(4):650–652.

Kang D, Duan R, Tan Y, Hong F, Wang B, Chen Z, Xu S, Lou X, Wei W, Yurke B, Xia F. Speeding up the self-assembly of a DNA nanodevice using a variety of polar solvents. *Nanoscale*. 2014;6(23):14153–14157.

Katz E, Bollella P. Fuel Cells and Biofuel Cells: From Past to Perspectives. *Isr. J. Chem*. 2021;61:68–84.

Ke Y, Liu Y, Zhang J, Yan H. A study of DNA tube formation mechanisms using 4-, 8-, and 12-helix DNA nanostructures. *J Am Chem Soc*. 2006 Apr 5;128(13):4414–4421.

Ke Y, Ong LL, Shih WM, Yin P. Three-dimensional structures self-assembled from DNA bricks. *Science*. 2012 Nov 30;338(6111):1177–1183.

Keller S, Marx A. The use of enzymes for construction of DNA-based objects and assemblies. *Chem Soc Rev*. 2011 Dec;40(12):5690–5697.

Kelly TR. Molecular motors: synthetic DNA-based walkers inspired by kinesin. *Angew Chem Int Ed Engl*. 2005 Jul 4;44(27):4124–4127.

Keren K, Berman RS, Buchstab E, Sivan U, Braun E. DNA-templated carbon nanotube field-effect transistor. *Science*. 2003 Nov 21;302(5649):1380–1382.

Keren K, Krueger M, Gilad R, Ben-Yoseph G, Sivan U, Braun E. Sequence-specific molecular lithography on single DNA molecules. *Science*. 2002 Jul 5;297(5578):72–75.

Khare HS, Gosvami NN, Lahouij I, Milne ZB, McClimon JB, Carpick RW. Nanotribological Printing: A Nanoscale Additive Manufacturing Method. *Nano Lett*. 2018 Nov 14;18(11):6756–6763.

Knudsen JB, Liu L, Bank, Kodal AL, Madsen M, Li Q, Song J, Woehrstein JB et al. Routing of individual polymers in designed patterns. *Nat Nanotechnol*. 2015 Oct;10(10):892–898.

Ko SH, Su M, Zhang C, Ribbe AE, Jiang W, Mao C. Synergistic self-assembly of RNA and DNA molecules. *Nat Chem*. 2010 Dec;2(12):1050–1055.

Kohll AX, Antkowiak PL, Chen WD, Nguyen BH, Stark WJ, Ceze L, Strauss K, Grass RN. Stabilizing synthetic DNA for long-term data storage with earth alkaline salts. *Chem Commun (Camb*. 2020 Mar 28;56(25):3613–3616.

Kotnala A, Zheng Y. Digital Assembly of Colloidal Particles for Nanoscale Manufacturing. *Part Part Syst Charact*. 2019 Aug;36(8):1900152.

Koumura N, Zijlstra RW, van, Delden RA, Harada N, Feringa BL. Light-driven monodirectional molecular rotor. *Nature*. 1999 Sep 9;401(6749):152–155.

Krissanaprasit A, Key CM, Pontula S, LaBean TH. Self-Assembling Nucleic Acid Nanostructures Functionalized with Aptamers. *Chem Rev*. 2021 Nov 24;121(22):13797–13868.

Kumar S, Milani G, Takatsuki H, Lana T, Persson M, Frasson C, te Kronnie G, Månsson A. Sensing protein antigen and microvesicle analytes using high-capacity biopolymer nano-carriers. *Analyst*. 2016 Feb 7;141(3):836–846.

Kumar S, ten, Siethoff L, Persson M, Lard M, te, Kronnie G, Linke H, Månsson A. Antibodies covalently immobilized on actin filaments for fast myosin driven analyte transport. *PLoS One*. 2012;7(10):e46298.

Kurzweil R. *The Age of Intelligent Machines*. MIT, Cambridge, MA, 1990.

LaBean TH, Winfree E, Reif JH. "Experimental Progress in Computational by Self-Assembly of DNA Tilings," in Winfree E, Gifford DK, eds., *DNA-Based Computers V*, American Mathematical Society, Providence, Rhode Island. 2000a;123–140.

LaBean TH, Yan H, Kopatsch J, Liu F, Winfree E, Reif JH et al. Construction, analysis, ligation, and self-assembly of DNA triple crossover complexes. *Journal of the American Chemical Society*. 2000b Mar 8;122(9):1848–1860.

Landon PB, Ramachandran S, Gillman A, Gidron T, Yoon D, Lal R. DNA zipper-based tweezers. *Langmuir*. 2012 Jan 10;28(1):534–540.

Lebrun A, Lavery R. Modelling extreme stretching of DNA. *Nucleic Acids Res*. 1996 Jun 15;24(12):2260–2267.

Lee GU, Chrisey LA, Colton RJ. Direct measurement of the forces between complementary strands of DNA. *Science*. 1994 Nov 4;266(5186):771–773.

Leng M. Left-handed Z-DNA. *Biochim Biophys Acta*. 1985 Aug 21;825(4):339–344.

Lewis, J. B., Smith, B., Krummenacker, M. "A proposed path from current biotechnology to a replicating assembler," in Krummenacker M, Lewis JB, eds., *Prospects in Nanotechnology: Toward Molecular Manufacturing*, Wiley, Hoboken, NJ, 1995; 123–135.

Li H, Wang Z, Huo F, Wang S. Dip- pen nanolithography (DPN): from micro/nano-patterns to biosensings. *Chem Res Chin Univ*. 2021;37(4):846–854.

Li J, Johnson-Buck A, Yang YR, Shih WM, Yan H, Walter NG. Exploring the speed limit of toehold exchange with a cartwheeling DNA acrobat. *Nat Nanotechnol*. 2018;13(8):723–729.

Li JJ, Tan W. A Single DNA Molecule Nanomotor. *Nano Letters*. 2002;2:315–318.

Li N, Shang Y, Han Z, Wang T, Wang ZG, Ding B. Fabrication of metal nanostructures on DNA templates. *ACS Appl Mater Interfaces*. 2019 Apr 17;11(15):13835–13852.

Li S, Tian T, Zhang, Cai X, Lin Y. Advances in biological applications of self-assembled DNA tetrahedral nanostructures. *Materials Today*. 2019;24:57–68.

Li Y, Breaker RR. Phosphorylating DNA with DNA. *Proc Natl Acad Sci U S A*. 1999 Mar 16;96(6):2746–2751.

Li Y, Liu Y, Breaker RR. Capping DNA with DNA. *Biochemistry*. 2000 Mar 21;39(11):3106–3114.

Li Y, Sen D. A catalytic DNA for porphyrin metallation. *Nat Struct Biol*. 1996;3(9):743–747.

Li Y, Tseng YD, Kwon SY, D'Espaux L, Bunch JS, McEuen PL, Luo D. Controlled assembly of dendrimer-like DNA. *Nat Mater*. 2004 Jan;3(1):38–42.

Li Z, Jin R, Mirkin CA, Letsinger RL. Multiple thiol-anchor capped DNA-gold nanoparticle conjugates. *Nucleic Acids Res*. 2002 Apr 1;30(7):1558–1562.

Liang C, Mislow K. On Borromean links. *Journal of Mathematical Chemistry*. 1994;16: 27–35.

Lilley DM. Structures of helical junctions in nucleic acids. *Q Rev Biophys*. 2000 May;33(2): 109–159.

Lim TS, Montague-Cardoso K. A DNA origami-based nanoscale molecular transport railway. *Commun Biol*. 2021 Sep 9;4(1):1075.

Liphardt J, Onoa B, Smith SB, Tinoco I Jr, Bustamante C. Reversible unfolding of single RNA molecules by mechanical force. *Science*. 2001 Apr 27;292(5517):733–737.

Liu D, Park SH, Reif JH, LaBean TH. DNA nanotubes self-assembled from triple-crossover tiles as templates for conductive nanowires. *Proc Natl Acad Sci U S A*. 2004 Jan 20;101(3):717–722.

Liu F, Zhang Y, Ou-Yang Z. Flexoelectric origin of nanomechanic deflection in DNA-microcantilever system. *Biosens Bioelectron*. 2003 May;18(5-6):655–660.

Liu J, Rinzler AG, Dai H, Hafner JH, Bradley RK, Boul PJ, Lu A, Iverson T, Shelimov K, Huffman CB, Rodriguez-Macias F, Shon YS, Lee TR, Colbert DT, Smalley RE. Fullerene pipes. *Science*. 1998 May 22;280(5367):1253–1256.

Liu M, Zhang Q, Li Z, Gu J, Brennan JD, Li Y. Programming a topologically constrained DNA nanostructure into a sensor. *Nat Commun*. 2016 Jun 23;7:12074.

Liu S, Shang Y, Jiao Y, Li N, Ding B. DNA-based plasmonic nanostructures and their optical and biomedical applications. *Nanotechnology*. 2021 Jul 15;32:402002.

Loakes D. Survey and summary: The applications of universal DNA base analogues. *Nucleic Acids Res*. 2001 Jun 15;29(12):2437–2447.

Loffler PMG, Ries O, Rabe A, Okholm AH, Thomsen, RP, Kjems J, Vogel SA. DNA-programmed liposome fusion cascade. *Angew. Chem. Int. Ed*. 2017;56:13228–13231.

Long Y, Ubych K, Jagu E, Neely RK. FRET-Based Method for Direct, Real-Time Measurement of DNA Methyltransferase Activity. *Bioconjug Chem*. 2021 Jan 20;32(1):192–198.

Loweth CJ, Caldwell WB, Peng X, Alivisatos AP, Schultz PG. DNA-Based Assembly of Gold Nanocrystals. *Angew Chem Int Ed Engl*. 1999 Jun 14;38(12):1808–1812.

Lu B, Vecchioni S, Ohayon YP, Sha R, Woloszyn K, Yang B, Mao C, Seeman NC. 3D Hexagonal Arrangement of DNA Tensegrity Triangles. *ACS Nano*. 2021 Oct 26;15(10):16788–16793.

Luhrs M, Eccleston K, Vesenka J. Construction and Examination of "G-wire" DNA, DNA-Based Molecular Construction, Intern. Workshop on DNA-based molecular construction, Jena, Germany 2002, Editor: W. Fritzsche, *AIP Conference Proceedings*. 2002;640:109–122.

Luo L, Liu J, Wang Z, Yang X, Dong S, Wang E. Fabrication of layer-by-layer deposited multilayer films containing DNA and its interaction with methyl green. *Biophys Chem*. 2001 Dec 11;94(1-2):11–22.

Luther A, Brandsch R, von Kiedrowski G. Surface-promoted replication and exponential amplification of DNA analogues. *Nature*. 1998 Nov 19;396(6708):245–248.

Luu MT, Berengut JF, Li J, Chen JB, Daljit Singh JK, Coffi Dit Glieze K et al. Reconfigurable nanomaterials folded from multicomponent chains of DNA origami voxels. *Sci Robot*. 2024 Nov 27;9(96):eadp2309.

Lymperopoulos K, Crawford R, Torella JP, Heilemann M, Hwang LC, Holden SJ, Kapanidis AN. Single-molecule DNA biosensors for protein and ligand detection. *Angew Chem Int Ed Engl*. 2010 Feb 8;49(7):1316–1320.

Ma N, Minevich B, Liu J, Ji M, Tian Y, Gang O. Directional assembly of nanoparticles by DNA shapes: towards designed architectures and functionality. *Top Curr Chem (Cham)*. 2020 Mar 27;378(2):36.

MacDonaill DA. "Tautomerism as a Constraint on the Composition of Alternative Nucleotide Alphabets," in Standish, Abbass, Bedau eds., *Artificial Life VIII*, MIT Press. 2002;106–110.

Mądzik MT, Asaad S, Youssry A, Joecker B, Rudinger KM, Nielsen E et al. Precision tomography of a three-qubit donor quantum processor in silicon. *Nature*. 2022 Jan;;601(7893):348–353.

Majumdar A. Bioassays based on molecular nanomechanics. *Dis Markers*. 2002;18(4): 167–174.

Månsson A, Sundberg M, Balaz M, Bunk R, Nicholls IA, Omling P, Tågerud S, Montelius L. In vitro sliding of actin filaments labelled with single quantum dots. *Biochem Biophys Res Commun*. 2004 Feb 6;;314(2):529–534.

Mao C, LaBean TH, Relf JH, Seeman NC. Logical computation using algorithmic self-assembly of DNA triple-crossover molecules. *Nature*. 2000 Sep 28;407(6803):493–496.

Mao C, Sun W, Seeman NC. Assembly of Borromean rings from DNA. *Nature*. 1997 Mar 13;386(6621):137–138.

Mao C, Sun W, Seeman NC. Designed two-dimensional DNA Holliday junction arrays visualized by atomic force microscopy. *Journal of the American Chemical Society*. 1999a Jun 16;121(23):5437–5443.

Mao C, Sun W, Shen Z, Seeman NC. A nanomechanical device based on the B-Z transition of DNA. *Nature*. 1999b Jan 14;397(6715):144–146.

Marsh TC, Henderson E. G-wires: self-assembly of a telomeric oligonucleotide, d(GGGGTTGGGG), into large superstructures. *Biochemistry*. 1994 Sep 6;33(35): 10718–10724.

Marsh TC, Vesenka J, Henderson E. A new DNA nanostructure, the G-wire, imaged by scanning probe microscopy. *Nucleic Acids Res*. 1995 Feb 25;23(4):696–700.

Maune HT, Han, SP, Barish, RD, Bockrath M, Iii WAG, Rothemund, PW, Winfree E. Self-assembly of carbon nanotubes into two-dimensional geometries using DNA origami templates. *Nat. Nanotechnol*. 2010;5:61–6.

Mbindyo JKN, Reiss BD, Martin BR, Keating CD, Natan MJ, Mallouk TE. DNA- directed assembly of gold nanowires on complementary surfaces. *Adv. Mater*. 2001;13:249–254.

McLaughlin LW, Stewart KM. Synthesis and self-assembly of metal-centered DNA lattices. Presented at: *11th Foresight Conference on Molecular Nanotechnology*. Palo Alto, CA. 2003.

McLeish MJ, Kenyon GL. Relating structure to mechanism in creatine kinase. *Crit Rev Biochem Mol Biol*. 2005 Jan-Feb;;40(1):1–20.

Mehta AD, Rock RS, Rief M, Spudich JA, Mooseker MS, Cheney RE. Myosin-V is a processive actin-based motor. *Nature*. 1999 Aug 5;400(6744):590–593.

Merkel RC, Nassoy P, Leung A, Ritchie K, Evans E. Energy landscapes of receptor-ligand bonds explored with dynamic force spectroscopy. *Nature*. 1999 Jan 7;397(6714):50–53.

Merkle RC. Biotechnology as a route to nanotechnology. *Trends Biotechnol*. 1999 Jul;;17(7): 271–274.

Michaelis J, Muschielok A, Andrecka J, Kügel W, Moffitt JR. DNA based molecular motors. *Phys Life Rev*. 2009;6(4):250–266.

Mirkin CA. Programming the assembly of two- and three-dimensional architectures with DNA and nanoscale inorganic building blocks. *Inorg Chem*. 2000 May 29;39(11):2258–2272.

Mirkin CA, Letsinger RL, Mucic RC, Storhoff JJ. A DNA-based method for rationally assembling nanoparticles into macroscopic materials. *Nature*. 1996 Aug 15;382(6592):607–609.

Mitchell GP, Mirkin CA, Letsinger RL. Programmed assembly of DNA functionalized quantum dots. *J Am Chem Soc*. 1999;121(35):8122–8123.

Mitchell JC, Yurke B, *DNA Based Computers VII*, No. 2340 in LNCS, N. Jonoska, N.C. Seeman, eds. Springer-Verlag, Heidelberg, 2002.

Moghaddam MJ, Taylor S, Gao M, Huang S, Dai L, McCall MJ. Highly efficient binding of DNA on the sidewalls and tips of carbon nanotubes using photochemistry. *Nano Lett*. 2004;4(1):89–93.

Morrison LE, Stols LM. Sensitive fluorescence-based thermodynamic and kinetic measurements of DNA hybridization in solution. *Biochemistry*. 1993 Mar 30;32(12): 3095–3104.

Mucic RC, Storhoff JJ, Mirkin CA, Letsinger RL. DNA-directed synthesis of binary nanoparticle network materials. *J. Am. Chem. Soc*. 1998;120(48):12674–12675.

Muir T, Morales E, Root J, Kumar I, Garcia B, Vellandi C, Jenigian D, Marsh T, Henderson E, Vesenka J. *J. Vac. Sci. Technol. A* 1998 May 1;16(3):1172–1177.

Murphy MC, Rasnik I, Cheng W, Lohman TM, Ha T. Probing single-stranded DNA conformational flexibility using fluorescence spectroscopy. *Biophys J*. 2004 Apr;86(4): 2530–2537.

Naik RR, Brott LL, Clarson SJ, Stone MO. Silica-precipitating peptides isolated from a combinatorial phage display peptide library. *J Nanosci Nanotechnol*. 2002 Feb;2(1):95–100.

Nakashima S, Matsuura N, Nagatsugi F, Maeda M, Sasaki S. Synthesis and evaluation of oligonucleotides incorporating novel artificial nucleobases for the selective formation of non-natural type triplexes. *Nucleic Acids Symp Ser*. 1997;(37):33–34.

Nayak AK, Mishra A, Jena BS, Mishra BK, Subudhi U. Lanthanum induced B-to-Z transition in self-assembled Y-shaped branched DNA structure. *Sci Rep.* 2016 May 31;6: 26855.

Nguyen BH, Takahashi CN, Gupta G, Smith JA, Rouse R, Berndt P, Yekhanin S, Ward DP, Ang SD, Garvan P, Parker HY, Carlson R, Carmean D, Ceze L, Strauss K. Scaling DNA data storage with nanoscale electrode wells. *Sci Adv.* 2021 Nov 26;7(48): eabi6714.

Nguyen CV, Delzeit L, Cassell AM, Li J, Han J, Meyyappan M. Preparation of Nucleic Acid Functionalized Carbon Nanotube Arrays. *Nano Lett.* 2002;2(10):1079–1081.

Nichols MH, Corces VG. A tethered-inchworm model of SMC DNA translocation. *Nat Struct Mol Biol.* 2018 Oct;25(10):906–910.

Nielsen PE. Targeting double stranded DNA with peptide nucleic acid (PNA). *Curr Med Chem.* 2001 Apr;8(5):545–550.

Nielsen PE, Egholm M. An introduction to peptide nucleic acid. *Curr Issues Mol Biol.* 1999;1(1-2):89–104.

Niemeyer CM, Bürger W, Peplies J. Covalent DNA-Streptavidin Conjugates as Building Blocks for Novel Biometallic Nanostructures. *Angew Chem Int Ed Engl.* 1998 Sep 4;37(16):2265–2268.

Niemeyer CM, Sano T, Smith CL, Cantor CR. Oligonucleotide-directed self-assembly of proteins: semisynthetic DNA–streptavidin hybrid molecules as connectors for the generation of macroscopic arrays and the construction of supramolecular bioconjugates. *Nucleic Acids Res.* 1994 Dec 25;22(25):5530–5539.

Nilsen TW, Grayzel J, Prensky W. Dendritic nucleic acid structures. *J Theor Biol.* 1997 Jul 21;187(2):273–284.

Numata M, Sugiyasu K, Hasegawa T, Shinkai S. Sol-gel reaction using DNA as a template: an attempt toward transcription of DNA into inorganic materials. *Angew Chem Int Ed Engl.* 2004 Jun 21;43(25):3279–3283.

Nummelin S, Shen B, Piskunen P, Liu Q, Kostiainen MA, Linko V. Robotic DNA Nanostructures. *ACS Synth Biol.* 2020 Aug 21;9(8):1923–1940.

Oak Ridge National Laboratory (ORNL) Frontier, 2022 https://www.ornl.gov/news/frontier-supercomputer-debuts-worlds-fastest-breaking-exascale-barrier (accessed September 20, 2025).

Obika S, Hari Y, Sekiguchi M, Imanishi T. Stable oligonucleotide-directed triplex formation at target sites with CG interruptions: strong sequence-specific recognition by 2′,4′-bridged nucleic-acid-containing 2-pyridones under physiological conditions. *Chemistry.* 2002 Oct 18;8(20):4796–4802.

Panyutin IG, Hsieh P. The kinetics of spontaneous DNA branch migration. *Proc Natl Acad Sci U S A.* 1994 Mar 15;91(6):2021–2025.

Parker J. Computing with DNA. *EMBO Reports.* 2003;4(1):7–10.

Parker N, Schneegurt M, Tu A, HT, Forster B, Lister P. *Microbiology.* Open Stax, Rice University, Houston, TX, 2016.

Paul A, Avci-Adali M, Ziemer G, Wendel HP. Streptavidin-coated magnetic beads for DNA strand separation implicate a multitude of problems during cell-SELEX. *Oligonucleotides.* 2009;19(3):243–254.

Pei H, Liang L, Yao G, Li J, Huang Q, Fan C. Reconfigurable three-dimensional DNA nanostructures for the construction of intracellular logic sensors. *Angew Chem Int Ed Engl.* 2012;51(36):9020–9024.

Pei R, Cui X, Yang X, Wang E. Assembly of alternating polycation and DNA multilayer films by electrostatic layer-by-layer adsorption. *Biomacromolecules.* 2001;2(2): 463–468.

Persson M, Gullberg M, Tolf C, Lindberg AM, Månsson A, Kocer A. Transportation of nanoscale cargoes by myosin propelled actin filaments. *PLoS One.* 2013;8(2):e55931.

Phan AT, Leroy JL, Guéron M. Determination of the residence time of water molecules hydrating B′-DNA and B-DNA, by one-dimensional zero-enhancement nuclear Overhauser effect spectroscopy. *J Mol Biol.* 1999 Feb 19;286(2):505–519.

Phoenix C. Private communication, 2004.

Pike AR, Lie L, Eagling RA, Ryder LC, Patole SN, Connolly BA, Horrocks BR, Houlton A. DNA on silicon devices: on-chip synthesis, hybridization, and charge transfer. *Angew. Chem. Int. Ed.* 2002;41:615–617.

Pomerantz RT, Anikin M, Zlatanova J, McAllister WT. RNA Polymerase as an Information-Dependent Molecular Motor. *11th Foresight Conference on Molecular Nanotechnology*, Palo Alto, CA, 9–12 Oct 2003.

Pu F, Wu L, Ran X, Ren J, Qu X. G-quartet-based nanostructure for mimicking light-harvesting antenna. *Angew Chem Int Ed Engl.* 2015 Jan 12;54(3):892–896.

Qian L, Winfree E, Bruck J. Neural network computation with DNA strand displacement cascades. *Nature.* 2011 Jul 20;475(7356):368–372.

Rakitin A, Aich P, Papadopoulos C, Kobzar Y, Vedeneev AS, Lee JS, Xu JM. Metallic conduction through engineered DNA: DNA nanoelectronic building blocks. *Phys Rev Lett.* 2001 Apr 16;86(16):3670–3673.

Ramachandran S, Ernst KH, Bachand GD, Vogel V, Hess H. Selective loading of kinesin-powered molecular shuttles with protein cargo and its application to biosensing. *Small.* 2006;2(3):330–334.

Rau Y, Chiang H. Bio-Inspired Self-Assembly Bridging System Based on Synthetic Peptides. *International Symposium on Nanoelectronic Circuits and Giga-Scale Systems (ISNCGS).* 2004.

Raza MT, Tandon A, Park S, Lee S, Nguyen TBN, Vu THN, Jo S, Nam Y, Jeon S, Jeong JH, Park SH. Demonstration of elementary functions via DNA algorithmic self-assembly. *Nanoscale.* 2021 Dec 2;13(46):19376–19384.

Reif JH. "The Design of Autonomous DNA Nanomechanical Devices: Walking and Rolling DNA," in Hagiya M, Ohuchi A. eds., *DNA Computing. DNA 2002. Lecture Notes in Computer Science*, Springer, Berlin, Heidelberg, 2003;2568.

Rhee S, Han, Zj, Liu K, Miles HT, Davies DR. Structure of a triple helical DNA with a triplex-duplex junction. *Biochemistry.* 1999 Dec 21;38(51):16810–16815.

Richter J, Seidel R, Kirsch R, Mertig M, Pompe W, Plaschke J, Schackert HK. Nanoscale Palladium Metallization of DNA. *Adv. Mater.* 2000;12:507–510.

Rief M, Clausen-Schaumann H, Gaub HE. Sequence-dependent mechanics of single DNA molecules. *Nat Struct Biol.* 1999 Apr;6(4):346–349.

Rittman M, Gilroy E, Koohya H, Rodger A, Richards A. Is DNA a worm-like chain in Couette flow? In search of persistence length, a critical review. *Sci Prog.* 2009;92(2):163–204.

Rothemund PW, Papadakis N, Winfree E. Algorithmic self-assembly of DNA Sierpinski triangles. *PLoS Biol.* 2004 Dec;2(12):e424.

Rothemund PW, Winfree E. The program-size complexity of self-assembled squares (extended abstract). In *Proceedings of the thirty-second annual ACM symposium on Theory of Computing.* ACM Press. 2000;459–468.

Rothemund PWK. Folding DNA to create nanoscale shapes and patterns. *Nature.* 2006;440:297–302.

Round AN, Berry M, McMaster TJ, Stoll S, Gowers D, Corfield AP, Miles MJ. Heterogeneity and persistence length in human ocular mucins. *Biophys J.* 2002 Sep;83(3):1661–1670.

Rousina-Webb A, Lachance-Brais C, Rizzuto FJ, Askari MS, Sleiman HF. Transition-metal-functionalized DNA double-crossover tiles: enhanced stability and chirality transfer to metal centers. *Angew Chem Int Ed Engl.* 2020 Mar 2;59(10):4091–4098.

Rubio-Sánchez R, Barker SE, Walczak M, Cicuta P, Michele LD. A modular, dynamic, DNA-based platform for regulating cargo distribution and transport between lipid domains. *Nano Lett.* 2021b Apr 14;21(7):2800–2808.

Rubio-Sánchez R, Fabrini G, Cicuta P, Di Michele L. Amphiphilic DNA nanostructures for bottom-up synthetic biology. *Chem Commun (Camb).* 2021a Nov 30;57(95):12725–12740.

Sameiyan E, Bagheri E, Ramezani M, Alibolandi M, Abnous K, Taghdisi SM. DNA origami-based aptasensors. *Biosens Bioelectron.* 2019 Oct 15;143:111662.

Sancar A, Hearst JE. Molecular matchmakers. *Science.* 1993 Mar 5;259(5100):1415–1420.

Saper G, Hess H. Synthetic Systems Powered by Biological Molecular Motors. *Chem Rev.* 2020 Jan 8;120(1):288–309.

Saran R, Wang Y, Li ITS. Mechanical Flexibility of DNA: A Quintessential Tool for DNA Nanotechnology. *Sensors (Basel).* 2020 Dec 8;20(24):7019.

Scala S, Bowler C. Molecular insights into the novel aspects of diatom biology. *Cell Mol Life Sci.* 2001 Oct;58(11):1666–1673.

Scheffler M, Dorenbeck A, Jordan S, Wüstefeld M, von Kiedrowski G. Self-Assembly of Trisoligonucleotidyls: The Case for Nano-Acetylene and Nano-Cyclobutadiene. *Angew Chem Int Ed Engl.* 1999 Nov 15;38(22):3311–3315.

Schembri MA, Kjaergaard K, Klemm P. Bioaccumulation of heavy metals by fimbrial designer adhesins. *FEMS Microbiol Lett.* 1999 Jan 15;170(2):363–371.

Schill J, Rosier BJ, Gumí Audenis B, Magdalena Estirado E, de Greef TF, Brunsveld L. Assembly of dynamic supramolecular polymers on a DNA origami platform. *Angew. Chem. Int.* 2021;60:7612–7616.

Scholl JA, García-Etxarri A, Koh AL, Dionne JA. Observation of quantum tunneling between two plasmonic nanoparticles. *Nano Lett.* 2013 Feb 13;13(2):564–569.

Schreiber R, Kempter S, Holler S, Schuller V, Schiffels D, Simmel SS, Nickels PC, Liedl T. DNA origami-templated growth of arbitrarily shaped metal nanoparticles. *Small.* 2011;7:1795–1799.

Schulman R, Winfree E. Controlling nucleation rates in algorithmic self-assembly. *SIAM J Comput.* 2010;39(4):1581–1616.

Seeman NC. Private communication. 2004.

Seeman NC. Nanotechnology and the double helix. *Sci Am.* 2004 Jun;290(6):64-69, 72–75.

Seeman NC. Structural DNA nanotechnology: an overview. *Methods Mol Biol.* 2005;303:143–166.

Seeman NC. Nanomaterials based on DNA. *Annu Rev Biochem.* 2010;79:65–87.

Seeman NC, Goddard WA, Vaidehi N, Winfree E. DNA based nanomechanical devices. Presented at: NSF Nanoscale Science and Engineering Grantees Conference; 2002 Dec 11–13; Arlington, VA.

Service RF. Nanotechnology. Biology offers nanotechs a helping hand. *Science.* 2002 Dec 20;298(5602):2322–2323.

Shao Z, Vollrath F. Surprising strength of silkworm silk. *Nature.* 2002 Aug 15;418(6899):741.

Sharpe WN, Turner KT, Edwards RL. Tensile testing of polysilicon. *Experimental Mechanics.* 1999;39:162–170.

Shchepinov MS, Mir KU, Elder JK, Frank-Kamenetskii MD, Southern EM. Oligonucleotide dendrimers: stable nanostructures. *Nucleic Acids Res.* 1999 Aug 1;27(15):3035–3041.

Shen Z, Yan H, Wang T, Seeman NC. Paranemic crossover DNA: a generalized Holliday structure with applications in nanotechnology. *J Am Chem Soc.* 2004 Feb 18;126(6):1666–1674.

Sheppard TL, Ordoukhanian P, Joyce GF. A DNA enzyme with N-glycosylase activity. *Proc Natl Acad Sci U S A.* 2000;97(14):7802–7.

Sherman WB, Seeman NC. A precisely controlled DNA biped walking device. *Nano Letters.* 2004;4(7):1203–1207.

Shi Y, Xu X, Yu H, Lin Z, Zuo H, Wu Y. Defined positive charge patterns created on DNA nanostructures determine cellular uptake efficiency. *Nano Lett.* 2022;22:5330–5338.

Shih WM, Quispe JD, Joyce GF. A 1.7-kilobase single-stranded DNA that folds into a nanoscale octahedron. *Nature.* 2004 Feb 12;427(6975):618–621.

Shimazaki T, Asai Y, Yamashita K. Theoretical rate constants of super-exchange hole transfer and thermally induced hopping in DNA. *J Phys Chem B*. 2005 Jan 27;109(3): 1295–1303.

Shin JS, Pierce NA. A synthetic DNA walker for molecular transport. *J Am Chem Soc*. 2004 Sep 8;126(35):10834–10835.

Shoji K, Kawano R. Recent advances in liposome-based molecular robots. *Micromachines (Basel)*. 2020 Aug 20;11(9):788.

Shu D, Guo P. A viral RNA that binds ATP and contains a motif similar to an ATP-binding aptamer from SELEX. *J Biol Chem*. 2003 Feb 28;278(9):7119–7125.

Sievers D, von, Kiedrowski G. Self-replication of complementary nucleotide-based oligomers. *Nature*. 1994 May 19;369(6477):221–224.

Silverman SK. Deoxyribozymes: DNA catalysts for bioorganic chemistry. *Org Biomol Chem*. 2004 Oct 7;2(19):2701–2706.

Simmel FC, Yurke B. "DNA-Based Nanodevices," in Nalwa HS, eds., *Encyclopedia of Nanoscience and Nanotechnology*, American Scientific Publishers, Stevenson Ranch, CA, 2004;2:495–504.

Sivasankari B, Varalakshmi P. Blockchain and IoT Technology in Healthcare: A Review. *Stud Health Technol Inform*. 2022 May 25;294:277–278.

Smith C. Diagnostic characters for common phytoplankton – Diatoms. Botany Dept, Univ Hawaii at Manoa. 2025. http://www.botany.hawaii.edu/BOT201/Algae/Phytoplankton%20lecture%20notes.htm (accessed September 20, 2025).

Smith DE, Tans SJ, Smith SB, Grimes S, Anderson DL, Bustamante C. The bacteriophage straight phi29 portal motor can package DNA against a large internal force. *Nature*. 2001 Oct 18;413(6857):748–752.

Smith SB, Cui Y, Bustamante C. Overstretching B-DNA: the elastic response of individual double-stranded and single-stranded DNA molecules. *Science*. 1996 Feb 9;271(5250):795–799.

Smith SS, Niu L, Baker DJ, Wendel JA, Kane SE, Joy DS. Nucleoprotein-based nanoscale assembly. *Proc Natl Acad Sci U S A*. 1997 Mar 18;94(6):2162–2167.

Soloveichik D, Winfree E. "Complexity of self-assembled scale-invariant shapes," in Ferretti C, Mauri G, Zandron C, eds., *DNA Computing*. 10th International Workshop on DNA; 2004 Jun 7–10, Springer, Milano, Italy. Berlin, 2005;344–354. (Lecture Notes in Computer Science; vol. 3384).

Sreedhara A, Li Y, Breaker RR. Ligating DNA with DNA. *J Am Chem Soc*. 2004 Mar 24;126(11):3454–3460.

Stachowiak JC, Yue M, Castelino K, Chakraborty A, Majumdar A. Chemomechanics of surface stresses induced by DNA hybridization. *Langmuir*. 2006 Jan 3;22(1):263–268.

Starikov EB, Quintilla A, Lee KH, Wenzel W. Conformational dependence of DNA ballistic conductivity. *J Chem Phys*. 2008 Oct 7;129(13):131101.

Stofer E, Chipot C, Lavery R. Free energy calculations of Watson–Crick base pairing in aqueous solution. *Journal of the American Chemical Society*. 1999;121(41):9503–9508.

Stoltenberg RM, Woolley AT. DNA-templated nanowire fabrication. *Biomed Microdevices*. 2004 Jun;6(2):105–111.

Stühmeier F, Welch JB, Murchie AI, Lilley DM, Clegg RM. Global structure of three-way DNA junctions with and without additional unpaired bases: a fluorescence resonance energy transfer analysis. *Biochemistry*. 1997 Nov 4;36(44):13530–13538.

Sun F, Dong Y, Ni M, Ping Z, Sun Y, Ouyang Q, Qian L. Mobile and self-sustained data storage in an extremophile genomic DNA. *Adv Sci (Weinh)*. 2023 Apr;10(10):e2206201.

Sun J, Li Y, Liu G, Chu F, Chen C, Zhang Y, Tian H, Song Y. Patterning a superhydrophobic area on a facile fabricated superhydrophilic layer based on an inkjet-printed water-soluble polymer template. *Langmuir*. 2020a Aug 25;36(33):9952–9959.

Sun W, Shen J, Zhao Z, Arellano N, Rettner C, Tang J et al. Precise pitch-scaling of carbon nanotube arrays within three-dimensional DNA nanotrenches. *Science*. 2020b May 22;368(6493):874–877.
Surana S, Bhat JM, Koushika SP, Krishnan Y. An autonomous DNA nanomachine maps spatiotemporal pH changes in a multicellular living organism. *Nat Commun*. 2011 Jun 7;2:340.
Svoboda K, Block SM. Force and velocity measured for single kinesin molecules. *Cell*. 1994 Jun 3;77(5):773–784.
Switzer CY, Moroney SE, Benner SA. Enzymatic incorporation of a new base pair into DNA and RNA. *J Am Chem Soc*. 1989;111:8322–8323.
Sygula A, Xu G, Marcinow Z, Rabideau PW. "Buckybowls" – introducing curvature by solution phase synthesis. *Tetrahedron*. 2001;57(17):3637–3644.
Tackett AJ, Morris PD, Dennis R, Goodwin TE, Raney KD. Unwinding of unnatural substrates by a DNA helicase. *Biochemistry*. 2001 Jan 16;40(2):543–548.
Takatsuki H, Tanaka H, Rice KM, Kolli MB, Nalabotu SK, Kohama K, Famouri P, Blough ER. Transport of single cells using an actin bundle-myosin bionanomotor transport system. *Nanotechnology*. 2011 Jun 17;22(24):245101.
Tanaka S, Takiguchi K, Hayashi M. Repetitive stretching of giant liposomes utilizing the nematic alignment of confined actin. *Communications Physics*. 2018;1(1):10.
Tang W, Wang H, Wang D, Zhao Y, Li N, Liu F. DNA tetraplexes-based toehold activation for controllable DNA strand displacement reactions. *J Am Chem Soc*. 2013 Sep 18;135(37):13628–13631.
Tang Y, Liu H, Wang Q, Qi X, Yu L, Šulc P, Zhang F, Yan H, Jiang S. DNA Origami Tessellations. *J Am Chem Soc*. 2023 Jun 28;145(25):13858–13868.
Taton TA, Mucic RC, Mirkin CA, Letsinger RL. The DNA-mediated formation of supramolecular mono and multilayered nanoparticle structures. *J. Am. Chem. Soc*. 2000 Jul 5;122(26):6305–6306.
Tian Y, He Y, Chen Y, Yin P, Mao C. A DNAzyme that walks processively and autonomously along a one-dimensional track. *Angew Chem Int Ed Engl*. 2005 Jul 11;44(28):4355–4358.
Tian Y, Mao C. Molecular gears: a pair of DNA circles continuously rolls against each other. *J Am Chem Soc*. 2004 Sep 22;126(37):11410–11411.
Tinland B, Pluen A, Sturm J, Weill G. Persistence length of single-stranded DNA. *Macromolecules*. 1997;30(19):5763–5765.
Tosello G. Micro/Nano Manufacturing. *Micromachines (Basel)*. 2017 Oct 2;8(10):297.
Travers A. *DNA-Protein Interactions*. Springer, Netherlands, 1993.
Travers A, Muskhelishvili G. DNA structure and function. *FEBS J*. 2015 Jun;282(12):2279–2295.
Turberfield AJ, Mitchell JC, Yurke B, Mills AP Jr, Blakey MI, Simmel FC. DNA fuel for free running nanomachines. *Phys Rev Lett*. 2003 Mar 21;90(11):118102.
Turberfield AJ, Yurke B, Mills AP Jr. DNA nanomechanical devices. In: Winfree E, Gifford DK, editors. DNA Based Computers V. Providence (RI): American Mathematical Society; 2000;171–182. (DIMACS Series in Discrete Mathematics and Theoretical Computer Science; vol. 54).
Uddin AH, Roman MA, Anderson JR, Damha MJ. A novel N3-functionalized thymidine linker for the stabilization of triple helical DNA. *Chem Commun (Camb)*. 1996;(2):171–172.
Uemura S, Higuchi H, Olivares AO, De La Cruz EM, Ishiwata S. Mechanochemical coupling of two substeps in a single myosin V motor. *Nat Struct Mol Biol*. 2004 Sep;11(9):877–883.
Valero J, Pal N, Dhakal S, Walter NG, Famulok M. A bio-hybrid DNA rotor-stator nanoengine that moves along predefined tracks. *Nat Nanotechnol*. 2018 Jun;13(6):496–503.
Valero J, Škugor M. Mechanisms, Methods of Tracking and Applications of DNA Walkers: A Review. *Chemphyschem*. 2020 Sep 2;21(17):1971–1988.

Ventra MD, Zwolak M. "DNA Electronics," *in Nalwa HS*, ed., *Encyclopedia of Nanoscience and Nanotechnology*, American Scientific Publishers, Stevenson Ranch, CA,2004;2:475–493.
Vesenka J, Marsh T, Miller R, Henderson E. Atomic force microscopy reconstruction of G-wire DNA. *J Vac Sci Technol B*. 1996 Mar 1;14(2):1413–1417.
Vittala SK, Saraswathi SK, Joseph J. Fullerene Cluster Assisted Self-Assembly of Short DNA Strands into Semiconducting Nanowires. *Chemistry*. 2017 Nov 7;23(62):15759–15765.
Vittala SK, Zhao Y, Han D. Programmed assembly of DNA templates by silver nanowires. *Chempluschem*. 2022 Jan;87(1):e202100478.
von, Delius M, Leigh DA. Walking molecules. *Chem Soc Rev*. 2011 Jul;40(7):3656–3676.
von Kiedrowski G. A self-replicating hexadeoxynucleotide. *Angew Chem Int Ed Engl*. 1986;25:932–935.
von Kiedrowski G. Molecular Principles of Artificial Self-Replication Biophys. *Acta – Biomembranes*. 2000;1464:151–164.
von Kiedrowski G, Eckardt L, Naumann K, Pankau W, Reimold M, Rein M. Toward replicable, multifunctional, nanoscaffolded machines. A chemical manifesto. *Pure Appl Chem*. 2003;75(5):609–619.
von Kiedrowski G, Wlotzka B, Helbing J. Sequence dependence of template-directed syntheses of hexadeoxynucleotide derivatives with 3′–5′ pyrophosphate linkage. *Angew. Chem. Int. Ed. Engl*. 1989;28:1235–1237.
von Kiedrowski G, Wlotzka B, Helbing J, Matzen M, Jordan S. Parabolic growth of a self-replicating hexadeoxynucleotide bearing a 3′-5′- phosphoamidate linkage. *Angew Chem Int Ed Engl*. 1991;30:423–426.
Wang C, Huang Z, Lin Y, Ren J, Qu X. Artificial DNA nano-spring powered by protons. *Adv Mater*. 2010;22(25):2792–2798.
Wang C, Tao Y, Song G, Ren J, Qu X. Speeding up a bidirectional DNA walking device. *Langmuir*. 2012 Oct 16;28(41):14829–14837.
Wang F, Willner B, Willner I. DNA-based machines. *Top Curr Chem*. 2014;354:279–338.
Wang G, Xu JJ, Chen HY. Interfacing cytochrome *c* to electrodes with a DNA – carbon nanotube composite film. *Electrochem. Commun*. 2002;4(6):506–509.
Wang H. Proving theorems by pattern recognition-II. *The Bell System Technical Journal*. 1961;40(1):1–41.
Wang H. *An Unsolvable Problem on Dominoes*. Harvard Computation Laboratory, Cambridge, Massachusetts. *Technical report BL-30*, 1962;1–5.
Wang MD, Schnitzer MJ, Yin H, Landick R, Gelles J, Block SM. Force and velocity measured for single molecules of RNA polymerase. *Science*. 1998 Oct 30;282(5390):902–907.
Wang W, Chen C, Vecchioni S, Zhang T, Wu C, Ohayon YP, Sha R, Seeman NC, Wei B. Reconfigurable two-dimensional DNA v. *Angew Chem Int Ed Engl*. 2021 Dec 1;60(49):25781–25786.
Wang Y, Wang H, Li Y, Yang C, Tang Y, Lu X et al. Chemically conjugated branched staples for super-DNA origami. *J Am Chem Soc*. 2024 Feb 14;146(6):4178–4186.
Wang Y, Zhang Y, Ong NP. Speeding up a single-molecule DNA device with a simple catalyst. *Phys Rev E Stat Nonlin Soft Matter Phys*. 2005;72(5):051918.
Wang ZG, Elbaz J, Willner I. DNA machines: bipedal walker and stepper. *Nano Lett*. 2011 Jan 12;11(1):304–309.
Watson JD, Crick FH. Molecular structure of nucleic acids: a structure for deoxyribose nucleic acid. *Nature*. 1953 Apr 25;171(4356):737–738.
Watson SM, Pike AR, Pate J, Houlton A, Horrocks BR. DNA-templated nanowires: morphology and electrical conductivity. *Nanoscale*. 2014 Apr 21;6(8):4027–4037.
Wen Q, Wei X, Jiang F, Lu J, Xu X. Focused ion beam milling of single-crystal sapphire with A-, C-, and M-orientations. *Materials (Basel)*. 2020 Jun 26;13(12):2871.

Weng Z, Yu H, Luo W, Guo Y, Liu Q, Zhang L, et al. Cooperative branch migration: a mechanism for flexible control of DNA strand displacement. *ACS Nano.* 2022 Feb 22;16(2):3135–3144.

Wettig SD, Li CZ, Long YT, Kraatz HB, Lee JS. M-DNA: a self-assembling molecular wire for nanoelectronics and biosensing. *Anal Sci.* 2003 Jan;19(1):23–26.

Whaley SR, English DS, Hu EL, Barbara PF, Belcher AM. Selection of peptides with semiconductor binding specificity for directed nanocrystal assembly. *Nature.* 2000 Jun 8;405(6787):665–668.

White O, Eisen JA, Heidelberg JF, Hickey EK, Peterson JD, Dodson RJ et al. Genome sequence of the radioresistant bacterium Deinococcus radiodurans R1. *Science.* 1999 Nov 19;286(5444):1571–1577.

Williams KA, Veenhuizen PT, de la Torre BG, Eritja R, Dekker C. Nanotechnology: carbon nanotubes with DNA recognition. *Nature.* 2002 Dec 19–26;420(6917):761.

Willner I, Patolsky F, Wasserman J. Photoelectrochemistry with controlled DNA-cross-linked CdS nanoparticle arrays. *Angew Chem Int Ed Engl.* 2001 May 18;40(10):1861–1864.

Winfree E. Simulations of Computing by Self-Assembly. Technical Report CaltechCSTR:1998.22. California Institute of Technology, Pasadena, CA, 1998.

Winfree E, Bekbolatov R. "Proofreading Tile Sets: Error Correction for Algorithmic Self-Assembly," in Chen, J., Reif, J. eds., *DNA Computing. DNA 2003. Lecture Notes in Computer Science*, Springer, Berlin, Heidelberg. 2004;2943.

Winfree E, Liu F, Wenzler LA, Seeman NC. Design and self-assembly of two-dimensional DNA crystals. *Nature.* 1998 Aug 6;394(6693):539–544.

Wu G, Ji H, Hansen K, Thundat T, Datar R, Cote R, Hagan MF, Chakraborty AK, Majumdar A. Origin of nanomechanical cantilever motion generated from biomolecular interactions. *Proc Natl Acad Sci U S A.* 2001 Feb 13;98(4):1560–1564.

Wu G, Li Y, Zhang J, Yun W, Xiong Z, Yang L. Simultaneous and ultra-sensitive detection of Cu^{2+} and Mg^{2+} in wine and beer based on dual DNA tweezers and entropy-driven three-dimensional DNA nanomachine. *Food Chem.* 2021 Oct 1;358:129835.

Wu Y, Fa M, Tae EL, Schultz PG, Romesberg FE. Enzymatic phosphorylation of unnatural nucleosides. *J Am Chem Soc.* 2002 Dec 11;124(49):14626–14630.

Xiong K, Erwin GS, Ansari AZ, Blainey PC. Sliding on DNA: from peptides to small molecules. *Angew Chem Int Ed Engl.* 2016 Nov 21;55(48):15110–15114.

Yakobson BI, Smalley RE. Fullerene nanotubes: C1,000,000 and beyond. *Am Sci.* 1997;85:324–337.

Yale University Department of Chemistry. Average Bond Energies Heats of Atomization. https://ursula.chem.yale.edu/~chem220/chem220js/STUDYAIDS/thermo/atomization.html#Single (accessed September 20, 2025).

Yamamoto T, Aoki K, Sugiyama A, Doi H, Kodama T, Shimizu Y, Kanai M. Design and synthesis of biotin analogues reversibly binding with streptavidin. *Chem Asian J.* 2015;10(4):1071–1078.

Yan H, LaBean TH, Feng L, Reif JH. Directed nucleation assembly of DNA tile complexes for barcode-patterned lattices. *Proc Natl Acad Sci U S A.* 2003 Jul 8;100(14): 8103–8108.

Yan H, Park SH, Finkelstein G, Reif JH, LaBean TH. DNA-templated self-assembly of protein arrays and highly conductive nanowires. *Science.* 2003 Sep 26;301(5641): 1882–1884.

Yan H, Zhang X, Shen Z, Seeman NC. A robust DNA mechanical device controlled by hybridization topology. *Nature.* 2002 Jan 3;415(6867):62–65.

Yang B, Li Q, Liu HJ, Fan CH. Recent progress of frame nucleic acids studies towards atomic fabrications. *Acta Phys. Sin.* 2021;70:026201.

Yang C, Moses D, Heeger AJ. Base-pair stacking in oriented films of DNA–surfactant complex. *Adv Mater.* 2003;15:1364–1367.

Yang W, Auciello O, Butler JE, Cai W, Carlisle JA, Gerbi JE, Gruen DM, Knickerbocker T, Lasseter TL, Russell JN Jr, Smith LM, Hamers RJ. DNA-modified nanocrystalline diamond thin-films as stable, biologically active substrates. *Nat Mater.* 2002 Dec;1(4):253–257.

Yang W, Butler JE, Russell JN Jr, Hamers RJ. Interfacial electrical properties of DNA-modified diamond thin films: intrinsic response and hybridization-induced field effects. *Langmuir.* 2004 Aug 3;20(16):6778–6787.

Yang X, Ebrahimi A, Li J, Cui Q. Fullerene-biomolecule conjugates and their biomedicinal applications. *Int J Nanomedicine.* 2014;9:77–92.

Yang Y, Wang J, Shigematsu H, Xu W, Shih WM, Rothman JE, Lin CX. Self-assembly of size-controlled liposomes on DNA nanotemplates. *Nat Chem.* 2016;8:476–483.

Yang Y, Wu Z, Wang L, Zhou K, Xia K, Xiong Q, Lin CX. Sorting sub-150-nm liposomes of distinct sizes by DNA-brick-assisted centrifugation. *Nat. Chem.* 2021;13: 335–342.

Yatsunyk LA, Piétrement O, Albrecht D, Tran PL, Renčiuk D, Sugiyama H, Arbona JM, Aimé JP, Mergny JL. Guided assembly of tetramolecular G-quadruplexes. *ACS Nano.* 2013 Jul 23;7(7):5701–5710.

Yin P, Yan H, Daniell XG, Turberfield AJ, Reif JH. A unidirectional DNA walker that moves autonomously along a track. *Angew Chem Int Ed Engl.* 2004 Sep 20;43(37): 4906–4911.

Yin Q, Zheng Y, Wang B, Zhang Q. Design of Constraint Coding Sets for Archive DNA Storage. *IEEE/ACM Trans Comput Biol Bioinform.* 2022;19(6):3384–3394.

You I, Pau G, Salerno VM, Sharma V. Special Issue: Internet of things for smart homes. *Sensors (Basel).* 2019 Sep 26;19(19):4173.

You M, Chen Y, Zhang X, Liu H, Wang R, Wang K, Williams KR, Tan W. An autonomous and controllable light-driven DNA walking device. *Angew Chem Int Ed Engl.* 2012 Mar 5;51(10):2457–2460.

You M, Peng L, Shao N, Zhang L, Qiu L, Cui C, Tan W. DNA "nano-claw": logic-based autonomous cancer targeting and therapy. *J Am Chem Soc.* 2014 Jan 29;136(4): 1256–1259.

Yurke B, Turberfield AJ, Mills AP Jr, Simmel FC, Neumann JL. A DNA-fuelled molecular machine made of DNA. *Nature.* 2000 Aug 10;406(6796):605–608.

Zadpoor AA. Meta-biomaterials. *Biomater Sci.* 2019 Dec 17;8(1):18–38.

Zaichuk T, Marko JF. Single-molecule micromanipulation studies of methylated DNA. *Biophys J.* 2021 Jun 1;120(11):2148–2155.

Zhang F, Jiang S, Li W, Hunt A, Liu Y, Yan H. Self-assembly of complex DNA tessellations by using low-symmetry multi-arm DNA tiles. *Angew Chem Int Ed Engl.* 2016 Jul 25;55(31):8860–8863.

Zhang H, Fu H, Shao X, Dehez F, Chipot C, Cai W. Changes in microenvironment modulate the B- to A-DNA transition. *J Chem Inf Model.* 2019 May 28;59(5):2324–2330.

Zhang J. Molecular biology methods to construct recombinant fibrous protein. *Methods Mol Biol.* 2021;2347:123–135.

Zhang J, Lang HP, Huber F, Bietsch A, Grange W, Certa U, McKendry R, Güntherodt HJ, Hegner M, Gerber C. Rapid and label-free nanomechanical detection of biomarker transcripts in human RNA. *Nat Nanotechnol.* 2006 Dec;1(3):214–220.

Zhang X, Wang R, Xue G. Programming macro-materials from DNA-directed self-assembly. *Soft Matter.* 2015 Mar 14;11(10):1862–1870.

Zhang Z, Yang Y, Pincet F, Llaguno MC, Lin CX. Placing and shaping liposomes with reconfigurable DNA nanocages. *Nat. Chem.* 2017;9:653–659.

Zhao M, Chen Y, Wang K, Zhang Z, Streit JK, Fagan JA et al. DNA-directed nanofabrication of high-performance carbon nanotube field-effect transistors. *Science.* 2020 May 22;368(6493):878–881.

Zheng M, Jagota A, Strano MS, Santos AP, Barone P, Chou SG et al. Structure-based carbon nanotube sorting by sequence-dependent DNA assembly. *Science*. 2003 Nov 28;302(5650):1545–1548.

Zhong R, Tang Q, Wang S, Zhang H, Zhang F, Xiao M, Man T, Qu X, Li L, Zhang W, Pei H. Self-assembly of enzyme-like nanofibrous G-molecular hydrogel for printed flexible electrochemical sensors. *Adv Mater.* 2018 Mar;30(12):e1706887.

Zhou L, Li W, Chen Z, Ju E, Ren J, Qu X. Growth of hydrophilic CuS nanowires via DNA-mediated self-assembly process and their use in fabricating smart hybrid films for adjustable chemical release. *Chemistry*. 2015 Feb 9;21(7):2930–2935.

Zhu G, Hannel M, Sha R, Zhou F, Ben, Zion MY, Zhang Y, Bishop K, Grier D, Seeman N, Chaikin P. Microchemomechanical devices using DNA hybridization. *Proc Natl Acad Sci U S A*.. 2021 May 25;118(21):e2023508118.

Zhu L, Lukeman PS, Canary JW, Seeman NC. Nylon/DNA: single-stranded DNA with a covalently stitched nylon lining. *J Am Chem Soc*. 2003 Aug 27;125(34):10178–10179.

5 The Next Big Leap in Human Health

Bridging the Gap between Life and Healthspan

Bryan Sykes

CONTENTS

5.1 Introduction: From Sickcare to Healthcare – A Paradigm Shift Whose Time Has Come 149
5.2 Part I: Salutogenesis – The Key to Prolonged Wellness 151
5.2.1 Defining Salutogenesis and Its Relevance in Modern Medicine 151
5.2.2 Aligning Salutogenesis with the 4P Medicine Model 152
5.2.3 Predictive Medicine – Comprehensibility 152
5.2.4 Preventive Medicine – Manageability 153
5.2.5 Personalized Medicine – Meaningfulness 153
5.2.6 Participatory Medicine – Coherence as a Whole 154
5.2.7 Salutogenesis as a Guiding Philosophy for Healthspan Optimization 154
5.3 Part II: Engineering Wellness: The Molecular Infrastructure of Future Health 155
5.3.1 Foundations of Molecular Manufacturing in Medicine 155
5.3.2 Precision Diagnostics: Early Detection at the Molecular Frontier 156
5.3.3 Nanoparticles and Regenerative Medicine 157
5.3.4 Engineering Salutogenesis: Operationalizing the 4Ps 157
5.3.5 Ethical and Equitable Implementation 158
5.3.6 Conclusion: Designing Health as a Right, Not a Privilege 158
References 159

5.1 INTRODUCTION: FROM SICKCARE TO HEALTHCARE – A PARADIGM SHIFT WHOSE TIME HAS COME

For decades, the dominant model of healthcare has centered around diagnosing and treating illness after it manifests, a model often labelled as "sickcare" rather than true healthcare. This approach, while vital in managing acute conditions and advancing clinical interventions, has increasingly proven inadequate in addressing the rising tide of chronic diseases, health disparities, and preventable conditions that strain healthcare resources and limit the well-being of populations (Holman, 2020). According

DOI: 10.1201/9781003724896-5

to the Centers for Disease Control and Prevention. (2024). Fast Facts: Health and Economic Costs of Chronic Conditions. https://www.cdc.gov/chronic-disease/data-research/facts-stats/index.html (accessed October 10, 2025). Approximately 90% of the nation's $4.9 trillion in annual healthcare expenditures are spent on chronic and mental health conditions. In response, thought leaders, public health experts, and policymakers have called for a transformative shift toward a more proactive, preventive, and holistic vision of healthcare – a shift from "sickcare" to proactive "healthcare."

This paradigm shift is not merely theoretical; it is undergirded by an expanding body of research that underscores the economic, social, and clinical value of prevention-centered health systems. Preventive care – ranging from vaccinations and screenings to lifestyle interventions – has demonstrated substantial cost-effectiveness and long-term health benefits across a range of settings (AbdulRaheem, 2023). For example, large-scale interventions targeting lifestyle-related risk factors such as tobacco use, poor diet, and physical inactivity have been linked to reductions in cardiovascular disease, diabetes, and the incidence of cancer (Fufaa, 2024). These gains are not just confined to individual outcomes; they generate ripple effects in reducing long-term healthcare expenditures and improving workforce productivity (Stenberg et al., 2019).

A striking example of this future-focused model is captured in the case of "Ana," a 38-year-old woman with a family history of breast cancer who used a consumer-grade genetic testing service linked to AI risk stratification. A polygenic risk score identified a high predisposition to BRCA2-related malignancy. Acting on this early insight, Ana partnered with her physician to begin routine imaging and implemented lifestyle modifications. Five years later, imaging detected a stage 0 carcinoma in situ, allowing for complete resolution with non-invasive intervention. This case illustrates how 4P medicine – predictive, preventive, personalized, and participatory – can pre-emptively shift the healthcare trajectory to before pathology begins to dominate (Alonso et al., 2019; Pack, 2016; Tian et al., 2012).

One of the central challenges in transitioning from reactive to proactive care lies in system design and incentive structures. Healthcare systems that primarily reimburse procedures and clinical treatments offer little motivation to invest in upstream health promotion activities. Studies highlight that payment reforms – such as value-based care models – can realign incentives toward prevention and patient-centered care (National Academy of Medicine, 2017). This shift also requires the rethinking of data infrastructures, workforce development, and community partnerships to sustain health at the population level (Welton et al., 1996). Moreover, transitioning to a healthcare model grounded in prevention necessitates a greater integration of behavioral health, environmental health, and social determinants of health (SDOH). Factors such as housing stability, food security, education, and transportation are now recognized as critical drivers of health outcomes, which often outweigh clinical care in their overall impact (Artiga & Hinton, 2018). For example, preventive interventions targeting health inequities through community-based engagement and culturally tailored messaging have improved participation in health screenings and chronic disease management in marginalized populations (Bazzano et al., 2023).

In addition to prevention, the future of healthcare must be supported by emerging technologies. Systems biology provides a framework for understanding the human body as a network of interconnected systems that respond dynamically to internal and external stimuli. Paired with hypothetical (for now) molecular manufacturing (MM) (envisaged to have the capacity to fabricate myriad materials, devices, and systems at molecular and even atomic resolution), the ability to manipulate biological matter at the nanoscale offers transformative potential in targeted diagnostics and regenerative therapies (Boisseau & Loubaton, 2011; Freitas, 2002). Artificial Narrow Intelligence (ANI) and, perhaps within the next decade, Artificial General Intelligence (AGI) and Artificial Superintelligence (ASI) will further amplify these advances through the integration of real-time data streams from wearables, environmental sensors, and digital health records to forecast risk and personalize treatments with precision (Hood & Friend, 2011).

Another critical dimension to this transformation is the growing divergence between lifespan and healthspan. While global average life expectancy has steadily increased, the number of years lived in full health has not kept pace, leaving many to endure prolonged periods of disability, disease, and diminished quality of life. This gap disproportionately affects low-income populations and communities of color (Kavanagh, 2016). Wellness, therefore, must be redefined not as a personal luxury but as a public mandate – a right to healthspan equity. Countries that have enshrined health as a constitutional right, such as Brazil and South Africa, demonstrate that structural commitment to wellness improves health outcomes, particularly when public policy integrates proactive care strategies (Kinney & Clark, 2004; Nampewo et al., 2022).

Technology also plays a critical role in accelerating this shift. Digital health platforms, wearable sensors, and AI are enabling earlier detection, personalized interventions, and continuous monitoring beyond traditional clinical settings (Willis & Edwards, 2022). Telemedicine, for instance, has expanded access to preventive counselling and mental health services, especially in underserved regions. Importantly, these technologies must be implemented with equity in mind to avoid widening the digital divide and reinforcing structural inequities (Richardson et al., 2022). Despite mounting evidence and technological enablers, cultural and institutional inertia remains a barrier. Many healthcare professionals are trained primarily in acute care rather than behavioral health coaching, public health strategies, or population health management (Schmutz et al., 2021). Achieving a true paradigm shift requires reimagining health professional education and aligning it with emerging needs and competencies. Ultimately, this chapter aims to explore the pragmatic, policy, and philosophical implications of transitioning from a sickcare to healthcare paradigm. Through a critical review of research evidence, innovative models, and global case studies, the author argues that such a shift is not only feasible – it is essential for the sustainability, equity, and effectiveness of modern health systems.

5.2 PART I: SALUTOGENESIS – THE KEY TO PROLONGED WELLNESS

5.2.1 Defining Salutogenesis and Its Relevance in Modern Medicine

Coined by Aaron Antonovsky in the 1970s, the term salutogenesis refers to a healthcare strategy that sets its focus on the origins of health rather than those of disease (Vinje et al., 1979). At the heart of salutogenesis is the concept of the "Sense of

Coherence" (SOC) – a framework that comprises comprehensibility, manageability, and meaningfulness, which enables individuals to navigate life's stressors while sustaining health (Lindström & Eriksson, 2006). A strong SOC equips individuals to adapt to stress and maintain functional well-being, even in the presence of biological or environmental risk factors. Unlike the pathogenic model, which centers on diagnosing and treating illness, the salutogenic model emphasizes adaptation, resilience, and the internal and external resources that assist individuals with maintaining or returning to a state of wellness (Herdiana et al., 2018; Mittelmark et al., 2017). In the context of rising chronic disease rates, aging populations, and the expanding gap between lifespan and healthspan, a salutogenic lens is not just beneficial – it is imperative (Bauer et al., 2020; Mittelmark et al., 2017).

This chapter situates salutogenesis as the philosophical and operational foundation of next-generation medicine. It proposes that the integration of the 4P model (predictive, preventive, personalized, and participatory medicine) provides a roadmap for the application of salutogenic thinking at scale, particularly when combined with innovations in systems biology, digital health, and MM (Alonso et al., 2019; Hood & Friend, 2011). The following sections explore how each "P" of the 4P framework extends and operationalizes salutogenesis.

5.2.2 Aligning Salutogenesis with the 4P Medicine Model

The emerging paradigm of 4P medicine – predictive, preventive, personalized, and participatory – offers a practical and technological framework to operationalize salutogenesis in modern healthcare. Each of the four Ps corresponds closely to one or more SOC elements, which together represent a systemic strategy for extending human wellness and optimizing individual health trajectories (Hood & Friend, 2011; Tian et al., 2012). The 4Ps support the salutogenic mandate by shifting attention upstream – toward early identification, personalized interventions, and active patient involvement.

5.2.3 Predictive Medicine – Comprehensibility

Predictive medicine identifies early risk trajectories long before symptoms emerge. This "forecasting" model mirrors the comprehensibility component of Antonovsky's SOC, which allows individuals and systems to make sense of complex health data that affects one's body to act accordingly (Hood & Price, 2014; Vinje et al., 1979). This clarity builds agency, as when an individual understands their personal health data, they can make more informed, pre-emptive choices; transforming uncertainty into empowerment (Alonso et al., 2019; Hood & Price, 2014). This is achieved through technologies such as polygenic risk scoring, AI-based health tracking, and early biomarker identification (Alum, 2025; Liu et al., 2025). Advancements in polygenic risk scores (PRS), digital phenotyping, and AI-driven modeling enable risk profiling for conditions like cancer, cardiovascular disease, and type 2 diabetes (Ho et al., 2023; Patel et al., 2023). For instance, the longitudinal tracking of gut microbiota and epigenetic markers can predict the onset of inflammation-correlated diseases (Hou et al., 2022) (Table 5.1).

TABLE 5.1
Tools of Predictive Medicine in a Salutogenic Framework

Tool	Predictive Function	Salutogenic Impact
Polygenic risk scores	Genetic predisposition	Risk awareness and early action
AI-based health monitors	Pattern recognition	Empowerment and personalization
Microbiome sequencing	Disease susceptibility	Lifestyle modulation

Sources: Alonso et al. (2019), Motahari-Nezhad et al. (2022), and Olawade et al. (2025).

5.2.4 Preventive Medicine – Manageability

Prevention is the most direct expression of salutogenesis in action, as it reflects the manageability aspect of SOC, which ensures that people and systems can access the resources required to respond to health risks (Lindström & Eriksson, 2006). Preventive interventions such as immunizations, early screenings, nutritional guidance, and stress reduction programs give individuals the tools they need to stay well. This proactive equipping corresponds with manageability, which Antonovsky described as possessing the internal and external resources to cope with life's challenges. Public health strategies like anti-smoking laws, cardiovascular screenings, and diabetes prevention programs showcase how policy and systems can provide collective manageability at scale (Evans et al., 2013; Purnell et al., 2016). When individuals feel that they have the means to manage risk, both adherence and resilience increase. Evidence supports the profound impact of anti-inflammatory diets (e.g., Mediterranean, plant-forward), regular physical activity, and toxin avoidance in reducing the risk of chronic disease (An et al., 2024; Neale et al., 2023; Sadiq, 2023). For example, public bans on BPA and trans fats have reduced hormone-disruptive conditions and cardiovascular events, respectively (Molina-López et al., 2023; Spurgeon, 2006).

Case Insight: Finland's "North Karelia Project" successfully reduced population-wide cardiovascular mortality by focusing on dietary reform and smoking cessation – a benchmark in population-level salutogenesis (Puska & Jaini, 2020).

5.2.5 Personalized Medicine – Meaningfulness

Personalized medicine respects that no two people have the same biology, context, or goals. By tailoring interventions to an individual's genomics, microbiome, environment, and values, healthcare becomes more relevant and meaningful (Yurkovich et al., 2024). Personalized medicine maps closely to meaningfulness, as it recognizes that people are more likely to stay healthy when they feel that their life and health choices matter. Personalized medicine also enhances cultural congruence, patient engagement, and shared decision-making, each of which is critical in chronic disease management (Magi et al., 2024). Rather than standardized regimens, treatments are tailored based on genomics, microbiomics, and metabolic phenotyping. Pharmacogenomic tools reduce adverse drug reactions and improve response rates (Zhou et al., 2024), while continuous glucose monitoring and microbiome-based diets allow for optimized nutritional interventions (Martinez-Garay & Djouder, 2023; Rein et al., 2022). Wearable technologies further enhance feedback loops for dynamic lifestyle adjustments (Wang et al., 2025).

5.2.6 Participatory Medicine – Coherence as a Whole

Participation reflects the full actualization of the SOC; creating systems where people find meaning, feel capable, and have the tools to act. Participatory medicine empowers individuals to co-create their care via access to data, digital tools, and decision-making roles. Health becomes a collaborative practice rather than a passive experience. When patients become partners, they develop a stronger SOC and overall blending with all three pillars of comprehensibility, manageability, and meaningfulness (Grosu, 2022). Research reveals that health literacy, when paired with access to actionable data, boosts adherence and engagement (Pitts & Freeman, 2021). Community-based health initiatives, gamified engagement platforms, wearable tech, and AI agent-enabled coaching systems are bringing this participatory ideal closer to reality. Digital health platforms, open-access patient records, and gamified wellness programs all exemplify this approach. Participatory programs in low-income communities increase screening uptake, reduce stigmatization, and build communal health resilience (Sofolahan-Oladeinde et al., 2015).

Case Vignette: "Ana," a 39-year-old woman with a family history of breast cancer, accessed her PRS, began lifestyle modifications, and avoided high-risk medications. Her annual screenings have since remained stable, demonstrating early risk reversal without intervention (Grosu, 2022; Johnson et al., 2016; Yuen et al., 2024).

5.2.7 Salutogenesis as a Guiding Philosophy for Healthspan Optimization

The core promise of salutogenesis is not just disease prevention; it is wellness preservation that optimizes each phase of the human health trajectory: lengthening periods of wellness and improving their quality, detecting and reversing transitions at the earliest possible moments, and greatly delaying (if not altogether avoiding) disease. This encompasses continuous monitoring and improving physical, mental, and emotional states before irreversible damage occurs. Advances in biosensors, machine learning, and digital twin technologies bridge a new era where continuous, personalized data collection supports the proactive management of wellness trajectories throughout life (Bagaria et al., 2020; Sulaiman et al., 2023). The 4P model supports this goal through real-time feedback loops, behavioral nudges, and integrative data platforms. Individually powerful, the 4Ps become transformative when viewed as a system of systems – aligned with the salutogenic model of human health. This integration:

- Predicts risk (comprehensibility).
- Manages resources (manageability).
- Builds meaning through co-creation (meaningfulness).

A systems biology framework connects these elements, which employs molecular and phenotypic data to identify inflection points in health trajectories (Price

TABLE 5.2
Mapping 4P Medicine to Sense of Coherence Domains

4P Medicine Component	Sense of Coherence Domain	Explanation
Predictive	Comprehensibility	Enhances understanding of health risks through data analytics, making health information more coherent.
Preventive	Manageability	Empowers individuals to manage health proactively, reducing uncertainty and enhancing control.
Personalized	Meaningfulness	Aligns healthcare with personal values and goals, increasing the relevance and purpose of interventions.
Participatory	All domains	Encourages active engagement, fostering a comprehensive sense of coherence.

Source: Veenstra et al. (2005).

et al., 2017). This enables the detection and reversal of pre-disease states, which empowers individuals to remain in high-functioning health zones. In this light, health is not the absence of disease, but an actively sustained state of coherence (Table 5.2). Salutogenesis becomes a human systems framework; a way of viewing individuals as adaptable ecosystems in constant dialogue with their environments.

As this chapter underscores, aligning salutogenesis with the principles of 4P medicine offers a visionary model for the optimization of human health. Nevertheless, even the most advanced healthcare frameworks will require equally sophisticated tools to actualize them. Enter MM, an emerging nanotechnological manufacturing frontier that is envisaged to enable ultraprecise, virtually non-invasive interventions at the cellular, organellar, molecular, and even atomic scales via (~Ø1 μm) autonomous nanomedical devices. By transforming health data into material solutions, MM will provide the infrastructure to operationalize salutogenesis across populations. What if health could be built – not just restored? In the next section, we explore how nanotechnology, biofabrication, and programmable molecular systems offer unprecedented potential to democratize wellness and extend healthspan, while redefining what is biologically possible.

5.3 PART II: ENGINEERING WELLNESS: THE MOLECULAR INFRASTRUCTURE OF FUTURE HEALTH

5.3.1 Foundations of Molecular Manufacturing in Medicine

MM refers to the precise fabrication of materials and devices at the molecular and/or atomic level, offering unprecedented control over matter. At its core, this process involves the use of molecular machines and nanoscale robotic systems to assemble materials with atomic precision, transforming everything from diagnostics to treatment delivery systems (Achal et al., 2018; Fahy, 1993; Forrest, 2001; Luo et al., 2025; Luu et al., 2024). While this domain originally emerged from concepts in physics and materials science, its most transformative applications have increasingly been realized in medicine, giving rise to what is now known as nanomedicine (Boehm, 2013; Boisseau & Loubaton, 2011; Feynman, 1960; Freitas, 2002).

In medicine, this translates to targeted nanoscale interventions and applications spanning diagnostics, therapeutics, tissue regeneration, biosensing, and nanoparticles that deliver drugs directly to cancer cells, to DNA origami that builds synthetic scaffolds for tissue repair (Boisseau & Loubaton, 2011; Datta & Jaitawat, 2006). The ultimate goal of MM in healthcare is to develop programmable molecular systems that are capable of detecting, repairing, and even preventing disease – effectively shifting medicine from intervention-based to maintenance-based care (Meetoo, 2009). Central components include:

- Nanocarriers: Liposomes, dendrimers, and exosomes engineered to carry specific drugs, RNA molecules, or immune signals (Güngör & Kahraman, 2021; Patra et al., 2018; Zhao et al., 2022; Zhu et al., 2025).
- Nanosensors: Diagnostic tools capable of detecting molecular signatures of disease at ultra-early stages (Arndt et al., 2020; Hang et al., 2021; Koohkansaadi et al., 2025; Palaniyandi et al., 2023).
- Bionanomachines: Synthetic molecular systems that can repair cellular damage or modulate gene expression (Arai, 2018; Inaba & Ueno, 2018; Majikes & Liddle, 2022; Wang & Nakano, 2023; Zhang et al., 2023).

The ability to generate therapeutic agents, diagnostic tools, and even artificial organs or tissue scaffolds at nanoscale dimensions creates myriad possibilities for customized, minimally invasive, and highly targeted interventions (Emerich & Thanos, 2006).

5.3.2 Precision Diagnostics: Early Detection at the Molecular Frontier

One of the most impactful prospective uses of MM may be in diagnostics. Nanoparticles and molecular devices can serve as biosensors for the detection of pathogens or biomarkers at ultra-low concentrations. For instance, medical nanosensors might provide pre-emptive disease detection (Datta & Jaitawat, 2006; Malik et al., 2023). Nanotechnology enables the detection of biomarkers at femtomolar concentrations, long before conventional imaging or serology could detect abnormalities (Kong et al., 2024

Case Study: Early Cancer Detection Using Quantum Dots: Quantum dot nanoparticles, decorated/functionalized with cancer-specific antibodies, are currently used to tag and visualize tumors in vivo (Hersh et al., 2022; Rasheed et al., 2024; Yan et al., 2023). This technique allows real-time detection of malignancies, especially in hard-to-reach tissues like the pancreas and brain (Chakraborty et al., 2022;

Hu et al., 2024; Liu et al., 2015, 2017; Zhao et al., 2022). Additionally, lab-on-a-chip nanodevices integrate microfluidics with AI to detect viral particles, inflammatory markers, or even circulating tumor DNA at the point of care, which makes them ideal for implementation in rural or resource-limited environments (Barakat et al., 2024; Clark et al., 2023; Mathekga et al., 2022; Rajalekshmi & Agrawal, 2025; Thwala et al., 2023).

5.3.3 Nanoparticles and Regenerative Medicine

Ultra-targeted treatments with fewer side effects might also be further empowered by MM. Nanoparticles can bypass the blood-brain barrier (Tang et al., 2019; Toader et al., 2025), precisely deliver chemotherapy (Li et al., 2022; Obijiofor & Novikov, 2025; Zhang et al., 2017), or even regulate immune responses in autoimmune diseases (Horwitz et al., 2021; Minardi, 2022; Zampieri et al., 2020). Gold nanoparticles (AuNPs) are being used to deliver siRNA to silence oncogenes (Child et al., 2015; Shaat et al., 2016), and smart hydrogels are being engineered to release insulin (Muñoz-Galán et al., 2024) or anti-inflammatory agents based on real-time physiological cues (Salehi et al., 2023). Stem cell-laden scaffolds, fabricated at the molecular level, enable neural, dermal, and cardiac tissue regeneration (Agarwal et al., 2020; Hu et al., 2022; Liu et al., 2019; Xu & Guan, 2016).

Callout Box: Nanoparticles in Diabetic Wound Healing: Chronic wounds, particularly in diabetes, resist healing due to impaired angiogenesis and infection. Nanoparticles that carry growth factors and antimicrobial agents are now used in smart bandages, which dramatically improve recovery rates (Azeltine et al., 2020; Chen et al., 2023; Farahani & Shafiee, 2021; Mostafalu et al., 2018).

5.3.4 Engineering Salutogenesis: Operationalizing the 4Ps

These applications (Table 5.3) map directly onto the salutogenic framework, which allows for real-time data collection, feedback, and interventions aligned with *comprehensibility, manageability,* and *meaningfulness* (Lindström & Eriksson, 2006; Vinje et al., 1979).

TABLE 5.3
Molecular Manufacturing Makes the 4P Framework *Actionable at Scale*. Here's How:

4P Element	Molecular Engineering Application
Predictive	Nanobiosensors that track biomarkers in sweat, blood, or breath
Preventive	Smart vaccines, detoxifying nanomaterials
Personalized	RNA-guided therapy, genomically tailored nanoformulations
Participatory	Wearable nanodevices, digital twin integration

5.3.5 Ethical and Equitable Implementation

The promise of MM raises key questions about equity, access, and planetary impact. Will only affluent societies benefit from designer wellness? Can low-resource regions adopt these tools affordably? Programs like the Nanotechnology Characterization Lab (NCL, 2025) and Millennium Development Goals (Salamanca-Buentello & Daar, 2021; Salamanca-Buentello et al., 2005) aim to ensure equitable dissemination. Furthermore, developments in open-source bioengineering and decentralized nanomanufacturing may enable communities to produce tailored diagnostics and therapies locally (NIH, 2023). As molecular medicine advances, ethical considerations become more prominent, particularly as they relate to gene editing and synthetic biology (Ayanoğlu et al., 2020). Gene-editing tools and synthetic biology raise ethical questions regarding human enhancement. Regulatory frameworks must balance innovation with risks, particularly for personalized therapies to address the safety, efficacy, and ethical implications of these new technologies.

5.3.6 Conclusion: Designing Health as a Right, Not a Privilege

The future of human health is no longer constrained by the limitations of reactive care or the inevitability of decline. Instead, we are entering an age where health and wellness can be designed, monitored, and sustained through the precise coordination of biological insights, molecular technologies, and participatory engagement. This is not merely a transformation of medical practice – it is a redefinition of what it means to live well, one grounded in the conviction that wellness is a human right rather than an honor reserved for the privileged few.

Globally, the right to health is enshrined in over 135 national constitutions, yet profound disparities persist between lifespan and healthspan, especially in low-income communities and under-resourced health systems (Backman et al., 2008; Gruskin et al., 2007). Countries such as Brazil, South Africa, and Thailand have implemented legal mandates to guarantee healthcare access as a constitutional right, which has resulted in expanded primary care, vaccination programs, and maternal health access (Forman et al., 2016). These nations demonstrate the feasibility of aligning public health infrastructures with constitutional commitments to health justice. However, to truly close the healthspan gap, a new vision is needed – one that integrates the right to wellness with the technological capacities of 21st-century medicine.

The convergence of salutogenic philosophy with 4P medicine provides an ethical and strategic blueprint for this future. By focusing on 4P approaches, healthcare systems can be reoriented around sustainability, adaptability, and dignity. This aligns not only with the World Health Organization's definition of health as "a state of complete physical, mental and social well-being" (WHO, 1946) but also with contemporary models of health citizenship, where individuals are empowered to engage meaningfully with their own biological futures (Rose & Novas, 2007).

This chapter has shown how salutogenesis has the potential to operationalize the flourishing of humanity – not just as the absence of disease, but as the cultivation of resilience, meaning, and coherence across lifespans. The addition of MM will provide the tools to actualize this vision at scale. With nanotechnologies that are capable of early disease detection, regenerative interventions, and personalized delivery systems, it is now technically possible to shift the healthcare paradigm from late-stage rescue to the design of continuous wellness (Lee, 2012; Wasti et al., 2023).

Yet, this transformation also demands new systems of accountability and inclusion. The deployment of AI-driven diagnostics and nanomedical tools cannot replicate the exclusions of the past, where novel technologies widened global inequities (O'Neil, 2016). Instead, regulatory frameworks must ensure equitable access to personalized disease prevention, with infrastructure investments that encompass rural, indigenous, and historically marginalized populations (Braveman & Gottlieb, 2014). Public-private partnerships, open-source biodesign platforms, and global knowledge commons will be essential for ensuring that salutogenic tools are not only available – but universally accessible and legally enforceable.

Importantly, wellness must be recognized not simply as a public good, but as an inalienable entitlement of the human condition. The United Nations' Sustainable Development Goal 3 – "Ensure healthy lives and promote well-being for all at all ages" – must be interpreted expansively, to include cognitive vitality, emotional well-being, and the capacity for lifelong thriving (UN, 2015). This holistic vision of health affirms that longevity without well-being is ethically inadequate.

Visionary public health frameworks already point the way. In Bhutan, the government's use of Gross National Happiness as a development metric integrates wellness, equity, and spiritual cohesion into national planning (Ura et al., 2012). In New Zealand, Wellbeing Budgets center health equity and indigenous knowledge in the allocation of healthcare resources (New Zealand Treasury, 2019). These models suggest that redefining progress in terms of wellness is not only possible – but already underway.

Going forward, medical science must embrace a dual mandate: to engineer the molecular basis of health, and to democratize its access. This calls for what we might term the bioethical infrastructures – policies, educational systems, and participatory platforms that ensure the right to wellness is robustly supported and safeguarded, not as a rhetorical ideal, but as a lived reality. From decentralized nanomanufacturing hubs in rural communities to open-access genomic literacy campaigns, the path toward health equity must be as innovative as the technologies that support it.

In this light, engineering wellness becomes the ethical imperative of our age. It is a project that merges precision molecular biology with planetary justice, predictive analytics with human dignity, and self-healing cells with collective empowerment. If the 20th century was defined by the conquest of infectious diseases, then the 21st must be defined by the democratization of health – ensuring that every individual, regardless of geography or income, has the capacity to live long, live well, and live meaningfully.

This is not merely the next frontier of medicine – it is the next frontier of human rights.

REFERENCES

AbdulRaheem Y. Unveiling the significance and challenges of integrating prevention levels in healthcare practice. *J Prim Care Community Health.* 2023 Jan–Dec;14: 21501319231186500.

Achal R, Rashidi M, Croshaw J, Churchill D, Taucer M, Huff T, et al. Lithography for robust and editable atomic-scale silicon devices and memories. *Nat Commun*. 2018 Jul 23;9(1):2778.

Agarwal G, Agiwal S, Srivastava A. Hyaluronic acid containing scaffolds ameliorate stem cell function for tissue repair and regeneration. *Int J Biol Macromol*. 2020 Dec 15;165(Pt A):388–401.

Alonso SG, de la Torre Díez I, Zapiraín BG. Predictive, personalized, preventive and participatory (4P) medicine applied to telemedicine and eHealth in the literature. *J Med Syst*. 2019 Apr 12;43(5):140.

Alum EU. AI-driven biomarker discovery: enhancing precision in cancer diagnosis and prognosis. *Discov Oncol*. 2025 Mar 13;16(1):313.

An X, Zhang Y, Sun W, Kang X, Ji H, Sun Y, et al. Early effective intervention can significantly reduce all-cause mortality in prediabetic patients: a systematic review and meta-analysis based on high-quality clinical studies. *Front Endocrinol (Lausanne)*. 2024 Mar 1;15:1294819.

Arai R. Hierarchical design of artificial proteins and complexes toward synthetic structural biology. *Biophys Rev*. 2018 Apr;10(2):391–410.

Arndt N, Tran HDN, Zhang R, Xu ZP, Ta HT. Different approaches to develop nanosensors for diagnosis of diseases. *Adv Sci (Weinh)*. 2020 Oct 28;7(24):2001476.

Artiga S, Hinton E. Beyond health care: the role of social determinants in promoting health and health equity. Kaiser Family Foundation (KFF) – Racial Equity and Health Policy. 2018. https://www.kff.org/racial-equity-and-health-policy/issue-brief/beyond-health-care-the-role-of-social-determinants-in-promoting-health-and-health-equity/ (accessed January 12, 2024).

Ayanoğlu FB, Elçin AE, Elçin YM. Bioethical issues in genome editing by CRISPR-Cas9 technology. *Turk J Biol*. 2020 Apr 2;44(2):110–120.

Azeltine M, Clark A, Zgheib C, Ghatak S. Nanotechnology in diabetic wound healing. In: Bagchi D, Das A, Roy S, editors. *Wound Healing, Tissue Repair, and Regeneration in Diabetes*. 1st ed. Academic Press, Cambridge (MA). 2020;417–437.

Backman G, Hunt P, Khosla R, Jaramillo-Strouss C, Fikre BM, Rumble C, et al. Health systems and the right to health: an assessment of 194 countries. *Lancet*. 2008 Dec 13;372(9655):2047–2085.

Bagaria N, Laamarti F, Badawi HF, Albraikan A, Martinez Velazquez R, et al. "Health 4.0: Digital Twins for Health and Well-Being," In El Saddik A, Hossain M, Kantarci B, eds., *Connected Health in Smart Cities*, Springer, Cham. 2020;143–152.

Barakat JM, Modica KJ, Lu L, Anujarerat S, Choi KH, Takatori SC. Surface topography induces and orients nematic swarms of active filaments: considerations for lab-on-a-chip devices. *ACS Appl Nano Mater*. 2024 May 8;7(10):12142–12152.

Bauer GF, Roy M, Bakibinga P, Contu P, Downe S, Eriksson M, et al. Future directions for the concept of salutogenesis: a position article. *Health Promot Int*. 2020 Apr 1;35(2):187–195.

Bazzano AN, Noel LA, Patel T, Dominique CC, Haywood C, Moore S, et al. Improving the engagement of underrepresented people in health research through equity-centered design thinking: qualitative study and process evaluation for the development of the grounding health research in design toolkit. *JMIR Form Res*. 2023 Feb 28;7:e43101.

Boehm F. *Nanomedical Device and Systems Design: Challenges, Possibilities, Visions, 1st ed.*, CRC Press, Boca Raton, FL. 2013.

Boisseau P, Loubaton B. Nanomedicine, nanotechnology in medicine. *Comptes Rendus. Physique*. 2011;12(7):620–636.

Braveman P, Gottlieb L. The social determinants of health: it's time to consider the causes of the causes. *Public Health Rep*. 2014 Jan–Feb;129(Suppl 2):19–31.

Chakraborty P, Das SS, Dey A, Chakraborty A, Bhattacharyya C, Kandimalla R, et al. Quantum dots: the cutting-edge nanotheranostics in brain cancer management. *J Control Release*. 2022;350:698–715.

Chen Y, Wang X, Tao S, Wang Q, Ma PQ, Li ZB, et al. Research advances in smart responsive-hydrogel dressings with potential clinical diabetic wound healing properties. *Mil Med Res*. 2023 Aug 23;10(1):37.

Child HW, Hernandez Y, Conde J, Mullin M, Baptista P, de la Fuente JM, et al. Gold nanoparticle-siRNA mediated oncogene knockdown at RNA and protein level, with associated gene effects. *Nanomedicine (Lond)*. 2015;10(16):2513–2525.

Clark IC, Mudvari P, Thaploo S, Smith S, Abu-Laban M, Hamouda M, et al. HIV silencing and cell survival signatures in infected T cell reservoirs. *Nature*. 2023 Feb;614(7947):318–325.

Datta R, Jaitawat SS. Nanotechnology – the new frontier of medicine. *Med J Armed Forces India*. 2006 Jul;62(3):263–268.

Emerich DF, Thanos CG. The pinpoint promise of nanoparticle-based drug delivery and molecular diagnosis. *Biomol Eng*. 2006 Sep;23(4):171–184.

Evans DB, Hsu J, Boerma T. Universal health coverage and universal access. *Bull World Health Organ*. 2013;91:546–546A.

Fahy GM. Molecular nanotechnology. *Clin Chem*. 1993 Sep;39(9):2011–2016.

Farahani M, Shafiee A. Wound healing: from passive to smart dressings. *Adv Healthc Mater*. 2021 Aug;10(16):e2100477.

Feynman R. There's plenty of room at the bottom. *Eng Sci*. 1960;23:22–36.

Forman L, Ooms G, Brolan C, Friedman E. Rights-based approaches to health: a key component of the post-2015 development agenda. *Health Hum Rights*. 2016;18(1):21–32.

Forrest DE. Molecular nanotechnology. *IEEE Instrum Meas Mag*. 2001;4(3):11–20.

Freitas RA Jr. The future of nanofabrication and molecular scale devices in nanomedicine. *Stud Health Technol Inform*. 2002;80:45–59.

Fufaa GD. The Role of Preventive Health Care in Preventing Chronic Diseases. *Public Health. IntechOpen*. 2024. http://dx.doi.org/10.5772/intechopen.1006694 (accessed May 29, 2025).

Grosu O. Predictive, preventive, personalized and participatory medicine (4P): Narrative review. *Bull Acad Sci Med*. 2022;74(3):3461.

Gruskin S, Mills EJ, Tarantola D. History, principles, and practice of health and human rights. *Lancet*. 2007 Aug 4;370(9585):449–455.

Güngör S, Kahraman E. Nanocarriers mediated cutaneous drug delivery. *Eur J Pharm Sci*. 2021 Mar 1;158:105638.

Hang X, He S, Dong Z, Minnick G, Rosenbohm J, Chen Z, et al. Nanosensors for single cell mechanical interrogation. *Biosens Bioelectron*. 2021 May 1;179:113086.

Herdiana I, Suryanto, Seger H. Family resilience: a conceptual review. In: Proceedings of the 3rd ASEAN Conference on Psychology, Counselling, and Humanities (ACPCH 2017). Atlantis Press, Paris; 2018:243–253.

Hersh AM, Alomari S, Tyler BM. Crossing the blood-brain barrier: advances in nanoparticle technology for drug delivery in neuro-oncology. *Int J Mol Sci*. 2022 Apr 9;23(8):4153.

Ho PJ, Tan IB, Chong DQ, Khor CC, Yuan JM, Koh WP, et al. Polygenic risk scores for the prediction of common cancers in East Asians: a population-based prospective cohort study. *Elife*. 2023 Mar 27;12:e82608.

Holman HR. The relation of the chronic disease epidemic to the health care crisis. *ACR Open Rheumatol*. 2020 Mar;2(3):167–173.

Hood L, Friend SH. Predictive, personalized, preventive, participatory (P4) cancer medicine. *Nat Rev Clin Oncol*. 2011 Mar;8(3):184–187.

Hood L, Price ND. Demystifying disease, democratizing health care. *Sci Transl Med.* 2014 Feb 26;6(225):225ed5.

Horwitz DA, Bickerton S, La Cava A. Strategies to use nanoparticles to generate CD4 and CD8 regulatory T cells for the treatment of SLE and other autoimmune diseases. *Front Immunol.* 2021 Jun 15;12:681062.

Hou K, Wu ZX, Chen XY, Wang JQ, Zhang D, Xiao C, et al. Microbiota in health and diseases. *Signal Transduct Target Ther.* 2022 Apr 23;7(1):135.

Hu W, Shi J, Lv W, Jia X, Ariga K. Regulation of stem cell fate and function by using bioactive materials with nanoarchitectonics for regenerative medicine. *Sci Technol Adv Mater.* 2022 Jun 22;23(1):393–412.

Hu Y, Wang X, Niu Y, He K, Tang M. Application of quantum dots in brain diseases and their neurotoxic mechanism. *Nanoscale Adv.* 2024 Jun 5;6(15):3733–3746.

Inaba H, Ueno T. Artificial bio-nanomachines based on protein needles derived from bacteriophage T4. *Biophys Rev.* 2018 Apr;10(2):641–658.

Johnson D, Deterding S, Kuhn KA, Staneva A, Stoyanov S, Hides L. Gamification for health and wellbeing: a systematic review of the literature. *Internet Interv.* 2016 Nov 2;6:89–106.

Kavanagh MM. The right to health: institutional effects of constitutional provisions on health outcomes. *St Comp Int Dev.* 2016;51:328–364.

Kinney ED, Clark, BA. Provisions for health and health care in the constitutions of the countries of the world. *Cornell International Law Journal.* 2004;37(2):Article 2 https://scholarship.law.cornell.edu/cilj/vol37/iss2/2/ (accessed September 20, 2025).

Kong S, Liu H, Zhang Y, Fan J, Huang W. Clinical applications of nanoprobes of high-resolution *in vivo* imaging. *iScience.* 2024 Dec 9;28(1):111459.

Koohkansaadi G, Tabean M, Mohagheghi A, Masoumi S, Jahanabad ZJ, Mobed A, et al. Aspirin nanosensors. *Clin Chim Acta.* 2025 Apr 15;571:120222.

Lee LM. Public health ethics theory: review and path to convergence. *J Law Med Ethics.* 2012;40(1):85–98.

Li B, Shao H, Gao L, Li H, Sheng H, Zhu L. Nano-drug co-delivery system of natural active ingredients and chemotherapy drugs for cancer treatment: a review. *Drug Deliv.* 2022 Dec;29(1):2130–2161.

Lindström B, Eriksson M. Contextualizing salutogenesis and Antonovsky in public health development. *Health Promot Int.* 2006 Sep;21(3):238–244.

Liu X, Braun GB, Qin M, Ruoslahti E, Sugahara KN. In vivo cation exchange in quantum dots for tumor-specific imaging. *Nat Commun.* 2017 Aug 24;8(1):343.

Liu Y, Hou W, Gao T, Yan Y, Wang T, Zheng C, et al. Influence and role of polygenic risk score in the development of 32 complex diseases. *J Glob Health.* 2025 Mar 10;15: 04071.

Liu J, Ruan J, Weir MD, Ren K, Schneider A, Wang P, et al. Periodontal bone-ligament-cementum regeneration via scaffolds and stem cells. *Cells.* 2019 Jun 4;8(6):537.

Liu H, Tang W, Li C, Lv P, Wang Z, Liu Y, et al. CdSe/ZnS quantum dots-labeled mesenchymal stem cells for targeted fluorescence imaging of pancreas tissues and therapy of type 1 diabetic rats. *Nanoscale Res Lett.* 2015 Dec;10(1):959.

Luo X, Ranno L, Sverko T, Lee JY, Sbalbi N, Jones A, et al. DNA origami directed integration of colloidal nanophotonic materials with silicon photonics. bioRxiv. 2025 Jan 23;2025.01.23.634416

Luu MT, Berengut JF, Li J, Chen JB, Singh Daljit JK, Coffi Dit Glieze K, et al. Reconfigurable nanomaterials folded from multicomponent chains of DNA origami voxels. *Sci Robot.* 2024 Nov 27;9(96):eadp2309.

Magi CE, Bambi S, Rasero L, Longobucco Y, El, Aoufy K, Amato C, et al. Health literacy and self-care in patients with chronic illness: a systematic review and meta-analysis protocol. *Healthcare (Basel).* 2024 Mar 31;12(7):762.

Majikes JM, Liddle JA. Synthesizing the biochemical and semiconductor worlds: the future of nucleic acid nanotechnology. *Nanoscale*. 2022 Nov 3;14(42):15586–15595.

Malik S, Muhammad K, Waheed Y. Nanotechnology: a revolution in modern industry. *Molecules*. 2023 Jan 9;28(2):661.

Mathekga BSP, Nxumalo Z, Thimiri Govinda Raj DB. Micro and nanofluidics for high throughput drug screening. *Prog Mol Biol Transl Sci*. 2022;187(1):93–120.

Meetoo D. Nanotechnology: the revolution of the big future with tiny medicine. *Br J Nurs*. 2009 Oct 22–Nov 11;18(19):1201–1206.

Minardi S. Editorial: Nano-biomaterials for the delivery of therapeutic and biological cues for regenerative medicine. *Front Bioeng Biotechnol*. 2022 Jan 13;9:832487.

Mittelmark MB, Sagy SM, Bauer GF, Pelikan JM, Lindström B, Geir Espnes GA, eds., *The Handbook of Salutogenesis*, Springer, Berlin, Germany. 2017.

Molina-López AM, Bujalance-Reyes F, Ayala-Soldado N, Mora-Medina R, Lora-Benítez A, Moyano-Salvago R. An overview of the health effects of bisphenol A from a one health perspective. *Animals (Basel)*. 2023 Jul 28;13(15):2439.

Mostafalu P, Tamayol A, Rahimi R, Ochoa M, Khalilpour A, Kiaee G, et al. Smart bandage for monitoring and treatment of chronic wounds. *Small*. 2018;14(33):e1703509.

Muñoz-Galán H, Enshaei H, Silva JC, Esteves T, Ferreira FC, Casanovas J, et al. Electroresponsive thiol-yne click-hydrogels for insulin smart delivery: tackling sustained release and leakage control. *ACS Appl Polym Mater*. 2024 Jul 11;6(14):8093–8104.

Nampewo Z, Mike JH, Wolff J. Respecting, protecting and fulfilling the human right to health. *Int J Equity Health*. 2022 Mar 15;21(1):36.

Finkelman EM, McGinnis JM, McClellan MB, Dzau VJ, eds. *Vital Directions for Health & Health Care: An Initiative of the National Academy of Medicine*, National Academy of Medicine: National Academies Press (US), Washington (DC). 2017.

NCL. https://www.cancer.gov/nano/research/ncl (accessed June 12, 2025).

Neale EP, Rosario VD, Probst Y, Beck E, Tran TB, Lambert K. Lifestyle interventions, kidney disease progression, and quality of life: a systematic review and meta-analysis. *Kidney Med*. 2023 Apr 18;5(6):100643.

New Zealand Treasury. Wellbeing Budget 2019. https://www.treasury.govt.nz/publications/wellbeing-budget/wellbeing-budget-2019-html (accessed May 29, 2025).

NIH. Cancer nanotechnology: Detection and diagnosis. National Cancer Institute. 2023. https://www.cancer.gov/nano/cancer-nanotechnology/detection-diagnosis (accessed April 10, 2024).

O'Neil C. *Weapons of Math Destruction: How Big Data Increases Inequality and Threatens Democracy*. Broadway Books, New York City, NY, 2016.

Obijiofor OC, Novikov AS. Exploring the role of density functional theory in the design of gold nanoparticles for targeted drug delivery: a systematic review. *J Mol Model*. 2025 Jun 10;31(7):186.

Olawade DB, Aderinto N, Clement David-Olawade A, Egbon E, Adereni T, Popoola MR, et al. Integrating AI-driven wearable devices and biometric data into stroke risk assessment: a review of opportunities and challenges. *Clin Neurol Neurosurg*. 2025 Feb;249:108689.

Pack AI. Application of personalized, predictive, preventative, and participatory (P4) medicine to obstructive sleep apnea. A roadmap for improving care? *Ann Am Thorac Soc*. 2016 Sep;13(9):1456–1467.

Palaniyandi T, Kanagavalli B, Prabhakaran P, Viswanathan S, Rahaman Abdul Wahab M, Natarajan S, et al. Nanosensors for the diagnosis and therapy of neurodegenerative disorders and inflammatory bowel disease. *Acta Histochem*. 2023 Feb;125(2):151997.

Patel AP, Wang M, Ruan Y, Koyama S, Clarke SL, Yang X, et al. A multi-ancestry polygenic risk score improves risk prediction for coronary artery disease. *Nat Med*. 2023;29(8):1741–1751.

Patra JK, Das G, Fraceto LF, Campos EVR, Rodriguez-Torres MDP, Acosta-Torres LS, et al. Nano based drug delivery systems: recent developments and future prospects. *J Nanobiotechnology.* 2018 Sep 19;16(1):71.

Pitts PJ, Freeman E. Health literacy: the common denominator of healthcare progress. *Patient.* 2021 Sep;14(5):455–458.

Price ND, Magis AT, Earls JC, Glusman G, Levy R, Lausted C, et al. A wellness study of 108 individuals using personal, dense, dynamic data clouds. *Nat Biotechnol.* 2017 Aug;35(8):747–756.

Purnell TS, Calhoun EA, Golden SH, Halladay JR, Krok-Schoen JL, Appelhans BM, et al. Achieving health equity: closing the gaps in health care disparities, interventions, and research. *Health Aff (Millwood).* 2016 Aug 1;35(8):1410–1415.

Puska P, Jaini P. The North Karelia project: prevention of cardiovascular disease in Finland through population-based lifestyle interventions. *Am J Lifestyle Med.* 2020 Mar 19;14(5):495–499.

Rajalekshmi R, Agrawal DK. Synergistic potential of stem cells and microfluidics in regenerative medicine. *Mol Cell Biochem.* 2025 Mar;480(3):1481–1493.

Rasheed PA, Ankitha M, Pillai VK, Alwarappan S. Graphene quantum dots for biosensing and bioimaging. *RSC Adv.* 2024 May 17;14(23):16001–16023.

Rein M, Ben-Yacov O, Godneva A, Shilo S, Zmora N, Kolobkov D, et al. Effects of personalized diets by prediction of glycemic responses on glycemic control and metabolic health in newly diagnosed T2DM: a randomized dietary intervention pilot trial. *BMC Med.* 2022 Feb 9;20(1):56.

Richardson S, Lawrence K, Schoenthaler AM, Mann D. A framework for digital health equity. *NPJ Digit Med.* 2022 Aug 18;5(1):119.

Rose N, Novas C. Chapter 23 "Biological Citizenship," in Ong A, Collier S, eds., *Global Assemblages: Technology, Politics, and Ethics as Anthropological Problems,* Blackwell, Oxford, UK. 2007.

Sadiq IZ. Lifestyle medicine as a modality for prevention and management of chronic diseases. *J Taibah Univ Med Sci.* 2023 Apr 15;18(5):1115–1117.

Salamanca-Buentello F, Daar AS. Nanotechnology, equity and global health. *Nat Nanotechnol.* 2021 Apr;16(4):358–361.

Salamanca-Buentello F, Persad DL, Court EB, Martin DK, Daar AS, Singer PA. Nanotechnology and the developing world. *PLoS Med.* 2005 May;2(5):e97.

Salehi S, Naghib SM, Garshasbi HR, Ghorbanzadeh S, Zhang W. Smart stimuli-responsive injectable gels and hydrogels for drug delivery and tissue engineering applications: a review. *Front Bioeng Biotechnol.* 2023 Feb 22;11:1104126.

Schmutz AMS, Jenkins LS, Coetzee F, Conradie H, Irlam J, Joubert EM, et al. Re-imagining health professions education in the coronavirus disease 2019 era: perspectives from South Africa. *Afr J Prim Health Care Fam Med.* 2021 Aug 10;13(1):e1–e5.

Shaat H, Mostafa A, Moustafa M, Gamal-Eldeen A, Emam A, El-Hussieny E, et al. Modified gold nanoparticles for intracellular delivery of anti-liver cancer siRNA. *Int J Pharm.* 2016 May 17;504(1–2):125–133.

Sofolahan-Oladeinde Y, Mullins CD, Baquet CR. Using community-based participatory research in patient-centered outcomes research to address health disparities in under-represented communities. *J Comp Eff Res.* 2015 Sep;4(5):515–523.

Spurgeon D. Banning trans fatty acids will reduce risk of heart attacks, study says. *BMJ.* 2006;332:931.

Stenberg K, Hanssen O, Bertram M, Brindley C, Meshreky A, Barkley S, et al. Guide posts for investment in primary health care and projected resource needs in 67 low-income and middle-income countries: a modelling study. *Lancet Glob Health.* 2019;7(11): e1500–e1510.

Sulaiman M, Håkansson A, Karlsen R. Fit-Twin: A Digital Twin of a User with Wearables and Context as Input for Health Promotion. *Proceedings of the 16th International Joint Conference on Biomedical Engineering Systems and Technologies – BIOSTEC*. Lisbon, Portugal. 2023;5:461–470.

Tang W, Fan W, Lau J, Deng L, Shen Z, Chen X. Emerging blood-brain-barrier-crossing nanotechnology for brain cancer theranostics. *Chem Soc Rev*. 2019 Jun 4;48(11): 2967–3014.

Thwala LN, Ndlovu SC, Mpofu KT, Lugongolo MY, Mthunzi-Kufa P. Nanotechnology-based diagnostics for diseases prevalent in developing countries: current advances in point-of-care tests. *Nanomaterials (Basel)*. 2023 Mar 31;13(7):1247.

Tian Q, Price ND, Hood L. Systems cancer medicine: towards realization of predictive, preventive, personalized and participatory (P4) medicine. *J Intern Med*. 2012 Feb;271(2):111–121.

Toader C, Serban M, Munteanu O, Covache-Busuioc RA, Enyedi M, Ciurea AV, et al. From synaptic plasticity to neurodegeneration: BDNF as a transformative target in medicine. *Int J Mol Sci*. 2025 Apr 30;26(9):4271.

UN. Transforming Our World: The 2030 Agenda for Sustainable Development. 2015. https://sdgs.un.org/2030agenda (accessed May 28, 2025).

Ura K, Alkire S, Zangmo T, Wangdi K. An extensive analysis of Gross National Happiness Index. Centre for Bhutan Studies and GNH Research; 2012.

Veenstra M, Moum T, Røysamb E. Relationships between health domains and sense of coherence: a two-year cross-lagged study in patients with chronic illness. *Qual Life Res*. 2005 Aug;14(6):1455–1465.

Vinje HF, Langeland E, Bull T. Aaron Antonovsky's development of salutogenesis, 1979 to 1994. In M. B. Mittelmark, G. F. Bauer, L. Vaandrager, J. M. Pelikan, S. Sagy, M. Eriksson, B. Lindström & C. Meier Magistretti (Eds.), *The Handbook of Salutogenesis*. Springer, Cham, Switzerland. 2016:25–40.

Wang S, Dong M, He J, Wu G, Li X, Pan X, et al. A multi-user waistband system for real-time health monitoring of respiration, ECG, and body temperature. *Microelectron Eng*. 2025;299:112346.

Wang J, Nakano T. A Biologically inspired model of collective bio-nanomachine rotation via chemical and physical interactions. *IEEE Trans Nanobioscience*. 2023 Jul;22(3):570–581.

Wasti S, Lee IH, Kim S, Lee JH, Kim H. Ethical and legal challenges in nanomedical innovations: a scoping review. *Front Genet*. 2023 May 12;14:1163392.

Welton WE, Kantner TA, Katz SM. Integrating our primary care and public health systems: a formula for improving community and population health. In: Donaldson MS, Committee on the Future of Primary Care, editors. Primary Care: America's Health in a New Era. National Academies Press, Washington (DC). 1996:341–368.

WHO. Constitution of the World Health Organization. 1946. https://www.who.int/about/governance/constitution (accessed May 29, 2025).

Willis M, Edwards A. Clinical applications of artificial intelligence: Understanding the barriers to adoption. *BMJ Health & Care Informatics*. 2022;29(1):e100495.

Xu Y, Guan J. Biomaterial property-controlled stem cell fates for cardiac regeneration. *Bioact Mater*. 2016 Apr 5;1(1):18–28.

Yan H, Wang Q, Wang J, Shang W, Xiong Z, Zhao L, et al. Planted graphene quantum dots for targeted, enhanced tumor imaging and long-term visualization of local pharmacokinetics. *Adv Mater*. 2023 Apr;35(15):e2210809.

Yuen E, Winter N, Savira F, Huggins CE, Nguyen L, Cooper P, et al. Digital health literacy and its association with sociodemographic characteristics, health resource use, and health outcomes: rapid review. *Interact J Med Res*. 2024 Jul 26;13:e46888.

Yurkovich JT, Evans SJ, Rappaport N, Boore JL, Lovejoy JC, Price ND, et al. The transition from genomics to phenomics in personalized population health. *Nat Rev Genet.* 2024;25(4):286–302.

Zampieri R, Brozzetti A, Pericolini E, Bartoloni E, Gabrielli E, Roselletti E, et al. Prevention and treatment of autoimmune diseases with plant virus nanoparticles. *Sci Adv.* 2020;6(19):eaaz0295.

Zhang D, Gorochowski TE, Marucci L, Lee HT, Gil B, Li B, et al. Advanced medical microrobotics for early diagnosis and therapeutic interventions. *Front Robot AI.* 2023 Jan 10;9:1086043.

Zhang J, Tang H, Liu Z, Chen B. Effects of major parameters of nanoparticles on their physical and chemical properties and recent application of nanodrug delivery system in targeted chemotherapy. *Int J Nanomedicine.* 2017 Nov 28;12:8483–8493.

Zhao X, Zhao R, Nie G. Nanocarriers based on bacterial membrane materials for cancer vaccine delivery. *Nat Protoc.* 2022 Oct;17(10):2240–2274.

Zhou Y, Peng S, Wang H, Cai X, Wang Q. Review of personalized medicine and pharmacogenomics of anti-cancer compounds and natural products. *Genes (Basel).* 2024 Apr 8;15(4):468.

Zhu Z, Lv Z, Fan Y, Wen H, Zhang Y, Guo Y, et al. A Neovascularization and EGFR Dual-targeting gold nanotheranostic system for non-small cell lung cancer treatment. *Adv Healthc Mater.* 2025 Aug;14(21):e2500854.

6 Subatomic Aggregates Are Us

Beyond Molecular Manufacturing

Angelika Domschke and Frank J Boehm

CONTENTS

6.1 Fundamental Units of Matter and Energy: Atomic Realms and Quantum Domains 168
6.1.1 From the First Atom to Periodic Table 169
6.1.2 Protons, Neutrons, and Electrons 171
6.1.3 Quantum Particles 172
6.1.3.1 Principles that Define the Quantum Domain 173
6.2 Functionalized Atomic/Subatomic Aggregates are Us 174
6.3 Self-Organization/Assembly Strategies 176
6.3.1 Self-Organization/Assembly in Living Matter 176
6.3.1.1 Adhesion Molecules 178
6.3.1.2 Cell-Matrix Interactions 179
6.3.1.3 Chemical Signaling 179
6.3.1.4 Mechanical Forces and Tension 179
6.3.1.5 Feedback Loops 180
6.3.1.6 Cellular Automata and Localized Rules 180
6.3.1.7 Chemical Reaction and Diffusion Systems 180
6.3.1.8 Genetic Regulation and Gene Networks 180
6.4 Myriad Roles of Water 180
6.4.1 A Broader Perspective of Conventional Definitions of Life and Their Implications 182
6.5 Demarcation of Progressively Higher State Structural Complexity Thresholds Correlated with Functionality 183
6.5.1 Distinguishing Features and Perceived Threshold Between Non-Living and Living Entities Infused with the Life Force 183
6.5.1.1 Distinguishing Features of the Life Force from a Philosophical or Spiritual Perspective 184
6.5.2 Distinguishing Features and Perceived Threshold Between Living Non-Brained and Brain-Endowed Organisms 185
6.5.2.1 Key Differences between Cognitive Thresholds 186
6.5.2.2 Evolutionary Perspective 186

DOI: 10.1201/9781003724896-6

6.5.2.3 Consciousness ... 186
6.5.2.4 Seat of Consciousness ... 187
6.5.3 Threshold and Distinguishing Features between Sentient Conscious Self-Aware Beings and the Unified Field: Quantum Universal Consciousness/Intelligence ... 188
6.5.3.1 Localized Consciousness ... 188
6.5.3.2 Unified Quantum/Universal Consciousness ... 189
6.5.3.3 Theories and Philosophical Parallels ... 189
6.5.3.4 Perceived Threshold ... 189
6.5.3.5 Speculative Correlations ... 189
6.5.3.6 Deduction ... 190
6.5.4 Connectedness and the Emergence of Higher Organisms ... 190
6.5.4.1 Gaia Hypothesis ... 190
6.5.4.2 Electromagnetic Fields as a Unifying Medium ... 190
6.5.4.3 Superorganism Theory ... 190
6.5.4.4 Collective Consciousness and Electromagnetic Fields ... 190
6.5.4.5 Technological Amplification ... 190
6.5.4.6 Panpsychism and Universal Consciousness ... 191
6.5.4.7 Global Connectivity/Collective Mind ... 191
6.5.4.8 Deduction ... 191
6.6 Beyond the Human Substrate: Ambient Resonance, Energy/Unified Consciousness Fields, Transcendence and Love ... 191
6.6.1 Mechanisms of EMF Generation ... 192
6.6.2 Exploring Energy Fields in Integrative Energetic Medicine ... 193
6.7 Ultimate Interconnectivity: Universal Mind and Interdimensionality ... 194
6.7.1 Interdimensionality ... 194
6.8 Conclusion ... 195
References ... 196

6.1 FUNDAMENTAL UNITS OF MATTER AND ENERGY: ATOMIC REALMS AND QUANTUM DOMAINS

In terms of the matter that comprises all in the physical universe, it might be broadly categorized (from our perception and the purpose of this chapter) as non-living inorganic matter (e.g., only compounds can be classified as inorganic; minerals, metals, water, rocks, and mountains) and living organic matter (e.g., cells, micro- and macro-organisms, and ecosystems).

The Big Bang theory proposes the expansion of space itself from an initial high-energy, high-density state that existed everywhere in the universe ~13.8 billion years ago. Inflationary cosmology explains the genesis of the universe as being devoid of atoms and exponentially expanding from an infinitesimal point, 100 billion times smaller (10^{-26} m)

TABLE 6.1
Earliest Atomic Nuclei Formed after the Big Bang

Atomic Nuclei Species	Proton/Neutron Configuration
Helium-4	Two protons/two neutrons
Deuterium	One proton and neutron
Helium-3	Two protons and one neutron
Tritium	One proton and two neutrons
Lithium-7	Three protons and four neutrons
Beryllium-7	Four protons and three neutrons

Source: Derived from Siegel (2023).

than a proton, to a meter in size within ~10^{-35} sec (Albrecht & Steinhardt, 1982; Guth, 1981; Linde, 1982). Subsequently, its velocity began to slow to the rate we observe today (Guth, 1997). During and immediately following the Big Bang, the universe was an extremely hot (hundreds of millions of degrees) and dense environment where subatomic particles were the primary forms of matter (NASA, Hubble Big Bang, 2024).

As the universe expanded further and cooled, these subatomic particles coalesced to form protons and neutrons, which later combined to create atomic nuclei. Eventually, electrons integrated with these nuclei to form neutral atoms. The first stable atomic nuclei were thought to have self-assembled from protons and neutrons within minutes (Table 6.1), while stable atoms (with the incorporation of electrons) would not begin to emerge for hundreds of thousands of years, leading to the formation of the first elements and the evolution of the universe as we know it today (Dieks, 2023; Siegel, 2023; Volpe et al., 2022; Weiss, 2006).

The evolution of inorganic matter and minerals on Earth began with the formation of the planet about 4.54 billion years ago. They were derived from the interactions between Earth's internal (e.g., volcanic activities and plate tectonics) and surface processes (e.g., weathering and biological activities), which led to the continual evolution and diversification of these materials. This process has required billions of years and continues today.

While the exact timing of the appearance of the first organic matter on Earth remains a topic of research, evidence suggests that organic molecules and potentially early forms of life appeared between 3.8 and 4.0 billion years ago. This period saw the transition from prebiotic chemistry to the emergence of the first living (life-force induced) organisms, which set the stage for the evolution of life on our planet (García-Ruiz et al., 2020; Magalon, 2023; Perez & Velazquez, 2009).

6.1.1 From the First Atom to Periodic Table

The history of the evolving elucidation of the atomic domain, the elements that comprise it, and the development of the periodic table of elements (Figure 6.1) spans centuries of scientific inquiry and discovery. These dates to the early atomic concepts in Ancient Greece led by Democritus (c. 460–c. 370 BC), who proposed that matter

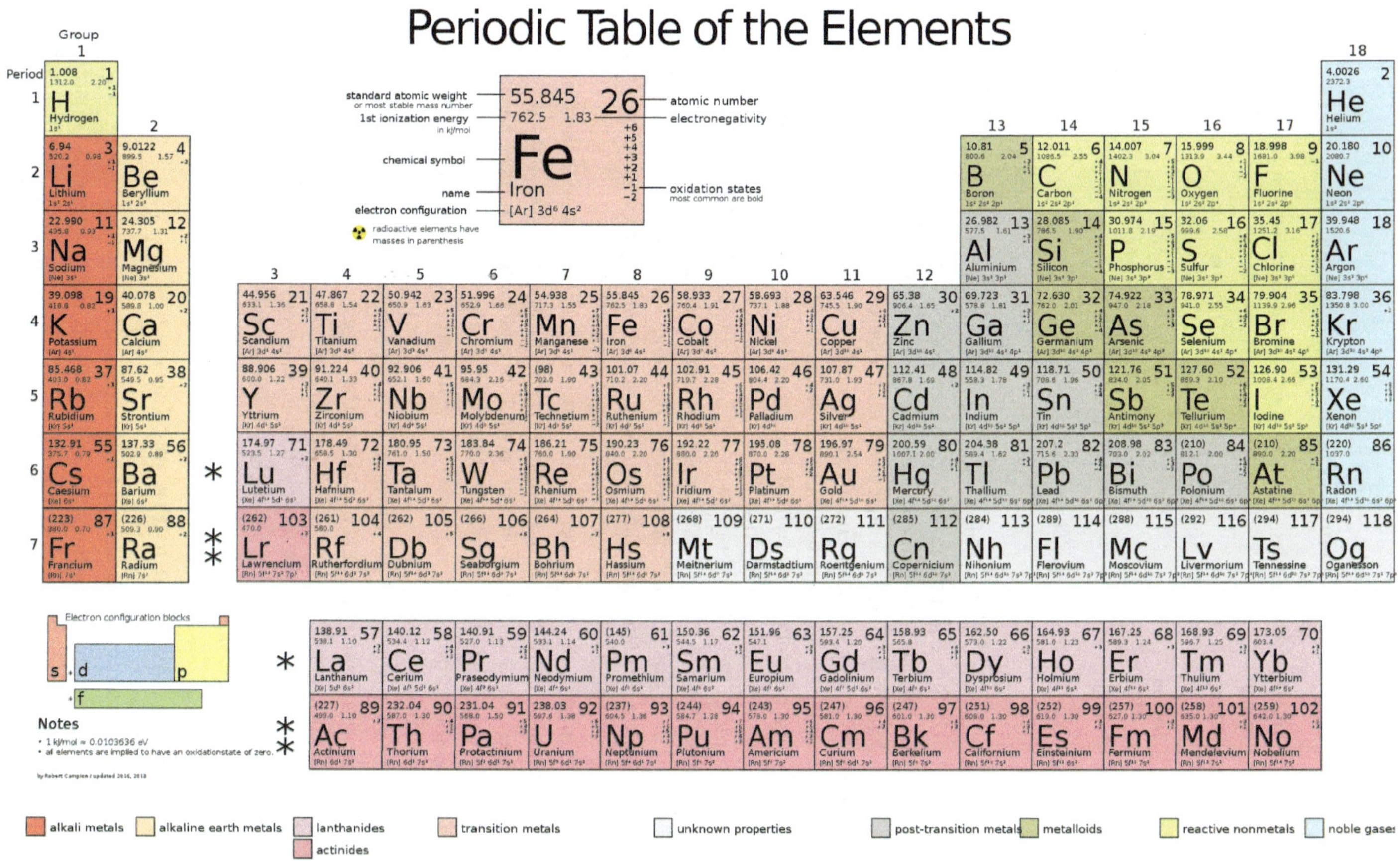

FIGURE 6.1 Periodic table of the elements (Reproduced from Wikimedia Commons. Creative Commons Attribution 3.0 Unported License.)

consists of small, indivisible particles referred to as "atomos" (meaning "uncuttable") (Magiorkinis et al., 2010; Democritus, 1949; Tsoucalas et al., 2012).

Key milestones toward the development of the Periodic Table of Elements were achieved by Dmitri Mendeleev (1834–1907), who developed the first widely recognized periodic table in 1869 (Edwards et al., 2020; Kaji, 2003; Romanoski & Mendelejew, 1954). Mendeleev arranged the elements by increasing atomic weights and grouped them according to similar chemical properties. His table left gaps for yet undiscovered elements and predicted their properties, many of which were subsequently confirmed.

The core contributors to the modern Periodic Table were Henry Moseley (1887–1915) (Egdell et al., 2020), who demonstrated in 1913 that the periodic properties of elements were better explained by their atomic number rather than atomic weight. The atomic number is defined as the number of protons in the nucleus of an atom. This led to the modern definition of the periodic table, which was based on the increasing atomic number. Glenn Seaborg (1912–1999) and Chapman (2020) contributed to the discovery of transuranium elements (those beyond uranium) and reconfigured the periodic table to include the actinide series.

In summary, our current understanding of the atomic realm and the elements it consists of evolved from early philosophical concepts to a detailed scientific framework. The development of the Periodic Table proceeded with the International Union of Pure and Applied Chemistry (IUPAC) release in November 2016, which reported the permanent names for the four latest elements, encompassing nihonium (113), moscovium (115), tennessine (117), and oganesson (118) (Cartlidge, 2018). This evolving framework incorporates fundamental subatomic particles, which themselves are comprised of even smaller constituents that reside within the quantum domain.

6.1.2 Protons, Neutrons, and Electrons

Protons, neutrons, and electrons comprise the fundamental subatomic particles that constitute atoms.

Protons: Reside within the atomic nuclei, are positively charged and define elements (Antognini et al., 2013) (Table 6.2).

TABLE 6.2
Proton Characteristics

Properties	Composition	Function
Charge Positive (+1 elementary charge). **Mass** Approximately 1.6726×10^{-27} kg (about 1836 times the mass of an electron). **Symbol** p or p^+	**Quarks** Protons are a type of baryon comprised of three quarks (two up quarks and one down quark), which are held together by the strong force and mediated by gluons.	**Atomic Nucleus** Protons reside within the atomic nucleus and determine the atomic number, which defines the element. **Electromagnetic Interaction** The positive charges of protons attract negatively charged electrons and keep them in orbit around the nucleus.

TABLE 6.3
Neutron Characteristics

Properties	Composition	Function
Charge Neutral (no charge) **Mass** Approximately 1.6750×10^{-27} kg (slightly more massive than a proton). **Symbol** n or n^0	**Quarks** Neutrons are also a type of baryon comprised of three quarks (one up quark and two down quarks), which are held together by the strong force and mediated by gluons.	**Atomic Nucleus** Neutrons are found within atomic nucleus, contribute to the mass of an atom, and provide stability to the nucleus by offsetting the electrostatic repulsions between protons. **Isotopes** The number of neutrons in the nucleus determines the isotope of an element.

Source: Derived from Smith (2010).

***Neutrons*:** Reside within the nucleus, are neutrally charged and contribute to the mass and stability of atoms (Smith, 2010) (Table 6.3).

***Electrons*:** Orbit the nucleus, are negatively charged and involved in the formation of chemical bonds and electric current (Sturm et al., 2014) (Table 6.4).

These subatomic particles interact through fundamental forces: protons and neutrons via the strong nuclear force, and electrons through electromagnetic interactions, which create the complex structures and behaviors observed in matter.

6.1.3 Quantum Particles

The quantum realm is defined by particles and interactions that follow the principles of quantum mechanics, which govern the behaviors of matter and energy at the

TABLE 6.4
Electron Characteristics

Properties	Composition	Function
Charge Negative (-1 elementary charge). **Mass** ~9.109×10^{-31} kg (~1/1836 the mass of a proton). **Symbol** e or e^-	**Elementary Particle** Since electrons are fundamental particles, they are not comprised of smaller constituents. **Lepton** Electrons belong to the lepton family of particles.	**Atomic Structure** Electrons orbit the atomic nucleus at various energy levels or shells to form electron clouds. **Chemical Bonds** Electrons are involved in chemical bonding, including ionic and covalent bonds, as they are shared/transferred between atoms. **Electricity and Magnetism** Electrons are the primary carriers of electric current in conductors and play a crucial role in the phenomena of electricity and magnetism.

Source: Derived from Sturm et al. (2014).

TABLE 6.5
Fundamental Quantum Particles

Fundamental Particles			
Quarks	**Leptons**	**Gauge Bosons**	**Higgs Boson**
Properties Quarks are elementary particles that combine to form protons and neutrons. They come in six "flavors": up, down, charm, strange, top, and bottom. **Interactions** Quarks interact via the strong force and are mediated by gluons.	**Properties** Leptons are a family of fundamental particles that include electrons, muons, tau particles, and their corresponding neutrinos (electron, muon, and tau neutrinos). **Interactions** Leptons interact via the weak force and electromagnetic force (except for neutrinos, which do not carry an electric charge and only weakly interact).	**Properties** Gauge bosons are force-carrying particles that mediate the fundamental forces. **Types** **Photon** (γ) Mediates the electromagnetic force. **W and Z bosons** Mediate the weak force. **Gluon (g)** Mediates the strong force. **Graviton (hypothetical)** Predicted to mediate the gravitational force but has not yet been observed.	**Properties** The Higgs boson is associated with the Higgs field, which conveys mass to other particles through a spontaneous symmetry breaking mechanism. **Interactions** It interacts with other particles via the Higgs field.

smallest scales. Several classes of particles are fundamental to the quantum realm (Dieks, 2020, 2023; Jaeger, 2021) (Table 6.5).

6.1.3.1 Principles that Define the Quantum Domain

Wave-Particle Duality: Particles exhibit both wave- and particle-like properties. For example, electrons can form interference patterns (wave behaviors) but also collide with surfaces as discrete particles (Carnio et al., 2019; Zhang et al., 2021).

Quantization: Certain properties (e.g., atomic energy levels) are quantized, meaning that they can only take on (snap between) specific discrete values (Carosso, 2022).

Uncertainty Principle: Formulated by Werner Heisenberg, this principle states that certain pairs of physical properties (e.g., position and momentum) cannot be simultaneously known to arbitrary precision (Cowley, 1949; Eastwood, 2017).

Superposition: Particles can exist in a superposition of states, where they simultaneously occupy multiple states or positions until measured (Aubrun et al., 2022; evan der Wal et al., 2000; Procopio et al., 2015).

Entanglement: Particles can become entangled, meaning their quantum states are interdependent regardless of the distance that separates them. Measuring one particle instantly affects the state of the other (Aubrun et al., 2022; Gisin, 2019; Levonian et al., 2022; Nakata & Murao, 2020).

These particles and principles form the foundation of quantum mechanics, which describes the behavior of matter and energy at subatomic scales, fundamentally shapes our understanding of the quantum realm.

6.2 FUNCTIONALIZED ATOMIC/SUBATOMIC AGGREGATES ARE US

From the perspective of atomic realms and quantum domains, humans may be considered as functionalized, animated, consciousness-infused, self-assembled aggregates comprised of atoms (and their constituents) that were cast at the birth of the universe. The typical adult human body, weighing 70 kg (150 lb), is composed of ~7×10^{27} (~7 octillion) atoms of various species in precisely arranged orientations that form molecules, proteins, organelles, cells, tissues, and organs (Dove, 2023; Hersey, 2024) (Figures 6.2 and 6.3).

Approximately 99% of the mass of the human body consists of just six elements, including oxygen (O), carbon (C), hydrogen (H), nitrogen (N), calcium (Ca), and phosphorus (P), with another 0.85% of macro-mineral/metal elements/trace elements, encompassing potassium (K), sulfur (S), sodium (Na), chlorine (Cl), and magnesium (Mg), iron (Fe), manganese (Mn), cobalt (Co), copper (Cu), zinc (Zn), molybdenum (Mo), selenium (Se), and iodine (I), which are essential for plant, animal, and human life (Zoroddu et al., 2019). Additional elements (e.g., vanadium (V), nickel (Ni), bromine (Br), silicon (Si), and tin (Sn)) continue to be investigated for their potential designation as essential (Figure 6.4) (Da Silva Williams, 2001; Helmenstine, 2015; Kaim & Schwederski, 1994; Mertz, 1981; Williams, 1991). Consequently, at the most fundamental level, far deeper than our distinctive superficial appearances and traits, we are essentially identical.

As a thought experiment, if each of the various atomic species listed above were simply measured in bulk, poured into a vat, heated to body temperature, and thoroughly mixed, we would certainly not expect to produce a human being. Even if all these atoms could somehow be positioned and bound in place and oriented exactly where they should be, would we still not be left with only a lifeless and woeful mound of mud? Thus, the critical question emerges … what might the underlying organizing drivers, principles, conditions, and protocols be that have the capacity to self-assemble us so exquisitely and dynamically? Further, once all these atoms are properly fixed in place to make up our each and every component, at what precise juncture, and how, are we initially imbued with the spark of life/the life force … and subsequently endowed with the ability to tap into what may well be ambient universal consciousness (Barlow, 2015; Barnden, 2020; Craddock et al., 2014; Hameroff, 2001; Hameroff & Penrose, 2014; Hameroff et al., 2014; Kanai & Fujisawa, 2024; Lohrey & Boreham, 2020; Torday, 2023).

To provide a simpler example, three-cell-thick *Trichoplax adhaerens* is currently designated as the simplest animal on Earth. It is thinner than a sheet of paper and devoid of a front or back, has no veins, muscles, mouth, or stomach and essentially functions as a "hungry suction cup" (Fox, 2018). As relates to our thought experiment, if it were possible to exactly replicate all the atoms that reside

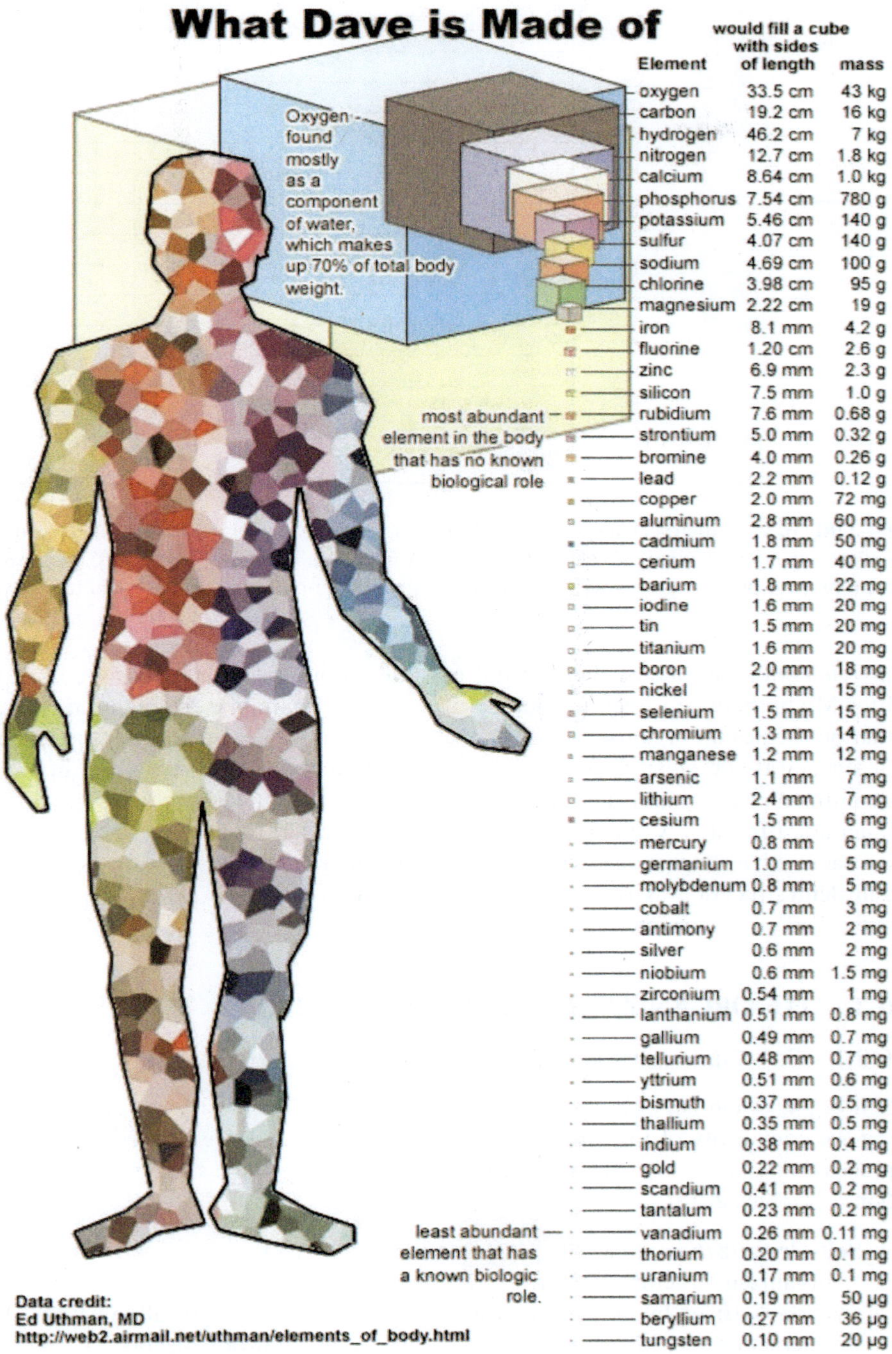

FIGURE 6.2 Composition of human body (atomic number and mass) with 56 elements shown at => 0.1 mg. (Data/Image credits: Ed Uthman, M.D.)

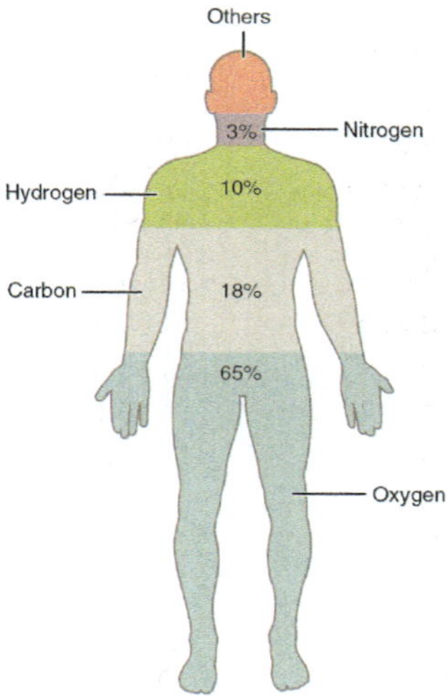

Element	Symbol	Percentage in Body
Oxygen	O	65.0
Carbon	C	18.5
Hydrogen	H	9.5
Nitrogen	N	3.2
Calcium	Ca	1.5
Phosphorus	P	1.0
Potassium	K	0.4
Sulfur	S	0.3
Sodium	Na	0.2
Chlorine	Cl	0.2
Magnesium	Mg	0.1
Trace elements include boron (B), chromium (Cr), cobalt (Co), copper (Cu), fluorine (F), iodine (I), iron (Fe), manganese (Mn), molybdenum (Mo), selenium (Se), silicon (Si), tin (Sn), vanadium (V), and zinc (Zn).		less than 1.0

FIGURE 6.3 Primary elements of the human body from the highest abundance to the lowest. OpenStax College (Creative Commons Attribution 3.0) https://openstax.org/books/anatomy-and-physiology/pages/2-1-elements-and-atoms-the-building-blocks-of-matter.

in *Trichoplax*, as well as their spatial positions, would it spontaneously come to life? Might there be something "other" from an (e.g., omnipresent quantum ether) that permeates an organism with life vitality once it exceeds a specific threshold of structural complexity? Alternately, might all atoms be inherently pre-infused as infinitesimal components of the life force, which cumulatively transition to a bona fide life force when precisely configured to be self-replicating/sustaining to interact with a supportive environment? Even a modicum of insightful elucidation and demystification as relates to these queries will be ample reward for having embarked on this chapter.

6.3 SELF-ORGANIZATION/ASSEMBLY STRATEGIES

6.3.1 Self-Organization/Assembly in Living Matter

Self-organization and self-assembly provide powerful frameworks for understanding the complexity of life and cast light on the very emergence of life itself. Both processes explain how collective order develops from "dynamic small-scale interactions" (Tibbits, 2013).

Self-Organization: Refers to a process where a system spontaneously forms organized structures or patterns without external guidance. This phenomenon is driven by the system's internal dynamics, which typically involves interactions between its components. Early on, self-organization in living tissues was described as the physical interactivities of molecules in a steady-state structure (Nicolis & Prigogine, 1977).

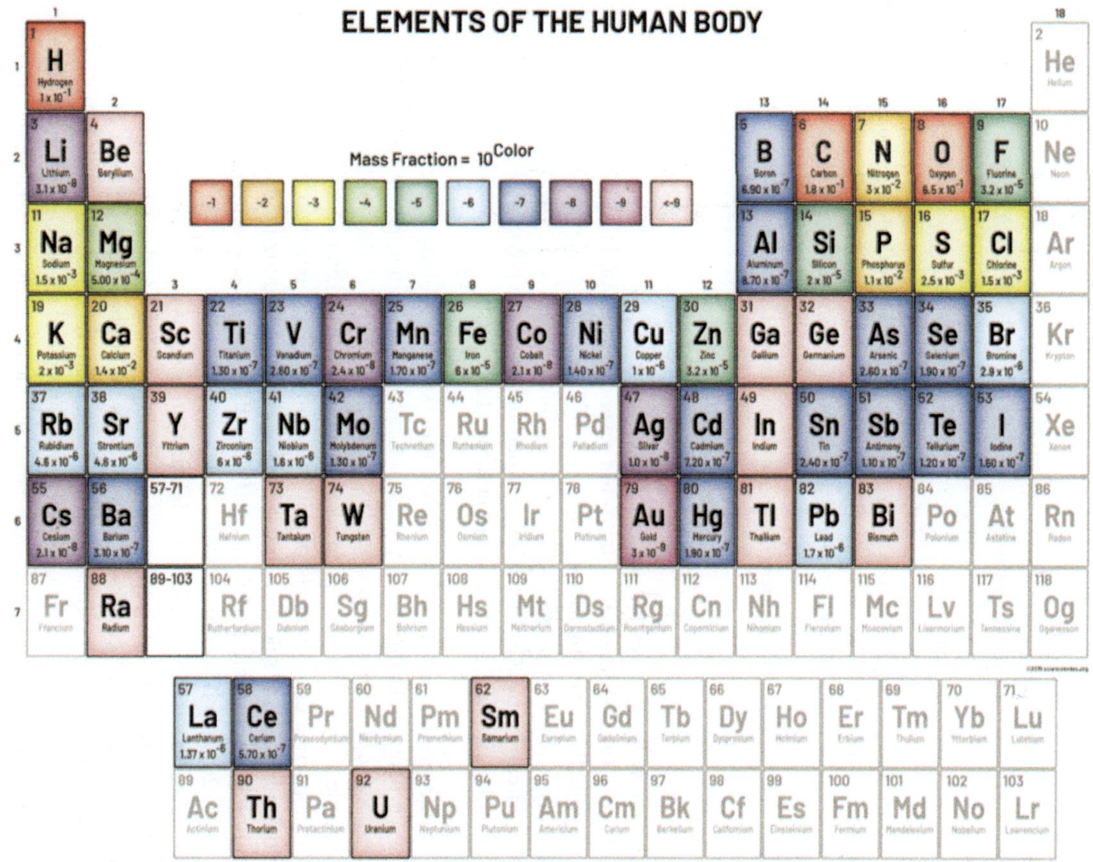

FIGURE 6.4 Periodic table of elements essential for life. (Reproduced from Helmenstine, 2015.)

TABLE 6.6
Key Features of Self-Organization and Self-Assembly

	Self-Organization
Localized Interactions	Components interact locally and follow simple rules, leading to complex overall behaviors.
Emergence	Complex structures and behaviors emerge from the collective dynamics of the system's components.
Adaptability	Self-organizing systems can adapt to changing environments by reorganizing their structures.
	Self-Assembly
Specifically Directed Interactions	Components and sub-components have specific affinities for each other, which facilitates their assembly.
Thermodynamic Stability	The resulting structures exist in a lower energy state, which gives them stability.
Scalability	Small-scale interactions can translate to large-scale structures.

Self-Assembly: Involves the autonomous organization of components into stable structures through specific, often reversible, interactions. This process is typically driven by the minimization of free energy, which can result in highly ordered structures. The self-assembly of molecules was acknowledged earlier by Kushner (1969), which involved the physical aggregation of molecules into equilibrium structures (Table 6.6).

The concepts of self-organization/assembly suggest that the generation of life forms is a process that is driven by precise interactions and emergent behaviors. From this perspective, life may have emerged when myriad molecules in the primordial soup exceeded a certain level of complexity and self-organized under specific conditions/in supportive environments into living entities. This extends Darwin's theory of evolution by natural selection and proposes that self-organization, selection, and probability are biospheric engines (Kauffman, 2014, 2020; Roli & Kauffman, 2020; Vasas et al., 2012; Wilds et al., 2008).

This view also has the potential to redefine what we consider as life, encompassing a more extensive array of both biological and artificial systems and phenomena. A closer examination of localized cell-to-cell interactions highlights the significant roles that self-organization and self-assembly play in living matter, particularly in highly developed organisms. In recent decades, numerous examples of these interactions – underpinning the strategies described above – have been identified and studied, revealing the remarkable extent to which they govern the fundamental processes of life (Athanasiou et al., 2013; Mueller et al., 2020; Xavier da Silveira Dos Santos & Liberali, 2019).

6.3.1.1 Adhesion Molecules

Cells adhere to each other using proteins such as cadherins and integrins (Dieterle et al., 2021; Weber et al., 2011), which enables them to tessellate and communicate.

This adhesion is vital for the development and maintenance of tissue structures. For example, epithelial cells adhere tightly to form continuous sheets of tissue.

Gap Junctions are channels that enable direct communications between neighboring cells, through which ions and small molecules may pass to enable coordinated cellular responses, such as the synchronized contractions of cardiac muscle cells (Kléber, 2011; Nielsen et al., 2012).

6.3.1.2 Cell-Matrix Interactions

***Extracellular Matrix (ECM) Binding*:** Cells interact with the extracellular matrix (ECM) (Karamanos et al., 2021; Theocharis et al., 2016), which is a network of proteins and polysaccharides that provides structural and biochemical support to surrounding cells. Cell surface resident integrins bind to ECM components (e.g., collagen or fibronectin), which influence cell shapes, migration, and differentiation.

***Mechanical Feedback*:** Cells have the capacity to sense and respond to the stiffness or elasticity of the ECM, which influences processes such as stem cell differentiation or tissue morphogenesis. This mechanotransduction is a type of localized interaction that guides self-organization based on the mechanical properties of the environment.

6.3.1.3 Chemical Signaling

***Diffusion Gradients*:** Cells release signaling molecules (morphogens) that diffuse through tissues to create concentration gradients and respond differently contingent on their position within the gradient, which translates to pattern formation. For example, during early embryonic development, gradients of morphogens such as BMP or Wnt (Henríquez et al., 2011; Yam & Charron, 2013) mediate the spatial organization of tissues.

***Autocrine and Paracrine Signaling*:** During autocrine signaling (Segers & De Keulenaer, 2021), cells respond to signaling molecules that they generate themselves, which reinforces certain behaviors. For paracrine signaling (Roy & Kornberg, 2015), cells secrete signaling molecules that influence others in proximity to coordinate their activities for wound healing or organ development.

6.3.1.4 Mechanical Forces and Tension

***Cytoskeletal Dynamics*:** The cytoskeleton (comprised of actin filaments, microtubules, and intermediate filaments) exerts forces within and between cells, which contribute to cell shapes, division, and movement (Geisterfer et al., 2021; Goldstein & Theriot, 2001; Pollard & Goldman, 2018). For instance, during mitosis, mitotic spindles exert forces to segregate chromosomes.

***Cellular Tension and Contractility*:** Cells generate tension through actomyosin networks, which can pull on neighboring cells or the ECM. This mechanical tension assists with tissue morphogenesis, such as the folding of epithelial sheets during embryonic development (Heer & Martin, 2017; Kattan et al., 2017).

6.3.1.5 Feedback Loops

Positive Feedback: In a positive feedback loop, an initial signal is amplified, which translates to a stronger response. For example, during blood clotting, the activation of one clotting factor triggers the cascading activation of others to rapidly form a clot (Green, 2006; Smith, 2009; Smith et al., 2015).

Negative Feedback: Negative feedback loops stabilize biological processes by reducing the outputs in response to an initial signal. For instance, if the body temperature increases, mechanisms are activated for its regulation by cooling the body down (Pigolotti et al., 2007; Prockop, 2013).

6.3.1.6 Cellular Automata and Localized Rules

Rules-Based Behaviors: Cells or molecules often observe simple rules based on localized conditions, which can translate to complex outcomes. For example, cells within a developing embryo might divide, become mobile, or differentiate based on their immediate environments, which results in the orderly development of tissues and organs (Lehotzky & Zupanc, 2019).

6.3.1.7 Chemical Reaction and Diffusion Systems

Reaction-Diffusion Systems: These systems involve interactions between substances that chemically react and spatially diffuse. A classic example is the Turing model, where activator and inhibitor molecules interact to create stable patterns such as spots or stripes on animal skin (Epstein & Xu, 2016; Krause et al., 2021; Nguindjel et al., 2022).

Enzyme Kinetics: Localized concentrations of substrates and enzymes can lead to spatial organization within cells. For example, certain metabolic pathways are localized within specific cellular compartments that optimize the efficacy of biochemical reactions (Gibson et al., 2021; Kastritis & Gavin, 2018; Midha et al., 2024).

6.3.1.8 Genetic Regulation and Gene Networks

Gene Regulatory Networks: Genes do not operate in isolation; they interact in networks where the expression of one gene influences the expressions of others. These exchanges can lead to self-organizing behaviors, such as the formation of gene expression patterns during development (McDonald & Reed, 2023; Wang & Wang, 2022).

Epigenetic Modification: The localized chemical modification of DNA or histones can influence gene expression, which contributes to cellular differentiation and tissue-specific patterns (Mishra et al., 2023; Sun et al., 2022; Zhao et al., 2020).

6.4 MYRIAD ROLES OF WATER

Water plays essential roles in the self-organizing processes of biological systems, including cells and tissues. The human body contains a significant quantity of water, with the approximate percentages in various tissues shown in Table 6.7. The unique physical and chemical properties of water are fundamental to the self-organization

TABLE 6.7
Percentages of Water in Tissues of the Human Body

Blood	~80%
Striated muscle	70.09%
Brain, spinal cord, nerve trunks	75.00%
Skin	57.71%
Liver	71.58%
Kidneys	70.58%
Heart	62.95%
Lungs (congested with blood)	77.28%
Spleen	78.79%
Pancreas	73.08%
Alimentary tract	77.40%
Skeleton	28.17%
Teeth	5.00%
Adipose tissue	23.02%
Average (weighted) All tissues	55.74%

Source: Derived from Forbes et al. (1953) and Beilin et al. (1966).

and functionality of biological systems (Giannetti et al., 2025; Schirò & Weik, 2019; Wiggins, 1990), which enable the formation, stability, and functionality of cellular and tissue structures. The primary contributions of water to self-organization include:

Solvent Properties: Water is a universal solvent, which means that it can dissolve an extensive range of substances. This confers the capacity for it to facilitate biochemical reactions and transport molecules (e.g., nutrients, ions, and metabolic waste products) within and between cells.

Hydration Shells: Water molecules surround charged or polar molecules to form hydration shells. This interaction assists with maintaining the structures and functions of proteins and nucleic acids by stabilizing their conformations and influencing their interactions (Laage et al., 2017; Penkova et al., 2021).

Cellular Organization: Water contributes to the organization of cellular components. For example, the aqueous environment within cells helps to stabilize and organize macromolecules such as proteins and nucleic acids into functional structures, such as enzymatic complexes and ribosomes (do Nascimento Vieira et al., 2020; Mardt et al., 2022; Vu et al., 2021).

Membrane Formation: The self-assembly of phospholipid bilayers, which form cell membranes, is driven by the hydrophobic effect. In an aqueous environment, the hydrophobic tails of phospholipids aggregate to avoid contact with water, while their hydrophilic heads interact with the surrounding water. This process leads to the formation of stable bilayer structures that are critical for cell integrity and functionality (Disalvo et al., 2015; Loitto et al., 2009).

Tissue Structure and Function: In tissues, water assists with maintaining the structure and function of the ECM. Water interacts with the ECM components, affects their mechanical properties, and facilitates cell-cell and cell-matrix interactions (Frantz et al., 2010; Georgi et al., 2019).

Thermal Regulation: Water has a high specific heat capacity, which helps it to regulate temperature within cells and tissues. This thermal stability is vital for maintaining the proper conditions for enzymatic reactions and overall cellular functions (Sladek & Johnson, 2013).

Dynamic Processes: The ability of water to form hydrogen bonds contributes to the dynamic nature of biological systems. This property enables water to participate in transient interactions and conformational changes in biomolecules, which are essential for processes like protein folding (Levy & Onuchic, 2006; Vajda & Perczel, 2014), enzyme catalysis (Ahsan et al., 2021; Pocker, 2000), and molecular signaling (Sharkey & Mawe, 2023).

In summary, these localized interactions (whether chemical, mechanical, or genetic) form the basis of self-organization/self-assembly in biological systems. Through these interactions, cells and molecules communicate, adapt, and organize into the complex structures and patterns that are the prerequisite for life. The emergent behavior that results from these interactions is critical for the development, maintenance, and functionalities of living organisms, which consequently reframes what the traditional definition of life is.

6.4.1 A Broader Perspective of Conventional Definitions of Life and Their Implications

Conventional definitions of life include attributes like metabolism, growth, reproduction, and responses to stimuli. However, self-organization and self-assembly challenge and broaden this definition by highlighting:

Emergent Complexity: Life might be understood as an emergent property of complex, dynamic systems rather than a set of specific, stand-alone biochemical processes (Weber, 2010).

Minimalist Life Forms: Synthetic biology and the creation of minimal cells or protocells reveal that life-like properties can emerge in simpler systems through self-organization and self-assembly (Samanta et al., 2024).

Non-Biological Life: The principles of self-organization and self-assembly suggest that life-like systems might exist in non-biological contexts, such as artificial life or even life on other planets, formed through distinct chemical pathways (Lawsin, 2023).

Flexibility: What it means to be alive might become more flexible, which may focus on the principles of organization and dynamics rather than specific molecular compositions (Skene, 2024).

Artificial Life: The creation of self-organizing/self-assembling systems in the lab blurs the line between living and non-living entities, which suggests that life could be an emergent property of certain systems under explicit conditions regardless of their origin (Baltieri et al., 2023; Gershenson, 2023; Mukhopadhyay, 2024).

Universal Life: A broader definition could accommodate diverse forms of life, potentially recognizing life forms that do not align with the Earth-centric biological mold. However, within certain parameters based on the constituents of matter laws of physics, there may be a universality to the attributes of life as we know them on Earth (Benner, 2010; Gómez-Márquez, 2021). As touched on above, even when all components are arranged in perfect order through self-organization and self-assembly to form living matter, many traditional cultural perspectives hold that a vital life-giving force must be bestowed for true life to emerge. Investigating progressively higher thresholds of structural complexity may offer insights into this viewpoint and its relation to the life force concept.

6.5 DEMARCATION OF PROGRESSIVELY HIGHER STATE STRUCTURAL COMPLEXITY THRESHOLDS CORRELATED WITH FUNCTIONALITY

6.5.1 Distinguishing Features and Perceived Threshold Between Non-Living and Living Entities Infused with the Life Force

The threshold and distinguishing features between non-living and living entities that are infused with what is typically referred to as the "life force" involve several key factors. Even though the concept of a life force is more commonly found in philosophical and spiritual contexts rather than mainstream science, we can explore how living entities are differentiated from those that are non-living based on conventional, observable, and scientific characteristics, which are often linked to their complexity. A few of the primary distinctions are listed below:

Organization and Complexity: Living entities exhibit a high degree of structural organization and complexity at atomic, molecular, cellular, and systemic levels. However, non-living entities, while they may possess complex structures, do not exhibit the same level of organization required for life functions.

Metabolism: Living organisms have metabolic processes that convert energy and matter to maintain their structure, growth, and reproduction. These processes include biochemical reactions such as respiration and digestion. Conversely, non-living entities do not possess metabolism or energy conversion systems.

Homeostasis: Living organisms maintain a stable internal environment through homeostasis, which involves several regulating factors (e.g., temperature, pH, and ion concentrations), while non-living entities do not have the capacity to self-regulate or maintain internal stability.

Growth and Development: Living entities have the capacity to grow and develop according to genetic instructions, which undergo specific stages of development from birth to maturity, whereas non-living entities do not grow or develop in this manner.

Reproduction: One of the defining features of living entities relates to their ability to reproduce and transfer genetic material to their offspring. Living biological organisms reproduce either sexually or asexually, while non-living entities do not reproduce, albeit they (e.g., crystals) can be designed to mimic the process (Kozlovskaya & Kharlampieva, 2022).

Responses to Stimuli: Living organisms respond to external stimuli through changes in their internal states or behaviors, while non-living entities do not react to stimuli in a way that involves biological processes.

Adaptation and Evolution: Living organisms can adapt to their environments over time through evolutionary processes, through changes in their genetic material across generations. Non-living entities do not evolve in this sense.

6.5.1.1 Distinguishing Features of the Life Force from a Philosophical or Spiritual Perspective

Vital Energy: In many philosophical and spiritual traditions, the "life force" or "vital energy" (e.g., chi, prana, or spirit) is believed to be the animating principle that distinguishes living from non-living entities. This force is thought to be responsible for vitality, consciousness, and the dynamic nature of living beings.

Consciousness and Awareness: Some traditions associate the life force with consciousness and awareness. They argue that living entities imbued with the life force possess a form of awareness or consciousness, which non-living entities are devoid of.

Spiritual or Holistic Perspective: From holistic or spiritual perspectives, the life force is considered as integral to the wholeness and interconnectedness of living systems. It is believed to be essential for harmony and balance within an organism, and as it relates to its environment.

Healing and Wellness: Many traditions believe that the life force is linked to health and wellness. Practices including acupuncture or energy healing aim to balance and enhance the life force to promote well-being.

From a scientific viewpoint, the distinction between non-living and living entities is based on measurable characteristics with high level of complexity, as listed above. From philosophical and spiritual perspectives, the life force is considered to be a universal non-material energy that permeates and animates all living entities regardless of their complexity, or even the presence of a brain.

While the cognitive faculties of sentient beings accommodate and enable more complex expressions of consciousness (e.g., thought, self-speak, and self-awareness), the life force itself is not contingent on the presence of a brain. Thus, it operates even within non-brained organisms such as plants, microbes, and the simplest of organisms (e.g., *Trichoplax adhaerens*). Certain spiritual traditions consider the brain as a channel or receiver of this energy rather than its source (Harp, 2012; McFadden, 2020; Mocombe, 2022). This alludes to an omnipresent consciousness continuum, manifest in myriad forms and intensities that spans all forms of life. From this view, the brain may serve to refine and focus consciousness and the life force, rather than generate them. Distilling this further, it may indeed be the case that consciousness and the life force are one in the same (Bernet, 2010; Dunne & Jahn, 2018; Schurger & Graziano, 2022).

In contrast to these philosophical and spiritual perceptions, scientific research increasingly points to the brain as the locus – or at least a critical enabler – of consciousness (Kitchener & Hales, 2022; Sarasso et al., 2021; Young & Pigott, 1999). To explore these seemingly divergent views, we delve deeper into the distinguishing features of brain-endowed versus non-brain organisms and re-examine the very definition of consciousness. This analysis aims to lay the groundwork for further discussion on the structural and functional thresholds that might be correlated with the emergence of consciousness and the life force (Kitchener & Hales, 2022; Sarasso et al., 2021; Young & Pigott, 1999).

6.5.2 Distinguishing Features and Perceived Threshold Between Living Non-Brained and Brain-Endowed Organisms

The distinction between brain-endowed and non-brained organisms primarily relates to the complexity of their nervous systems and capacity to process information, control behaviors, and adapt to their environments (Table 6.8).

TABLE 6.8
Key Thresholds and Distinguishing Features between Brain-Endowed and Non-Brained Organisms

	Brain-Endowed Organisms	Non-Brained Organisms
Nervous System Complexity	Host a centralized nervous system, where the brain serves as the command center for the processing of sensory information, making decisions, and controlling complex behaviors. Examples encompass vertebrates like mammals, birds, and reptiles.	Typically possess a decentralized nervous system or a simple network of neurons (e.g., jellyfish and hydra). They rely on localized networks of nerve cells to coordinate actions, such as movement or basic responses to stimuli.
Behavioral Complexity	Exhibit relatively sophisticated behaviors, including learning, memory, and decision-making. They can engage in more adaptive and complex actions, like problem-solving and social interactions.	Exhibit relatively crude reflexive behaviors. Their actions are often stereotypical and driven by immediate environmental cues with little capacity for learning or memory.
Adaptability and Learning	Have a high capacity for learning and adaptation, allowing them to modify their behavior based on past experiences or environmental changes.	Have a limited ability to adapt to new situations. They generally respond in a fixed way to specific stimuli.
Sensory Processing	Sensory information is centrally processed in the brain, allowing for integration from multiple senses and more complex perception.	Sensory information is processed locally or distributed throughout the organism's body (e.g., in the nerve nets of jellyfish).
Motor Control	Fine motor control, coordinated movements, and more intricate actions are possible, including voluntary and planned movements.	Simple, often repetitive movements are controlled by reflex arcs or local neural circuits.

In summary, the thresholds between brain-endowed and non-brained organisms reside in the complexities of information processing, behavior, and adaptability. Brain-endowed organisms exhibit centralized control over their bodies (or parts of them as in the case of the octopus) (Nödl et al., 2015; Sivitilli et al., 2022; Olson & Ragsdale, 2023), along with advanced sensory processing and the ability to learn and adapt. In contrast, non-brained organisms depend on simpler, decentralized nervous systems and largely reflexive behaviors.

6.5.2.1 Key Differences between Cognitive Thresholds

Simple brain-endowed creatures possess the basic neural structures required to process information and react to stimuli. However, they may not experience subjective feelings or have the higher cognitive functions that define sentience, consciousness, or self-awareness.

Sentience comprises the first threshold of subjective experience. While all brained creatures can process stimuli, sentient beings can actually *feel* and experience the world, which is often linked to more complex brain functions like those of vertebrates and some invertebrates.

Consciousness adds a further layer of cognitive complexity, which enables beings to not only feel and react, but to integrate sensory data into a unified experience of their environments/the world, to function within it, facilitated by a degree of understanding and decision-making.

Self-awareness demarcates the highest cognitive threshold, where beings not only perceive the world and themselves but can reflect on their own thoughts, plan for the future, and recognize/ponder their distinct individuality.

6.5.2.2 Evolutionary Perspective

From an evolutionary perspective, these thresholds likely progressively evolved as the complexity of nervous systems and brains deepened. A simple brain can support basic survival functions; however, more complex brains (e.g., those with specialized regions for memory, social interactions, and planning) appear to be the prerequisite for consciousness and self-awareness. Thus, the distinction between brain-endowed creatures and sentient, conscious, self-aware beings is likely manifest in the degree of neural complexity and depth of cognitive processing. Sentient, conscious, and self-aware beings require more specialized and integrated neural architectures to support subjective experiences and introspective thought.

6.5.2.3 Consciousness

Consciousness is a complex and multifaceted concept that, from a non-philosophical or non-spiritual perspective, often refers to the state of being aware of and able to ponder one's own existence, thoughts, and surroundings (Kanaev, 2022; Konen & Zirlinger, 2024; Kuhn, 2024; Mashour, 2024).

This definition encompasses several key aspects:

Awareness of internal states (thoughts, feelings) and external stimuli (sensory inputs) allows individuals to perceive, interpret, and respond to their environments.

Self-Awareness is the ability to recognize oneself as an individual who is distinct from others and involves the reflection on one's own mental state and understanding one's own identity.

Experience includes the capacity to have subjective experiences or qualia, which are the personal, internal experiences of sensations, emotions, and thoughts that are unique for everyone.

Attention often involves the ability to focus attention on specific stimuli or thoughts while filtering out others. This selective focus assists with the processing of relevant information and arriving at decisions.

Intentionality is the capacity for intentional thought and action, which enables individuals to set goals, make plans, and purposefully direct their behavior.

Integration of various aspects of experience into a coherent narrative or stream of thought. This integration facilitates the creation of a continuous sense of self and understanding of one's place in the world and universe. These key aspects of consciousness help catalyze the expansion of other scientific and philosophical disciplines.

Psychology: Psychologists examine consciousness in terms of cognitive processes, including perception, memory, and attention. They investigate how conscious and unconscious processes influence behavior and mental states (Bayne et al., 2024; Walach et al., 2005).

Artificial Intelligence: In the realm of AI, researchers consider whether machines might ever truly be imbued with consciousness, or merely sentience as a simulative aspect, and what the implications would be for a computer/humanoid robot to be conscious (Aru et al., 2023; Baluška & Reber, 2019; Chella, 2023; Hildt, 2019, 2023; McFadden, 2023).

Philosophy: Philosophers explore the nature of consciousness, its origins, and its relationship to the mind and body. Key questions include the nature of subjective experience and the "hard problem" of explaining how and why consciousness arises from physical processes (Kozuch, 2024; León & Zahavi, 2023).

6.5.2.4 Seat of Consciousness

Overall, consciousness remains one of the most profound and challenging areas of study, as it touches on fundamental questions relating to the nature of reality, identity, and the human experience. The "seat of consciousness" is a concept that refers to the physical or neurological core of/basis for consciousness (Turner, 1955). While no single location in the body has been definitively identified as such, current scientific understanding places the basis of conscious experience primarily in the brain, specifically in several key regions:

Cerebral Cortex: The cerebral cortex, particularly the regions involved in higher-order functions, is critical for the functionality of consciousness. Areas such as the prefrontal cortex (involved in decision-making, self-awareness, and executive functions) and the parietal cortex (involved in sensory perception and spatial awareness) play significant roles in conscious experience (Eickhoff et al., 2018; Fernández et al., 2016; Gogolla, 2017; McConnell, 1995; Swindale, 1990).

Thalamus: The thalamus serves as a relay station for sensory and motor signals to the cerebral cortex. It is involved in regulating states of consciousness and wakefulness. Damage to the thalamus can result in altered states of consciousness (Habas et al., 2019; Schlag-Rey & Schlag, 1989; Taber et al., 2004).

Brainstem: The brainstem, including the reticular activating system, regulates wakefulness and alertness. It is essential for maintaining a fundamental level of consciousness and arousal (Angeles Fernández-Gil et al., 2010; Cai et al., 2024; Smith & DeMyer, 2003).

Default Mode Network (DMN): This network of brain regions, which includes portions of the medial prefrontal cortex, posterior cingulate cortex, and the angular gyrus, is active when the mind is at rest and not focused on the external environment. The DMN is thought to be involved in self-referential thoughts and the sense of self (Menon, 2023; Smallwood et al., 2021; Yeshurun et al., 2021).

Global Workspace Theory: According to this theory, consciousness arises from the interactions between multiple brain regions working conjunctively to integrate information into a unified experience. This suggests that consciousness is not localized to a single site, but rather emerges from the collective activity of various brain regions (Baars, 2005; Baars et al., 2021; Black, 2020).

Integrated Information Theory: This theory posits that consciousness corresponds to the ability of a system to integrate information, which emphasizes the importance of the complexity and connectivity of neural processes across different brain regions. Overall, consciousness is understood as the product of complex interactions within the brain's neural networks rather than being localized to a single "seat." This involves the coordinated activities of multiple brain regions that work cooperatively to produce awareness, perception, and self-reflection (Desmond & Huneman, 2022; Mediano et al., 2022; Tononi et al., 2016).

6.5.3 Threshold and Distinguishing Features between Sentient Conscious Self-Aware Beings and the Unified Field: Quantum Universal Consciousness/Intelligence

The boundary between individually conscious self-aware beings and a perceived unified universal or quantum consciousness demarcates a profound shift in the definition of the nature of consciousness itself. This, as it transitions from localized residence and individual experience within a biological substrate to a holistic, nonlocalized decentralized form that is often associated with speculative scientific and philosophical frameworks.

6.5.3.1 Localized Consciousness

Definition: Arises from neural networks in biological beings, such as humans and animals, which enable subjective experience, self-awareness, and environmental interactions.

***Characteristics*:** Localized within the brain or equivalent structures. Dependent on biological processes. Bound by physical limits (e.g., time and space).
***Mechanism*:** Seen as an emergent property of complex neural activity.

6.5.3.2 Unified Quantum/Universal Consciousness

***Definition*:** Non-localized consciousness that is possibly fundamental to reality, which transcends individual organisms and biological systems.
***Characteristics*:** Not confined to physical boundaries or individual brains. Speculatively tied to phenomena such as quantum entanglement (Silvestri et al., 2024; Tamulis & Grigalavicius, 2014) or aforementioned wave-particle duality. Aligns with metaphysical ideas like cosmic consciousness or the "ground of being" (Wildman, 1994–2025).

6.5.3.3 Theories and Philosophical Parallels

***Eastern Philosophy*:** Concepts like Brahman (Chaudhuri, 1954) or Sunyata (Barua, 2016) describe a unified awareness beyond the self.
***Quantum Mind Theories*:** Suggest that consciousness might emerge from quantum processes in the brain (e.g., via microtubules) (Avila et al., 2024; Craddock et al., 2014, 2015; Hameroff, 2022; Hameroff et al., 2014; Hameroff & Penrose, 2014) or exists as a dynamic universal unified field (Evans, 2005–2006; Zhang & Zou, 2025).

6.5.3.4 Perceived Threshold

The perceived transition between these manifested forms of consciousness might involve the following attributes:

***Localization vs. Non-Localization*:** Individual consciousness is confined, while universal consciousness is expansive and interconnected.
***Materialist vs. Non-Materialist Views*:** Neuroscience attributes consciousness to neural activities confined to the brain or other physiological substrates, while quantum theories suggest it as a fundamental property of the universe.
***Emergent vs. Fundamental Nature*:** Biological consciousness emerges from complexity, whereas universal consciousness is inherent to existence.
***Temporal Boundaries*:** Individual consciousness is limited by lifespans, while universal consciousness exists beyond temporal and spatial domains.

6.5.3.5 Speculative Correlations

***Quantum Mysticism*:** Explores consciousness as an aspect of quantum reality.
***Panpsychism*:** Suggests all matter has a form of consciousness, implying a universal unified field.
***Holographic Universe*:** Proposes the universe as a reflective system, with consciousness being potentially intrinsic to its structure.

6.5.3.6 Deduction

Traversing this consciousness threshold boundary represents an interfacial transition from localized, biological consciousness frameworks to speculative notions of an interconnected cosmic awareness. While universal conscious intelligence remains unproven, it opens possibilities for the elucidation of consciousness as an integral part of the universe's fabric, which transcends localized limits.

6.5.4 Connectedness and the Emergence of Higher Organisms

The notion that energetic connections between humans, animals, and non-living entities (e.g., cities or infrastructure) might manifest as a higher organism aligns with systems theory and concepts like the Gaia hypothesis, panpsychism, and biocentrism. These frameworks suggest that interconnected networks of living and non-living entities may contribute to a collective consciousness or unified system.

6.5.4.1 Gaia Hypothesis

Proposed by James Lovelock, the Gaia hypothesis views Earth as a self-regulating organism, where interactions between living beings and non-living systems maintain conditions necessary for life. Electromagnetic fields might serve as a medium for these interactions, linking organisms and environments in a dynamic, self-sustaining system (Doolittle, 2017; Free & Barton, 2007; Gajić, 2011; Lovelock, 1989; Lovelock & Margulis, 1974).

6.5.4.2 Electromagnetic Fields as a Unifying Medium

Humans, animals, and plants generate electromagnetic fields that may interact with their environments, creating networks of energy. In urban settings, these fields could form collective energy webs that influence the "vibe" of cities and contribute to their bioenergetic signatures. Smart cities, with their interconnected technologies, may further amplify this effect, evolving into higher-order systems with dynamic sensory and regulatory functions.

6.5.4.3 Superorganism Theory

Superorganisms, akin to ant colonies or beehives, demonstrate how individual components can function collectively as a single entity. Similarly, urban environments, with their dense populations and complex infrastructures, could evolve into superorganisms, where humans, animals, and technologies evolve integrated cooperative networks (Chang et al., 2021; Rock & Wills, 2025; Wilson & Sober, 1989).

6.5.4.4 Collective Consciousness and Electromagnetic Fields

Speculatively, electromagnetic waves serve as the carriers of information, which foster a collective consciousness. Dense electromagnetic fields in urban centers, resulting from human activities, technology, and communication networks, might synchronize individual minds and energy fields, which may influence societal behaviors and decisions (Hunt et al., 2024; Kiouvrekis et al., 2024; McFadden, 2002).

6.5.4.5 Technological Amplification

Modern technologies amplify electromagnetic connectivities between living beings and non-living entities. As human societies integrate more intimately with technology, this interconnectedness might lead to a hybrid cybernetic organism, where human energy fields merge with technological systems to create higher-order intelligences (Chemingui et al., 2024; Degtyarevskaya & Vokhmintsev, 2024; Ismail & Azis, 2024; Lai & Levitt, 2023).

6.5.4.6 Panpsychism and Universal Consciousness

Panpsychism posits that consciousness is a fundamental feature of all matter, which suggests that even inanimate objects contribute to a universal consciousness. Electromagnetic fields might serve as a medium for distributing this consciousness across interconnected networks of living and non-living components (Butler, 2010; Chalmers, 2003; Strawson, 2016).

6.5.4.7 Global Connectivity/Collective Mind

As societies become increasingly globally interconnected, the notion of global consciousness emerges; a collective mind awareness influenced by shared electromagnetic fields, technologies, and experiences. Concepts like Teilhard de Chardin's noosphere describe this as a "thinking layer" around Earth, driving the evolution of a planetary or cosmic consciousness (Čartolovni, 2017; de Chardin, 1955; Vidal, 2021; Zybailov et al., 2024). Recently, one of the co-authors (Boehm) and a consortium of co-authoring researchers, has proposed that a hypothetical (for now) nanomedically enabled brain/cloud interface may (contingent on its ensured safety, security, tech-trust, and appropriate level of acceptability/implementation) evolve as a personalized cognitive prosthetic through which any individual on the planet (and perhaps, in the further future, even beyond it) might engage with an "always on" global mind (Martins et al., 2019).

6.5.4.8 Deduction

The idea that energetic connectedness might give rise to higher organisms is speculative yet rooted in scientific and philosophical precepts. This hypothetically interconnected system, characterized by energetic coherence and global consciousness, may represent an evolution toward a more integrated, synchronized future for life on Earth.

6.6 BEYOND THE HUMAN SUBSTRATE: AMBIENT RESONANCE, ENERGY/UNIFIED CONSCIOUSNESS FIELDS, TRANSCENDENCE AND LOVE

How can we begin to comprehend the possibility that individuals might interconnect with, access, or participate in a broader cosmic consciousness? From the perspective that humans are functionalized aggregates of atomic and subatomic particles; vibrating in harmony with the universe at large, and comprised of the same fundamental elements, it follows that our biological systems may synergistically resonate with broader energetic patterns. Moreover, the human body generates electromagnetic fields (EMFs) through physiological processes driven by the ionic activities of cells, tissues, and organs, which suggests a potential mechanism for interactions with the ambient environment on an energetic level. This raises the possibility that electromagnetic phenomena may play a role in mediating our far more intimate connection with the world, and perhaps even with the cosmos itself.

6.6.1 Mechanisms of EMF Generation

Nerve impulses create electric fields through neuronal action potentials that are quantifiable through an electroencephalogram (EEG) (Müller-Putz, 2020). The heart generates the body's most potent EMF, which is detectable using an electrocardiogram (ECG) (Sporton, 2001). Further, muscle contractions generate fields that can be observed through electromyography (EMG) (Dondelinger, 2010). Cellular resting potentials form weak, steady bioelectric fields, whereas thermal emissions unrelated to electrical activities, interact with electromagnetic spectra.

***Medical Measurements and Interactions*:** Diagnostic techniques like EEG, ECG, EMG, and magnetic resonance imaging (MRI) leverage various electrophysiological and nuclear magnetic resonance fields, respectively, to assess bodily functions. Bioelectric fields also interact with environmental EMFs, which influence natural processes such as tissue repair and healing.

***Bioelectric Fields in Healing*:** Injury currents guide keratinocytes and immune cells in wound healing. The piezoelectric properties of bones stimulate osteoblast activities and fracture repair (Khare et al., 2020). Bioelectric gradients direct axonal regrowth and synaptic repair in neural recovery (Qin et al., 2023). Further, electric fields attract immune cells, which aids in inflammation resolution and wound healing (Yang et al., 2022; Zhao et al., 2022).

***Therapeutic Applications*:** Pulsed Electromagnetic Field Therapy (PEMF) promotes bone healing (Dolkart et al., 2021), osteogenesis (Popović et al., 2023), and reduces inflammation (Kubat et al., 2015). Transcranial Magnetic Stimulation (TMS) enhances neuroplasticity, aiding in stroke recovery and depression treatment (Burke et al., 2019). Electrical stimulation using microcurrents accelerates chronic wound healing by improving cell migration and angiogenesis (Konstantinou et al., 2020).

***Scientific Applications*:** Scientific tools have quantified the emotional energies of specific organs, such as the heart and brain. The HeartMath Institute (HeartMath Institute, 2025) investigates heart-brain coherence and emotional energy using techniques that include:

Heart Rate Variability (HRV): It was found that positive emotions produced smooth, coherent HRV patterns, while negative emotions created erratic patterns. Biofeedback devices, such as Inner Balance (Hernandez et al., 2016) or emWave (Whited et al., 2014), can monitor HRV in real-time.

***Heart-Brain Coherence*:** Positive emotions synchronize the heart, brain, and nervous system, which optimizes physiological functions. HeartMath's tools use algorithms to measure coherence through HRV data.

***Electromagnetic Fields*:** The heart generates a measurable electromagnetic field that reflects emotional states, with positive emotions amplifying the field and influencing interpersonal and environmental energy dynamics. HeartMath suggests that emotional energy not only impacts individuals but also contributes to a collective energetic field. These therapeutic and scientific applications demonstrate the capacity of the human energy field to interact with external influences, which emphasizes a dynamic connection between the human body and its environment in medical science.

6.6.2 Exploring Energy Fields in Integrative Energetic Medicine

Integrative Energetic Medicine (IEM) serves as an umbrella term for a subset of recognized energy-based healing modalities. The capacity of the human energy field to interact with external influences forms the very foundation of numerous recognized ancient and contemporary energy medicine practices. The premise of IEM is that health and healing involve more than just the physical body. They are also contingent on the balance and flow of subtle energies within and around an individual. IEM combines conventional medical practices with energy-based approaches from various traditions to treat the whole person: body, mind, emotion, and spirit. Energy healing can extend beyond the constraints of physical proximity that also aligns with the holistic views that see individuals as interconnected within a pervasive universal energy system.

Recognized complementary IEM healing modalities (Rogers et al., 2021) include Reiki, Healing Touch, and Therapeutic Touch, all of which involve working with the body's energy field for healing, with varying origins and often connections to traditional Eastern medicine (Table 6.9).

TABLE 6.9
Recognized Complementary IEM Healing Modalities

IEM Modalities	Origin	Modus Operandi	Techniques
Reiki	Developed in Japan in the early 20th century by Mikao Usui.	Reiki is rooted in Japanese spiritual practices and draws from Buddhist and Shinto traditions. It shares concepts with Qi (Chi) in Chinese medicine and Prana in Indian traditions – ideas about a universal life energy that can be channeled for healing.	Hands are placed lightly on or near the body to channel healing energy.
Healing Touch	Developed in the late 20th century by Janet Mentgen, a nurse in the U.S.	Healing Touch incorporates energy-based concepts similar to Eastern ideas. Healing Touch is influenced by both Western nursing practices and some Eastern energy theories like chakras and meridians.	Practitioners use hands-on or near-body techniques to balance the energy field.
Therapeutic Touch	Created in the 1970s by Dolores Krieger (a nurse) and Dora Kunz (a theosophist).	Therapeutic Touch is inspired by Eastern philosophies and the concept of life energy (like Qi). It is more structured and scientific in approach compared to Reiki.	Involves consciously directing energy with the hands, usually without touching the body.

This widespread and long-established application of the human energy field in ancient medical traditions, as well as recognized current complementary healing modalities further substantiates its existence and suggests its potential to access, interconnect with, or participate in a larger energetic or universal field.

6.7 ULTIMATE INTERCONNECTIVITY: UNIVERSAL MIND AND INTERDIMENSIONALITY

The Universal Mind refers to a metaphysical concept which suggests the existence of an infinite, all-encompassing consciousness or intelligence that underlies and connects all things in the universe. This concept is rooted in various spiritual and philosophical traditions, including:

***Eastern Philosophy*:** Concepts like the *Akashic Records* (Lores, n.d.) in Hinduism and the Tao in Taoism (Nweke & Obienyem, 2020) align with the idea of a Universal Mind.

***Western Mysticism*:** In theosophy (Faivre, 1994) and Hermeticism (Chandler, n.d.), the Universal Mind is viewed as the source of all creation and thought.

***Modern Contexts*:** Some interpretations link the Universal Mind with quantum physics (Zhi & Xiu, 2023) suggesting that it might represent a field of infinite potential or the interconnected nature of existence. In this framework, individuals are often considered fragments of this Universal Mind, meaning that human consciousness is intrinsically interfaced with a greater whole.

6.7.1 INTERDIMENSIONALITY

Interdimensionality refers to the concept of the existence of, or interactions between, multiple dimensional domains beyond the three spatial dimensions (and one temporal dimension) we normally experience. It suggests that:

***Higher Dimensions*:** Beyond our perceivable reality, there may exist additional dimensions that are not directly observable but influence our universe. This is explored in theoretical physics (e.g., string theory) (Maccaferri et al., 2025) and metaphysics (Arroyo & Morgant, 2025).

***Spiritual Realms*:** In metaphysical contexts, interdimensionality often refers to spiritual planes or realities inhabited by beings, energies, or consciousness that are beyond human perception.

***Interdimensional Beings or Experiences*:** Some spiritual and paranormal traditions describe beings (e.g., angels, extraterrestrials, or spirits) as interdimensional entities that traverse or exist in multiple dimensions. Experiences such as out-of-body journeys, near-death experiences, or meditative states are sometimes interpreted as interdimensional interactions.

The two concepts are often interwoven in spiritual discussions. For instance, the Universal Mind may be thought of as operating across all dimensions, connecting the seen and unseen aspects of existence. Interdimensional experiences might be seen as glimpses or connections with the Universal Mind or other levels of reality. In both cases, these ideas emphasize interconnectedness, transcendence, and the exploration of realities beyond the physical realm.

From the perspective of IEM, the fact that energy healing can extend beyond the constraints of physical proximity suggests that it is conceivable that energetic connections can be made that transcend space and time. This aligns with quantum principles such as entanglement and non-locality, whereby particles can influence each other instantaneously, regardless of distance. Such phenomena reflect the possibility of energy or information exchange that operates beyond traditional boundaries. The Zero-Point Field, which is a quantum field that exists even in a vacuum, might serve as a medium through which these exchanges take place.

6.8 CONCLUSION

This chapter has endeavored to prospectively look well beyond the envisaged hypothetical capabilities of molecular manufacturing. This, to explore how, or indeed whether, precisely arranged patterns of molecules or atoms of sufficient complexity might serve as substrates that spontaneously give rise to the life force or gain the capacity to tap into an omnipresent universal consciousness. We explore various structural complexity thresholds that encompass inorganic non-living (from our perspective) entities to progressively more complex organic/biological organisms. The aim being to glean any modicum of insight into what prerequisite conditions might be involved in the transition from lifeless structured substrates to those that somehow become imbued with the life force and consciousness (which may indeed be one in the same). Further, structural complexity might not, in and of itself, set the stage for the mysterious infusion of the life force or consciousness. For example, extremely structurally complex and atomically precise computer chips do not spontaneously come to life! Consequently, something much deeper may be in operation that remains perhaps still far beyond the grasp of our current comprehension. As but only one of the amazing myriad species that inhabit this amazing and beautiful Earth, we ceaselessly strive to define our place and purpose on our home planet, and ultimately, where we fit within this inconceivably immense, enigmatic universe.

Historically, humankind has, for the most part, distinguished and set itself apart as the "other" who are somehow distinct, separate from nature and (in a truly oversimplified and isolating sense) superior to our fellow creatures and lifeforms. Through this chapter the authors encourage readers to pause, if only briefly, to set their gazes inward. This is where we may find the reality that we are precious, radiant beings, our essences permeated with intricate and unique vibrations and waveforms that intimately entangle with and contribute to the vast, dynamic ocean of all energetically bound lifeforms on Earth.

Everything in existence on this planet emanates from the quantum realm. Peering into the deepest cores of our true natures, we are by design empowered to choose and express our energies to all of life around us, and beyond. Unique energetic blueprints interweave our every cell, tissue, and organ to be experienced by the sensory faculties of our whole being. By gazing lovingly inward and connecting to our senses, we may experience the benevolent life-giving force that connects us with every life form and nonliving substrate on Earth, which likely stretches far beyond time and space, entangling us with the entire universe through all. The emerging message for humanity rings increasingly clear and true …

We are One.

REFERENCES

Ahsan M, Pindi C, Senapati S. Hydrogen bonding catalysis by water in epoxide ring opening reaction. *J Mol Graph Model.* 2021 Jun;105:107894.

Albrecht A, Steinhardt PJ. Cosmology for grand unified theories with radiatively induced symmetry breaking. *Phys Rev Lett.* 1982 Apr 26;48:1220–1223.

Angeles Fernández-Gil M, Palacios-Bote R, Leo-Barahona M, Mora-Encinas JP. Anatomy of the brainstem: a gaze into the stem of life. *Semin Ultrasound CT MR.* 2010 Jun;31(3):196–219.

Antognini A, Nez F, Schuhmann K, Amaro FD, Biraben F, Cardoso JM, et al. Proton structure from the measurement of 2S-2P transition frequencies of muonic hydrogen. *Science.* 2013;339(6118):417–420.

Arroyo R, Morgant M. Naturalistic metaphysics and the parity thesis: why scientific realism doesn't lead to realism about metaphysics. *Synthese.* 2025. https://doi.org/10.48550/arXiv.2505.07751 (accessed July 08, 2025).

Aru J, Larkum ME, Shine JM. The feasibility of artificial consciousness through the lens of neuroscience. *Trends Neurosci.* 2023 Dec;46(12):1008–1017.

Athanasiou KA, Eswaramoorthy R, Hadidi P, Hu JC. Self-organization and the self-assembling process in tissue engineering. *Annu Rev Biomed Eng.* 2013;15:115–136.

Aubrun G, Lami L, Palazuelos C, Plávala M. Entanglement and superposition are equivalent concepts in any physical theory. *Phys Rev Lett.* 2022 Apr 22;128(16):160402.

Avila J, Marco J, Plascencia-Villa G, Bajic VP, Perry G. Could there be an experimental way to link consciousness and quantum computations of brain microtubules? *Front Neurosci.* 2024 Jun 24;18:1430432.

Baars. Global workspace theory of consciousness: toward a cognitive neuroscience of human experience. *Prog Brain Res.* 2005;150:45–53.

Baars BJ, Geld N, Kozma R. Global workspace theory (GWT) and prefrontal cortex: recent developments. *Front Psychol.* 2021 Nov 10;12:749868.

Baltieri M, Iizuka H, Witkowski O, Sinapayen L, Suzuki K. Hybrid life: integrating biological, artificial, and cognitive systems. *Wiley Interdiscip Rev Cogn Sci.* 2023 Nov–Dec;14(6):e1662.

Baluška F, Reber A. Sentience and consciousness in single cells: how the first minds emerged in unicellular species. *Bioessays.* 2019 Mar;41(3):e1800229.

Barlow PW. The natural history of consciousness, and the question of whether plants are conscious, in relation to the Hameroff-Penrose quantum-physical "Orch OR" theory of universal consciousness. *Commun Integr Biol.* 2015 Jul 9;8(4):e1041696.

Barnden JA. The meta-dynamic nature of consciousness. *Entropy (Basel).* 2020 Dec 18;22(12):1433.

Barua P. The Concept of Śūnyatā in Madhyamaka Philosophy. 2016. ID-20131100, BA-4. https://www.academia.edu/30777585/The_Concept_of_%C5%9A%C5%ABnyat%C4%81_in_Madhyamaka_Philosophy_by_Pranab_Barua#loswp-work-container (accessed July 06, 2025).

Bayne T, Seth AK, Massimini M, Shepherd J, Cleeremans A, Fleming SM, et al. Tests for consciousness in humans and beyond. *Trends Cogn Sci.* 2024 May;28(5):454–466.

Beilin LJ, Knight GJ, Munro-Faure AD, Anderson J. The sodium, potassium, and water contents of red blood cells of healthy human adults. *J Clin Invest.* 1966 Nov;45(11):1817–1825.
Benner SA. Defining life. *Astrobiology.* 2010 Dec;10(10):1021–1030.
Bernet R. "Bergson on the Driven Force of Consciousness and Life," in Kelly MR, ed., *Bergson and Phenomenology*, London/New York, Palgrave Macmillan. 2010;42–62.
Black D. The global workspace theory, the phenomenal concept strategy, and the distribution of consciousness. *Conscious Cogn.* 2020 Sep;84:102992.
Burke MJ, Fried PJ, Pascual-Leone A. Transcranial magnetic stimulation: neurophysiological and clinical applications. *Handb Clin Neurol.* 2019;163:73–92.
Butler C. Panpsychism and the dissolution of dispositional properties. *Southwest Philos Rev.* 2010;26(2):87–108.
Cai J, Wang Y, McKeown MJ. Advances in functional and structural imaging of the brainstem: implications for disease. *Curr Opin Neurol.* 2024 Aug 1;37(4):361–368.
Carnio EG, Breuer HP, Buchleitner A. Wave-particle duality in complex quantum systems. *J Phys Chem Lett.* 2019 May 2;10(9):2121–2129.
Carosso A. Quantization: history and problems. *Stud Hist Philos Sci.* 2022Dec;96:35–50.
Cartlidge E. The elemental fight behind the periodic table's latest additions. *Nature.* 2018 Jun;558(7709):175–176.
Čartolovni A. Teilhard de Chardin's oeuvre within an ongoing discussion of a gene drive release for public health reasons. *Life Sci Soc Policy.* 2017 Dec 20;13(1):18.
Chalmers DJ. "Panpsychism and panprotopsychism," in Hameroff SR, Kaszniak AW, Scott AC, eds., *Consciousness and its Place in Nature,* Imprint Academic, Exeter, UK, 2003:217–250.
Chandler KA. Hermes and Hermeticism: A Historical Introduction (n.d.). https://www.academia.edu/34992032/Hermes_and_Hermeticism_A_Historical_Introduction (accessed July 08, 2025).
Chang J, Ge Y, Wu Z, Du Y, Pan K, Yang G, et al. Modern cities modelled as "super-cells" rather than multicellular organisms: implications for industry, goods and services. *Bioessays.* 2021;43(7):e2100041.
Chapman K. The transuranic elements and the island of stability. *Philos Trans A Math Phys Eng Sci.* 2020 Sep 18;378(2180):20190535.
Chaudhuri H. The concept of Brahman in Hindu philosophy. *Philos East West.* 1954;4(1): 47–66.
Chella A. Artificial consciousness: the missing ingredient for ethical AI? *Front Robot AI.* 2023 Nov 21;10:1270460.
Chemingui M, Elzanaty A, Tafazolli R. *EMF-Efficient MU-MIMO Networks: Harnessing Aerial RIS Technology.* 2024;arXiv:2412.05925.
Cowley WL. Uncertainty principle. *Nature.* 1949 Mar 26;163(4143):492.
Craddock TJ, Hameroff SR, Ayoub AT, Klobukowski M, Tuszynski JA. Anesthetics act in quantum channels in brain microtubules to prevent consciousness. *Curr Top Med Chem.* 2015;15(6):523–33.
Craddock TJ, Priel A, Tuszynski JA. Keeping time: could quantum beating in microtubules be the basis for the neural synchrony related to consciousness? *J Integr Neurosci.* 2014;13(2):293–311.
Da Silva, Williams RJP. *The Biological Chemistry of the Elements: The Inorganic Chemistry of Life*, Oxford University Press, Oxford, England, 2001.
de Chardin PT. *The Phenomenon of Man*, Harper, New York, 1955.
Degtyarevskaya T, Vokhmintsev A. Effects of electromagnetic field frequency on the behavior and mortalities of living organisms. *Online J Biol Sci.* 2024;24(2):244–254.
Democritus on the atom. *Sci Am.* 1949 Nov;181(5):48.
Desmond H, Huneman P. The integrated information theory of agency. *Behav Brain Sci.* 2022 Mar 23;45:e45.

Dieks D. Identical quantum particles, entanglement, and individuality. *Entropy (Basel)*. 2020 Jan 23;22(2):134.

Dieks D. Emergence and identity of quantum particles. *Philos Trans A Math Phys Eng Sci*. 2023 Sep 18;381(2255):20220107.

Dieterle MP, Husari A, Rolauffs B, Steinberg T, Tomakidi P. Integrins, cadherins and channels in cartilage mechanotransduction: perspectives for future regeneration strategies. *Expert Rev Mol Med*. 2021 Oct 27;23:e14.

Disalvo EA, Pinto OA, Martini MF, Bouchet AM, Hollmann A, Frías MA. Functional role of water in membranes updated: a tribute to Träuble. *Biochim Biophys Acta*. 2015;1848(7):1552–1562.

do Nascimento, Vieira A, Kleinermanns K, Martin WF, Preiner M. The ambivalent role of water at the origins of life. *FEBS Lett*. 2020 Sep;594(17):2717–2733.

Dolkart O, Kazum E, Rosenthal Y, Sher O, Morag G, Yakobson E, Chechik O, Maman E. Effects of focused continuous pulsed electromagnetic field therapy on early tendon-to-bone healing. *Bone Joint Res*. 2021 May;10(5):298–306.

Dondelinger RM. Electromyography–an overview. *Biomed Instrum Technol*. 2010;44(2):128–31.

Doolittle WF. Darwinizing Gaia. *J Theor Biol*. 2017;434:11–19.

Dove LL. How many atoms are in a person? 2023. https://science.howstuffworks.com/atoms-in-person.htm (accessed September 21, 2025).

Dunne B, Jahn R, eds. *Being & Biology: Is Consciousness the Life Force?* ICRL Press, Princeton, NJ, 2018.

Eastwood MA. Heisenberg's uncertainty principle. *QJM*. 2017 May 1;110(5):335–336.

Edwards PP, Egdell RG, Fenske D, Yao B. The periodic law of the chemical elements: "*The new system of atomic weights which renders evident the analogies which exist between bodies*." *Philos Trans A Math Phys Eng Sci*. 2020 Sep 18;378(2180):20190537.

Egdell RG, Bruton E. Henry Moseley, X-ray spectroscopy and the periodic table. *Philos Trans A Math Phys Eng Sci*. 2020 Sep 18;378(2180):20190302.

Eickhoff SB, Constable RT, Yeo BTT. Topographic organization of the cerebral cortex and brain cartography. *Neuroimage*. 2018 Apr 15;170:332–347.

Epstein IR, Xu B. Reaction-diffusion processes at the nano- and microscales. *Nat Nanotechnol*. 2016;11(4):312–319.

evan der Wal CH, ter Haar AC, Wilhelm FK, Schouten RN, Harmans CJ, Orlando TP, et al. Quantum superposition of macroscopic persistent-current states. *Science*. 2000 Oct 27;290(5492):773–777.

Evans MW. *Generally Covariant Unified Field Theory: The Geometrization of Physics* (Vols. 1–5), Arima Publishing, Suffolk, UK, 2005–2006.

Faivre A. Theosophy and the Esoteric Currents in the Modern West. *J Hist Ideas*. 1994;55(4):621–640.

Fernández V, Llinares-Benadero C, Borrell V. Cerebral cortex expansion and folding: what have we learned? *EMBO J*. 2016 May 17;35(10):1021–1044.

Forbes RM, Cooper AR, Mitchell HH. The composition of the adult human body as determined by chemical analysis. *J Biol Chem*. 1953 Jul;203(1):359–366.

Fox D. Living Mysteries: Meet Earth's Simplest Animal. *Science News Explores*. April 5, 2018. https://www.snexplores.org/article/living-mysteries-meet-earths-simplest-animal (accessed June 24, 2025).

Frantz C, Stewart KM, Weaver VM. The extracellular matrix at a glance. *J Cell Sci*. 2010 Dec 15;123(Pt 24):4195–4200.

Free A, Barton NH. Do evolution and ecology need the Gaia hypothesis? *Trends Ecol Evol*. 2007;22(11):611–619.

Gajić V. [Dr James Lovelock and story about GAIA hypothesis]. *Med Pregl*. 2011;64(11–12):607–611.

García-Ruiz JM, van Zuilen MA, Bach W. Mineral self-organization on a lifeless planet. *Phys Life Rev*. 2020 Dec;34–35:62–82.

Geisterfer ZM, Guilloux G, Gatlin JC, Gibeaux R. The cytoskeleton and its roles in self-organization phenomena: insights from *Xenopus* egg extracts. *Cells*. 2021 Aug 26;10(9):2197.

Georgi J, Metere R, Jäger C, Morawski M, Möller HE. Influence of the extracellular matrix on water mobility in subcortical gray matter. *Magn Reson Med*. 2019 Feb;81(2): 1265–1279.

Gershenson C. Emergence in artificial life. *Artif Life*. 2023 May 1;29(2):153–167.

Giannetti G, Matsumura F, Caporaletti F, Micha D, Koenderink GH, Ilie IM, et al. Water and collagen: a mystery yet to unfold. *Biomacromolecules*. 2025 Apr 10;26(5):2784–2799.

Gibson CR, Wang YH, Varkhede N, Ma B. Variability in human in vitro enzyme kinetics. *Methods Mol Biol*. 2021;2342:443–479.

Gisin N. Entanglement 25 years after quantum teleportation: testing joint measurements in quantum networks. *Entropy (Basel)*. 2019 Mar 26;21(3):325.

Gogolla N. The insular cortex. *Curr Biol*. 2017 Jun 19;27(12):R580–R586.

Goldstein LS, Theriot JA. Cytoskeleton. *Curr Opin Cell Biol*. 2001 Feb;13(1):17–18.

Gómez-Márquez J. What is life? *Mol Biol Rep*. 2021 Aug;48(8):6223–6230.

Green D. Coagulation cascade. *Hemodial Int*. 2006 Oct;10 Suppl 2:S2–4.

Guth AH. Inflationary universe: a possible solution to the horizon and flatness problems. *Phys Rev D*. 1981 Jan 15;23:347.

Guth AH. *The Inflationary Universe: The Quest for a New Theory of Cosmic Origins*, Addison-Wesley, Reading, MA, 1997.

Habas C, Manto M, Cabaraux P. The cerebellar thalamus. *Cerebellum*. 2019;18(3):635–648.

Hameroff S. Consciousness, the brain, and spacetime geometry. *Ann N Y Acad Sci*. 2001;929:74–104.

Hameroff S. Consciousness, cognition and the neuronal cytoskeleton – a new paradigm needed in neuroscience. *Front Mol Neurosci*. 2022 Jun 16;15:869935.

Hameroff S, Penrose R. Consciousness in the universe: a review of the "Orch OR" theory. *Phys Life Rev*. 2014 Mar;11(1):39–78.

Hameroff SR, Craddock TJ, Tuszynski JA. Quantum effects in the understanding of consciousness. *J Integr Neurosci*. 2014;13(2):229–252.

Harp SS. The brain as a receiver and consciousness as a fundamental. *J Conscious Explor Res*. 2012;3(11):1145–1148.

HeartMath Institute. https://www.heartmath.org/ (accessed July 08, 2025).

Heer NC, Martin AC. Tension, contraction and tissue morphogenesis. *Development*. 2017;144(23):4249–4260.

Helmenstine A. Elements in the Human Body and What They Do. *Science Notes*. May 20, 2015. https://sciencenotes.org/elements-in-the-human-body-and-what-they-do/ (accessed April 19, 2025).

Henríquez JP, Krull CE, Osses N. The Wnt and BMP families of signaling morphogens at the vertebrate neuromuscular junction. *Int J Mol Sci*. 2011;12(12):8924–8946.

Hernandez C, Daly K, Mehta A, Verduin M. A pilot study examining biofeedback and structured napping to promote medical student wellbeing. *MedEdPublish* (2016). 2019 May 27;8:110.

Hersey G. 35 Facts About Atom. Facts.net. Oct 14, 2024. https://facts.net/science/35-facts-about-atom/ (accessed November 19, 2024).

Hildt E. Artificial intelligence: does consciousness matter? *Front Psychol*. 2019 Jul 2;10: 1535.

Hildt E. The prospects of artificial consciousness: ethical dimensions and concerns. *AJOB Neurosci*. 2023 Apr–Jun;14(2):58–71.

Hunt T, Jones M, McFadden J, Delorme A, Hales CG, Ericson M et al. Editorial: electromagnetic field theories of consciousness: opportunities and obstacles. *Front Hum Neurosci*. 2024;17:1342634.

Ismail I, Azis RS. A review of magnetic nanocomposites for EMI shielding: synthesis, properties, and mechanisms. *J Mater Sci*. 2024;59:5293–5329.

Jaeger G. The elementary particles of quantum fields. *Entropy (Basel)*. 2021 Oct 28;23(11):1416.

Kaim W, Schwederski B, eds. *Bioinorganic Chemistry: Inorganic Elements in the Chemistry of Life: An Introduction and Guide*, Wiley, Chichester, 1994.

Kaji M. Social background of the discovery and the reception of the periodic law of the elements: recognizing the contributions of Dmitri Ivanovich Mendeleev and Julius Lothar Meyer. *Ann N Y Acad Sci*. 2003 May;988:302–306.

Kanaev IA. Evolutionary origin and the development of consciousness. *Neurosci Biobehav Rev*. 2022;133:104511.

Kanai R, Fujisawa I. Toward a universal theory of consciousness. *Neurosci Conscious*. 2024 May 31;2024(1):niae022.

Karamanos NK, Theocharis AD, Piperigkou Z, Manou D, Passi A, Skandalis SS, et al. A guide to the composition and functions of the extracellular matrix. *FEBS J*. 2021Dec;288(24):6850–6912.

Kastritis PL, Gavin AC. Enzymatic complexes across scales. *Essays Biochem*. 2018 Oct 26;62(4):501–514.

Kattan WM, Alarfaj SF, Alnooh BM, Alsaif HF, Alabdul Karim HS, Al-Qattan NM, et al. Myofibroblast-mediated contraction. *J Coll Physicians Surg Pak*. 2017;27(1):38–43.

Kauffman S. Answering Schrödinger's "What is life?" *Entropy (Basel)*. 2020 Jul 25;22(8):815.

Kauffman SA. Prolegomenon to patterns in evolution. *Biosystems*. 2014 Sep;123:3–8.

Khare D, Basu B, Dubey AK. Electrical stimulation and piezoelectric biomaterials for bone tissue engineering applications. *Biomaterials*. 2020 Nov;258:120280.

Kiouvrekis Y, Givisis I, Panagiotakopoulos T, Tsilikas I, Ploussi A, Spyratou E, et al. A comparative analysis of explainable artificial intelligence models for electric field strength prediction over eight European cities. *Sensors (Basel)*. 2024;25(1):53.

Kitchener PD, Hales CG. What neuroscientists think, and don't think, about consciousness. *Front Hum Neurosci*. 2022 Feb 24;16:767612.

Kléber AG. Gap junctions and conduction of cardiac excitation. *Heart Rhythm*. 2011;8(12):1981–1984.

Konen C, Zirlinger M. *Consciousness. Neuron*. 2024 May 15;112(10):1519.

Konstantinou E, Zagoriti Z, Pyriochou A, Poulas K. Microcurrent stimulation triggers MAPK signaling and TGF-β1 release in fibroblast and osteoblast-like cell lines. *Cells*. 2020 Aug 19;9(9):1924.

Kozlovskaya V, Kharlampieva E. Anisotropic particles through multilayer assembly. *Macromol Biosci*. 2022 Jan;22(1):e2100328.

Kozuch B. Better bridges: integrating the neuroscience and philosophy of consciousness. *Conscious Cogn*. 2024 Nov;126:103774.

Krause AL, Klika V, Maini PK, Headon D, Gaffney EA. Isolating patterns in open reaction-diffusion systems. *Bull Math Biol*. 2021 Jun 4;83(7):82.

Kubat NJ, Moffett J, Fray LM. Effect of pulsed electromagnetic field treatment on programmed resolution of inflammation pathway markers in human cells in culture. *J Inflamm Res*. 2015;8:59–69.

Kuhn RL. A landscape of consciousness: toward a taxonomy of explanations and implications. *Prog Biophys Mol Biol*. 2024 Aug;190:28–169.

Kushner DJ. Self-assembly of biological structures. *Bacteriol Rev*. 1969 Jun;33(2):302–345.

Laage D, Elsaesser T, Hynes JT. Water dynamics in the hydration shells of biomolecules. *Chem Rev*. 2017 Aug 23;117(16):10694–10725.

Lai H, Levitt BB. Cellular and molecular effects of non-ionizing electromagnetic fields. *Rev Environ Health*. 2023;39(3):519–529.

Lawsin J. *Autognorics: The Science of Engineered Lifeforms*. Lulu.com. August 8, 2023.

Lehotzky D, Zupanc GKH. Cellular automata modeling of stem-cell-driven development of tissue in the nervous system. *Dev Neurobiol*. 2019 May;79(5):497–517.

León F, Zahavi D. Consciousness, philosophy, and neuroscience. *Acta Neurochir (Wien).* 2023;165(4):833–839.
Levonian DS, Riedinger R, Machielse B, Knall EN, Bhaskar MK, Knaut CM, et al. Optical entanglement of distinguishable quantum emitters. *Phys Rev Lett.* 2022 May 27;128(21):213602.
Levy Y, Onuchic JN. Water mediation in protein folding and molecular recognition. *Annu Rev Biophys Biomol Struct.* 2006;35:389–415.
Linde AD. Coleman-Weinberg theory and the new inflationary universe scenario. *Physics Letters B.* 1982 Aug 12;114(6):431–435.
Lohrey A, Boreham B. The nonlocal universe. *Commun Integr Biol.* 2020 Oct 11;13(1):147–159.
Loitto VM, Karlsson T, Magnusson KE. Water flux in cell motility: expanding the mechanisms of membrane protrusion. *Cell Motil Cytoskeleton.* 2009 May;66(5):237–247.
Lores E. The Akashic Records. (n.d.). https://www.academia.edu/36287444/The_Akashic_Records (accessed July 08, 2025).
Lovelock JE. Geophysiology, the science of Gaia. *Rev Geophys.* 1989;27(2):215–222. https://www.jameslovelock.org/geophysiology-the-science-of-gaia/ (accessed July 07, 2025).
Lovelock JE, Margulis L. Atmospheric homeostasis by and for the biosphere: the Gaia hypothesis. *Telus Magazine.* 1974;XXVI(1–2):2–10.
Maccaferri C, Marino F, Valsesia B. Introduction to string theory. *SciPost Physics Lecture Notes.* 2025;90. https://doi.org/10.21468/SciPostPhysLectNotes.90 (accessed July 08, 2025).
Magalon A. History of maturation of prokaryotic molybdoenzymes – a personal view. *Molecules.* 2023 Oct 20;28(20):7195.
Magiorkinis E, Beloukas A, Diamantis A. The philosophic and biological views of the "atomic" philosophers, Leucippus and Democritus. *Hell J Nucl Med.* 2010 May–Aug;13(2):111–117.
Mardt A, Gorriz RF, Ferraro F, Ulrich P, Zahran M, Imhof P. Effect of a U:G mispair on the water around DNA. *Biophys Chem.* 2022 Apr;283:106779.
Martins NRB, Angelica A, Chakravarthy K, Svidinenko Y, Boehm FJ, Opris I, Lebedev MA, Swan M, Garan SA, Rosenfeld JV, Hogg T and Freitas RA Jr. Human Brain/Cloud interface. *Front Neurosci.* 2019;13:112.
Mashour GA. Anesthesia and the neurobiology of consciousness. *Neuron.* 2024 May 15;112(10):1553–1567.
McConnell SK. Constructing the cerebral cortex: neurogenesis and fate determination. *Neuron.* 1995 Oct;15(4):761–768.
McDonald JMC, Reed RD. Patterns of selection across gene regulatory networks. *Semin Cell Dev Biol.* 2023 Aug;145:60–67.
McFadden J. Synchronous firing and its influence on the Brain's electromagnetic field evidence for an electromagnetic field theory of consciousness. *J Conscious Stud.* 2002;9:23–50.
McFadden J. Integrating information in the brain's EM field: the cemi field theory of consciousness. *Neurosci Conscious.* 2020 Sep 22;2020(1):niaa016.
McFadden J. Consciousness: matter or EMF? *Front Hum Neurosci.* 2023 Jan 18;16:1024934.
Mediano PAM, Rosas FE, Bor D, Seth AK, Barrett AB. The strength of weak integrated information theory. *Trends Cogn Sci.* 2022 Aug;26(8):646–655.
Menon V. 20 years of the default mode network: a review and synthesis. *Neuron.* 2023 Aug 16;111(16):2469–2487.
Mertz W. The essential trace elements. *Science.* 1981;213(4514):1332–1338.
Midha T, Kolomeisky AB, Igoshin OA. Insights into error control mechanisms in biological processes: copolymerization and enzyme-kinetics revisited. *J Phys Chem B.* 2024 Jun 13;128(23):5612–5622.
Mishra S, Raval M, Kachhawaha AS, Tiwari BS, Tiwari AK. Aging: epigenetic modifications. *Prog Mol Biol Transl Sci.* 2023;197:171–209.
Mocombe PC. The nature of consciousness: a channel of a frequency wavelength. *J Clin Case Rep Stu.* 2022;3(2).

Mueller M, Rasoulinejad S, Garg S, Wegner SV. The importance of cell-cell interaction dynamics in bottom-up tissue engineering: concepts of colloidal self-assembly in the fabrication of multicellular architectures. *Nano Lett.* 2020 Apr 8;20(4):2257–2263.

Mukhopadhyay RD. A template for artificial life. *Nat Chem.* 2024 Aug;16(8):1214–1216.

Müller-Putz GR. Electroencephalography. *Handb Clin Neurol.* 2020;168:249–262.

Nakata Y, Murao M. Generic entanglement entropy for quantum states with symmetry. *Entropy (Basel).* 2020 Jun 19;22(6):684.

NASA, Hubble Big Bang. https://science.nasa.gov/mission/hubble/science/science-behind-the-discoveries/hubble-big-bang/ October 22, 2024 (accessed September 21, 2025).

Nguindjel AC, de Visser PJ, Winkens M, Korevaar PA. Spatial programming of self-organizing chemical systems using sustained physicochemical gradients from reaction, diffusion and hydrodynamics. *Phys Chem Phys.* 2022 Oct 12;24(39):23980–24001.

Nicolis G, Prigogine I. *Self-Organization in Nonequilibrium Systems*, John Wiley & Sons, Inc., New York, 1977;512.

Nielsen MS, Axelsen LN, Sorgen PL, Verma V, Delmar M, Holstein-Rathlou NH. Gap junctions. *Compr Physiol.* 2012 Jul;2(3):1981–2035.

Nödl MT, Fossati SM, Domingues P, Sánchez FJ, Zullo L. The making of an octopus arm. *Evodevo.* 2015;6:19.

Nweke CC, Obienyem V. On the philosophy of Taoism. *J Philos Cult Relig.* 2020;50:1–8.

Olson CS, Ragsdale CW. Toward an understanding of octopus arm motor control. *Integr Comp Biol.* 2023 Dec 29;63(6):1277–1284.

Penkova NA, Sharapov MG, Penkov NV. Hydration shells of DNA from the point of view of terahertz time-domain spectroscopy. *Int J Mol Sci.* 2021 Oct 14;22(20):11089.

Perez Velazquez JL. Finding simplicity in complexity: general principles of biological and nonbiological organization. *J Biol Phys.* 2009 Aug;35(3):209–221.

Pigolotti S, Krishna S, Jensen MH. Oscillation patterns in negative feedback loops. *Proc Natl Acad Sci U S A.* 2007 Apr 17;104(16):6533–6537.

Pocker Y. Water in enzyme reactions: biophysical aspects of hydration-dehydration processes. *Cell Mol Life Sci.* 2000 Jul;57(7):1008–1017.

Pollard TD, Goldman RD. Overview of the cytoskeleton from an evolutionary perspective. *Cold Spring Harb Perspect Biol.* 2018 Jul 2;10(7):a030288.

Popović T, Matavulj M, Nežić L, Radulović TN, Škrbić R. Pulsed electromagnetic field attenuates bone fragility in estrogen-deficient osteoporosis in rats. *Technol Health Care.* 2023;31(4):1343–1353.

Prockop DJ. Concise review: two negative feedback loops place mesenchymal stem/stromal cells at the center of early regulators of inflammation. *Stem Cells.* 2013 Oct;31(10):2042–2046.

Procopio LM, Moqanaki A, Araújo M, Costa F, Alonso Calafell I, Dowd EG, et al. Experimental superposition of orders of quantum gates. *Nat Commun.* 2015 Aug 7;6:7913.

Qin C, Qi Z, Pan S, Xia P, Kong W, Sun B, et al. Advances in conductive hydrogel for spinal cord injury repair and regeneration. *Int J Nanomedicine.* 2023 Dec 6;18:7305–7333.

Rock T, Wills MA. Quantifying the configurational complexity of biological systems in multivariate "complexity space." *J R Soc Interface.* 2025 Jan;22(222):20240558.

Rogers L, Phillips K, Cooper NN. Energy healing therapies: a systematic review and critical appraisal. *Health Psychol Rev.* 2021;2(3):162–170.

Roli A, Kauffman SA. Emergence of Organisms. *Entropy (Basel).* 2020 Oct 16;22(10):1163.

Romanoski H, Mendelejew D. 1834–1907 [Dmitri Ivanovich Mendeleev, 1834–1907]. *Farm Pol.* 1954 Feb;10(2):52–54.

Roy S, Kornberg TB. Paracrine signaling mediated at cell-cell contacts. *Bioessays.* 2015 Jan;37(1):25–33.

Samanta A, Baranda Pellejero L, Masukawa M, Walther A. DNA-empowered synthetic cells as minimalistic life forms. *Nat Rev Chem.* 2024 Jun;8(6):454–470.

Sarasso S, Casali AG, Casarotto S, Rosanova M, Sinigaglia C, Massimini M. Consciousness and complexity: a consilience of evidence. *Neurosci Conscious.* 2021 Aug 30;2021(2):niab023.

Schirò G, Weik M. Role of hydration water in the onset of protein structural dynamics. *J Phys Condens Matter.* 2019 Nov 20;31(46):463002.

Schlag-Rey M, Schlag J. The central thalamus. *Rev Oculomot Res.* 1989;3:361–390.

Schurger A, Graziano M. Consciousness explained or described? *Neurosci Conscious.* 2022 Jan 21;2022(2):niac001.

Segers VFM, De Keulenaer GW. Autocrine signaling in cardiac remodeling: a rich source of therapeutic targets. *J Am Heart Assoc.* 2021 Feb 2;10(3):e019169.

Sharkey KA, Mawe GM. The enteric nervous system. *Physiol Rev.* 2023 Apr 1;103(2):1487–1564.

Siegel E. What was it like when the first atoms formed? *Big Think.* December 07, 2023. https://bigthink.com/starts-with-a-bang/when-first-atoms-formed/ (accessed June 11, 2024).

Silvestri R, Yu H, Strömberg T, Hilweg C, Peterson RW, Walther P. Experimental observation of Earth's rotation with quantum entanglement. *Sci Adv.* 2024 Jun 14;10(24):eado0215.

Sivitilli DM, Smith JR, Gire DH. Lessons for robotics from the control architecture of the octopus. *Front Robot AI.* 2022 Jul 18;9:862391.

Skene KR. Systems theory, thermodynamics and life: integrated thinking across ecology, organization and biological evolution. *Biosystems.* 2024 Feb;236:105123.

Sladek CD, Johnson AK. Integration of thermal and osmotic regulation of water homeostasis: the role of TRPV channels. *Am J Physiol Regul Integr Comp Physiol.* 2013;305(7):R669–678.

Smallwood J, Bernhardt BC, Leech R, Bzdok D, Jefferies E, Margulies DS. The default mode network in cognition: a topographical perspective. *Nat Rev Neurosci.* 2021 Aug;22(8):503–513.

Smith SA. The cell-based model of coagulation. *J Vet Emerg Crit Care (San Antonio).* 2009 Feb;19(1):3–10.

Smith TP. The anatomy of a neutron. *American Scientist.* 2010;98(6):478–485.

Smith LH, DeMyer WE. Anatomy of the brainstem. *Semin Pediatr Neurol.* 2003Dec; 10(4):235–240.

Smith SA, Travers RJ, Morrissey JH. How it all starts: initiation of the clotting cascade. *Crit Rev Biochem Mol Biol.* 2015;50(4):326–336.

Sporton S. The place of the electrocardiogram in modern cardiology. *Hosp Med.* 2001Apr;62(4):196–197.

Strawson G. "Mind and being: the primacy of panpsychism," in Brüntrup G, Jaskolla L, eds., *Panpsychism: Contemporary perspectives*, Oxford University Press, Oxford, 2016;85–125.

Sturm S, Köhler F, Zatorski J, Wagner A, Harman Z, Werth G, et al. High-precision measurement of the atomic mass of the electron. *Nature.* 2014 Feb 27;506(7489):467–470.

Sun L, Zhang H, Gao P. Metabolic reprogramming and epigenetic modifications on the path to cancer. *Protein Cell.* 2022 Dec;13(12):877–919.

Swindale NV. Is the cerebral cortex modular? *Trends Neurosci.* 1990 Dec;13(12):487–492.

Taber KH, Wen C, Khan A, Hurley RA. The limbic thalamus. *J Neuropsychiatry Clin Neurosci.* 2004;16(2):127–132.

Tamulis A, Grigalavicius M. Quantum entanglement in photoactive prebiotic systems. *Syst Synth Biol.* 2014 Jun;8(2):117–140.

Theocharis AD, Skandalis SS, Gialeli C, Karamanos NK. Extracellular matrix structure. *Adv Drug Deliv Rev.* 2016 Feb 1;97:4–27.

Tibbits S. The emergence of "4D printing." TED Talk. February 2013. https://www.ted.com/talks/skylar_tibbits_the_emergence_of_4d_printing?subtitle=en (accessed November 04, 2024).

Tononi G, Boly M, Massimini M, Koch C. Integrated information theory: from consciousness to its physical substrate. *Nat Rev Neurosci.* 2016 Jul;17(7):450–461.

Torday JS. The holism of cosmology and consciousness. *Prog Biophys Mol Biol.* 2023Jan;177:181–184.

Tsoucalas G, Karamanou M, Kousoulis AA, Androutsos G. Democritus (460–370 BC) on embryology, anatomy and pediatrics: the unknown aspects of the Greek atomic scientist. *Ital J Anat Embryol.* 2012;117(3):199–204.

Turner E. The seat of consciousness. *Lancet.* 1955 Dec 24;269(6904):1305–1307.

Vajda T, Perczel A. Role of water in protein folding, oligomerization, amyloidosis and miniprotein. *J Pept Sci.* 2014 Oct;20(10):747–759.

Vasas V, Fernando C, Santos M, Kauffman S, Szathmáry E. Evolution before genes. *Biol Direct.* 2012 Jan 5;7:1; discussion 1.

Vidal C. Pierre Teilhard de Chardin: a visionary in controversy. *Hist Philos Life Sci.* 2021 Nov 26;43(4):125.

Volpe G, Bechinger C, Cichos F, Golestanian R, Löwen H, Sperl M, Volpe G. Active matter in space. *NPJ Microgravity.* 2022 Nov 24;8(1):54.

Vu QV, Jiang Y, Li MS, O'Brien EP. The driving force for co-translational protein folding is weaker in the ribosome vestibule due to greater water ordering. *Chem Sci.* 2021 Aug 3;12(35):11851–11857.

Walach H, Kohls N, Belschner W. Transpersonale Psychologie – Psychologie des Bewusstseins: Chancen und Probleme [Transpersonal psychology – psychology of consciousness: chances and problems]. *Psychother Psychosom Med Psychol.* 2005 Sep–Oct;55(9–10):405–415.

Wang Y, Wang Z. Inference on the structure of gene regulatory networks. *J Theor Biol.* 2022 Apr 21;539:111055.

Weber BH. What is life? Defining life in the context of emergent complexity. *Orig Life Evol Biosph.* 2010 Apr;40(2):221–229.

Weber GF, Bjerke MA, DeSimone DW. Integrins and cadherins join forces to form adhesive networks. *J Cell Sci.* 2011 Apr 15;124(Pt 8):1183–1193.

Weiss A. Big Bang Nucleosynthesis: Cooking up the first light elements. in: *Einstein Online*, Band 02 (2006), 02-1017. https://www.einstein-online.info/en/spotlight/bbn/ (accessed September 21, 2025).

Whited A, Larkin KT, Whited M. Effectiveness of emWave biofeedback in improving heart rate variability reactivity to and recovery from stress. *Appl Psychophysiol Biofeedback.* 2014 Jun;39(2):75–88.

Wiggins PM. Role of water in some biological processes. *Microbiol Rev.* 1990Dec; 54(4):432–449.

Wildman W. Paul Tillich Resources. 1994–2025. https://people.bu.edu/wwildman/tillich/index.htm (accessed July 06, 2025).

Wilds R, Kauffman SA, Glass L. Evolution of complex dynamics. *Chaos.* 2008;18(3):033109.

Williams RJP. The chemical elements of life. *J Chem Soc Dalton Trans.* 1991;(S):539–546.

Wilson DS, Sober E. Reviving the superorganism. *J Theor Biol.* 1989 Feb 8;136(3):337–356.

Xavier da Silveira Dos Santos A, Liberali P. From single cells to tissue self-organization. *FEBS J.* 2019;286(8):1495–1513.

Yam PT, Charron F. Signaling mechanisms of non-conventional axon guidance cues: the Shh, BMP and Wnt morphogens. *Curr Opin Neurobiol.* 2013 Dec;23(6):965–973.

Yang J, Liu X, Wang W, Chen Y, Liu J, Zhang Z, et al. Bioelectric fields coordinate wound contraction and re-epithelialization process to accelerate wound healing via promoting myofibroblast transformation. *Bioelectrochemistry.* 2022 Dec;148:108247.

Yeshurun Y, Nguyen M, Hasson U. The default mode network: where the idiosyncratic self meets the shared social world. *Nat Rev Neurosci.* 2021 Mar;22(3):181–192.

Young GB, Pigott SE. Neurobiological basis of consciousness. *Arch Neurol.* 1999;56(2): 153–157.

Zhang B, Ran D, Ianconescu R, Friedman A, Scheuer J, Yariv A, Gover A. Quantum wave-particle duality in free-electron-bound-electron interaction. *Phys Rev Lett.* 2021 Jun 18;126(24):244801.

Zhang X, Zou S. Unified Field Theory. *viXra.org e-Print archive.* 2025. viXra:2501.0132. https://vixra.org/pdf/2501.0132v1.pdf (accessed July 06, 2025).

Zhao M, Rolandi M, Isseroff RR. Bioelectric signaling: role of bioelectricity in directional cell migration in wound healing. *Cold Spring Harb Perspect Biol.* 2022;14(10):a041236.

Zhao LY, Song J, Liu Y, Song CX, Yi C. Mapping the epigenetic modifications of DNA and RNA. *Protein Cell.* 2020 Nov;11(11):792–808.

Zhi G, Xiu R. Quantum theory of consciousness. *J Appl Math Phys.* 2023;11(9):2652–2670.

Zoroddu MA, Aaseth J, Crisponi G, Medici S, Peana M, Nurchi VM. The essential metals for humans: a brief overview. *J Inorg Biochem.* 2019 Jun;195:120–129.

Zybailov BL, Kosovsky GY, Glazko GV, Glazko VI, Skobel OI. Evolutionary perspectives on human-artificial intelligence convergence. *Acta Naturae.* 2024 Jul–Sep;16(3):4–17.

Index

Note: Pages in **bold** and *italics* refer to tables and figures, respectively.

3D printers/printing, 5, 16–18, *17*
in biomedical industry, 23–24
camera printed by, 18
in construction industry, 23
in consumer goods sector, 20–21
in energy sector, 19–20
jet engines printed by, 18, *19*
and off-world manufacturing, 77
ratchet wrench printed by, 18
in transportation sector, 20, *21*
4P medicine model, salutogenesis and, 152–154, **155**

A

Actin, 103
Adhesion molecules, 178–179
AGI (Artificial General Intelligence), 7, 151
ALANA (humanoid robot), 23
ANI (Artificial Narrow Intelligence), 7, 151
Antonovsky, Aaron, 151
Artificial intelligence (AI), 5, 31, 67
Artificial lattices, 58–59
Artificial life, 182
ASI (Artificial Superintelligence), 7, 151
Assembly, molecular manufacturing hardware, 51
bio-MEMS/NEMS, 51
biomolecular condensates, 52–53
MEMS/NEMS, 51
molecular assemblers, 52
nanoparticles, 51–52
self-assembled drug delivery platforms, 53
Asset Store, 68
Atomically precise manufacturing, 48–49, *49*
Atomic Force Microscopy (AFM), 106
Atomic precision advanced manufacturing (APAM), 49–50
Atomic scale fabrication
atomically precise manufacturing, 48–49, *49*
atomic precision advanced manufacturing, 49–50
digital methods and autonomous robotic nanofabrication, 50
Atomic spin qubits, 50
Atoms, 47, 169, 171
ATP (adenosine triphosphate), 103
Autocrine signaling, 179
Autonomous nanomedical robots, *15*, 15–16
Autonomous robotic nanofabrication, 50

B

Batish, Shashwat, 23
B-DNA, 101, 127
Berkeley Humanoid Lite, 23
Big Bang theory, 168–169
BioBlockchain, 72
Bioelectric fields, 192
Biological liquid crystals, 55–56
Biology, molecular manufacturing and, 75–77
Chern-Simons theory, 76–77
human brain, 75
neuroscience physics, 76, **76**
Biomedical industry, 3D printers/printing in, 23–24
Bio-MEMS/NEMS, 51
Biomolecular condensates, 52–53
Bionanomachines, 156
Biotechnology, 104
Biotin, 111
Bitcoin, 70
Blade supercar, 20, *21*
Bloch theorem, 59
Blockchain
adoption, status of, 71
applications in manufacturing industries, 72–74
cryptoeconomics, 70–71
definition and status, 69–72
mining nodes, 70
non-fungible token intellectual property registries, 71–72
peer-to-peer networks, 74
science, 72
Blood-brain barrier (BBB), 52
Boltzmann machine, 69
Born machine, 69
Bosonic codes, 65
Bosons, 61
BQ, 21
Brain-endowed organisms *vs.* non-brained organisms, 185–188
Brainstem, 188
Branch point migration, 106
B-Z transition device, 106–107

C

Camera, printed by 3D printers, 18
Catalytic reactions, 110
Cellular automata, 180
Cerebral cortex, 187
Chemical reaction, 180
Chemical signaling, 179
Chern-Simons theory, 76–77
Classical machine learning, 68–69
Cloning and amplification technology (CAT), 115
Conceptual autonomous nanomedical robot, *15*, 15–16
Conductivity, DNA, 126–128
Consciousness, 186–187
  localized, 188–189
  seat of, 187–188
  unified quantum/universal, 189, 191
Construction industry, 3D printers/printing in, 23
Consumer goods sector, 3D printers/printing in, 20–21
Consumer-grade virtual reality headsets, 68
Cooperative branch migration (CBM), 106
Cryptoeconomics, 70–71
Crystals, 59
Cytoskeleton, 179

D

Decentralized finance (DeFi), 70
Default Mode Network (DMN), 188
de Grey, Aubrey, 7
Dendrimers, 121–122
Deoxyribozymes, 110
Desktop units, F@H, dimensional constraints of, 28
Diffusion gradients, 179
Diffusion systems, 180
Digital twins, 66–67
Discrete time crystals, 59
Divergent Technologies, 20
DNA, 47
  adhesion to dendrimers, 121–122
  adhesion to surfaces, 122, *123*
  binding to nanotubes, 120–121
  data storage, 102–103
  fractal triangular Sierpinski pattern, 125
  function, 101–102
  manipulation of proteins using, 113
  manufacturing strategy, 128–129
  nanodevices based on, 103–106, *105*
  nanoelectronics, 126–128
  replication of structural elements, 118–119
  structure of, 101–102
  thiolated, 121
  three state device, 110–111
  Velcro strips, 122
DNA-based actuation devices, 107–108
DNA-based manufacturing complex, 114–117
  potential computational requirements for, 123–125
DNA-based molecular motors, 111–113
DNA chirality inversion, 55–56
DNA crossover junction-based device, 111
DNA G-quartet device, 111
DNA Guided Assembly of Proteins (DGAP), 113
DNA nanotechnology, 53–54
  topological molecules, 54–55
DNA origami technology, *119*, 119–120
DNA scissors, 109–110
DNA tweezers, 108–109, *109*
DNA walkers, 113–114, *114*
Downs, Molly, 4
Drexler, Eric, 5, 25
Drones, 3D printers/printing for creation of, 20
Drug delivery platforms, self-assembled, 53
Dynein, 103

E

EcoWorld Magazine, 3
Electricity decentralization, F@H system and, 27
Electromagnetic field (EMF), 190, 191
  generation, mechanisms of, 192–193
Electron engineering, 58
Electrons, 172, **172**
Electrospinning, AI-guided, 31–32
Electrospraying, AI-guided, 31–32
Energy
  fields, nanomedicine and, 6–8
  fundamental units of, 168–169, **169**
Energy sector, 3D printing in, 19–20
Engines of Creation (Drexler), 5
Entanglement, quantum property, 47
Epigenetics, 14
Extracellular matrix (ECM) binding, 179

F

Factory-at-Home (F@H) system, 5, 13–14
  advantages of, 14–16
  desktop units, dimensional constraints, 28
  and electricity decentralization, 27
  future of, 28–29
  iterations, price:performance ratio and, 30
  potential for abuse and regulatory prudence, 29–30
  and poverty eradication, 26–27
  3D printing *vs.*, 24
Fashion industry, 3D printers/printing in, 20–21

Feedback loops, 180
Fermions, 61
Feynman, Richard, 25
First-generation F@H system, 30–31, *31*
Flexible Electronics Laboratory, 19
Floquet circuits, 59–60
Floquet (periodic time) engineering, 59
  discrete time crystals, 59
  Floquet circuits, 59–60
  Floquet tri-unitary time circuits, 60
Floquet tri-unitary time circuits, 60
Fluorescence resonance energy transfer (FRET), 107
Food industry, 3D printers/printing in, 21
Food Ink (international restaurant chain), 21
Fossil fuels, 2–3
Fractal triangular Sierpinski pattern, 125
Fraunhofer Institute for Solar Energy Systems ISE, Freiburg, Germany, 3
Freitas, Robert A., Jr., 16
Fuel cells, 3D printing for creation of, 20

G

Gaia hypothesis, 190
Game engine platforms, 67–68
Gap junctions, 179
Giant unilamellar vesicle (GUV), 103
GKP (Gottesman, Kitaev, Preskill) codes, 65
Global Blockchain Survey (2019), 71
Global Health Care Equivalency (GHCE) platform, 15
Global workspace theory:, 188
Google Quantum AI, 63
G-quartets, 111
GTP (guanosine triphosphate), 103

H

Hardware, quantum nanoscience technologies
  artificial lattices, 58–59
  electron engineering, 58
  Floquet (periodic time) engineering, 59
  molecular coding
    Bosonic codes, 65
    quantum error correction, 64, **64**
    stabilizer codes, 65
  quantum dots (semiconductor nanoparticles), 56–57
  quantum matter (topological materials), 60–64, **61**
    long-range entanglement, 62–63
    quantum spin liquids and topological insulators, 61, **61**
    short-range protected topological order, 63–64
  spin engineering, 57–58
Hardware platforms, molecular manufacturing, 47–48, *48*
  assembly, 51
    bio-MEMS/NEMS, 51
    biomolecular condensates, 52–53
    MEMS/NEMS, 51
    molecular assemblers, 52
    nanoparticles, 51–52
    self-assembled drug delivery platforms, 53
  atomic scale fabrication
    atomically precise manufacturing, 48–49, *49*
    atomic precision advanced manufacturing, 49–50
    digital methods and autonomous robotic nanofabrication, 50
  DNA nanotechnology, 53–54
    topological molecules, 54–55
Headsets, virtual reality, 68
Healthcare systems, 149–151
Heart rate variability (HRV), 192
Herbertsmithite, 62
Holliday junction, 106
Hubble Space Telescope, 45
Human body mass, elements in, 174–176, *175*, *176*
Human brain, 75
Humanity, current state of, 2–3

I

ICOs (Initial Coin Offerings), 70
Industrial Internet of Things (IIOT), 31
InP nanowire-based solar cell, 3
Inria, 21–22
Institution of Mechanical Engineers, UK, 2
Integrated information theory, 188
Integrative Energetic Medicine (IEM), 192–194, **193**
Intellectual property registration, non-fungible token, 71–72
Intelligent matter, 67
Interdimensionality, 194–195
Interference, quantum property, 47
International Union of Pure and Applied Chemistry (IUPAC), 171
IQ system, 67

J

James Webb Space Telescope, 45
Jet engines, printed by 3D printers, 18, *19*

K

Kinesin, 103
Korea Institute of Energy Research, 3

L

LIDAR mapping, 20
Life, conventional definitions of, 182–183
Life 1.0 (biology), 67
Life 2.0 (culture), 67
Life 3.0 (technology), 67
Lipofuscin, 7
Liquid crystals, 55–56
Lithography, atomically precise, 48
Living entities, non-living entities *vs.*, 183–184
  from philosophical and spiritual perspectives, 184–185
Localized consciousness, 188–189
Lund University, Sweden, 3

M

Machine learning
  classical, 68–69
  quantum, 68–69
Many-body localization (MBL), 61
Matter, fundamental units of, 168–169, **169**
M-DNA, 127
Mendeleev, Dmitri, 171
Mesoscopic scale, 47
Microelectromechanical systems (MEMS), 24–25, 51
Microscopy methods, 48–49
Mining nodes, blockchain, 70
Molecular assemblers, 25, 52
Molecular coding
  Bosonic codes, 65
  quantum error correction, 64, **64**
  stabilizer codes, 65
Molecular knotting, 54
Molecular manufacturing (MM), 3, 4–6, 8–9, 25–26
  application areas
    biology, 75–77
    space, 77
  hardware platforms, *See* Hardware platforms, molecular manufacturing
  in medicine, 155–156
  scale of, 45–47, **46**
  software platforms, *See* Software platforms, molecular manufacturing
  stack of, *46*
Molecular weaving, 55
Molecules, 45, 47
  adhesion, 178–179
  dimensions of, 47
  topological, 54–55
Monash Centre for Additive Manufacturing (MCAM), 18
Monte Carlo approaches, 69
Moore's law, 49
Moseley, Henry, 171
Moynihan, Michael, 4
Myosin, 103

N

NAND gates, 67
Nanocarriers, 156
Nanocrystalline diamond (NCD), 127–128
Nanoelectromechanical systems (NEMS), 25, 51
Nanoelectronics, DNA, 126–128
  conductivity, 126–128
  nanowires, 126
Nanofactory, conceptual, 16
Nanogels, 53
Nanomedicine, and energy fields, 6–8
Nanomotors, 25
Nanoparticles, 51–52
  and regenerative medicine, 157
  spin engineering, 57–58
Nanosensors, 156
Nanotechnology, 3, 4–6
Nanotubes, DNA binding to, 120–121
Nanowires, DNA, 126
NASA, 21, 45
Neurons, 67
Neuroscience physics, 76, **76**
Neutrons, 172, **172**
Non-biological life, 182
Non-fungible token (NFT) intellectual property registries, 71–72
Non-living entities *vs.* living entities, 183–184
  from philosophical and spiritual perspectives, 184–185

O

Octahedron, 118–119
Off-world manufacturing, 77

P

Panpsychism, 191
Paracrine signaling, 179
Participatory medicine, 154
Periodic table of elements, 169–174, *170*
Personalized medicine, 153–154
PharmaBlockchains, 73–74
Pharmaceutical manufacturing, 73–74
Photovoltaic cells, 19
PIB (printable intelligent bot), 22–23
Plasma-based printing system, 77
Polygenic risk score (PRS), 150, 152
Polymerases, 116
Poppy Humanoid (robot), 22, *22*
Poverty eradication, F@H system and, 26–27
Predictive medicine, 152, **153**

Preventive care, 150
Preventive medicine, 153
Price:performance ratio, 30
Proteins, manipulation using DNA, 113
Protons, 171, **171**
Pulsed Electromagnetic Field Therapy (PEMF), 192

Q

Quantum chips, manufacturing of, 50
Quantum dots (semiconductor nanoparticles), 56–57
  early cancer detection using, 156–157
Quantum error correction (QEC), 64, **64**
Quantum Hall states, 62
Quantum machine learning, 68–69
Quantum matter (topological materials), 60–64, **61**
  long-range entanglement, 62–63
  quantum spin liquids and topological insulators, 61, **61**
  short-range protected topological order, 63–64
Quantum nanoscience technologies, 48, 69
  hardware
    artificial lattices, 58–59
    electron engineering, 58
    Floquet (periodic time) engineering, 59
    quantum dots (semiconductor nanoparticles), 56–57
    quantum matter (topological materials), 60–64, **61**
    spin engineering, 57–58
Quantum particles, 172, **173**, 173–174
Quantum scale, 47
Quantum spin liquids, 62–63
Quantum tunneling, 49

R

Radical Abundance (Drexler), 5
Ribosomes, 25
Robots/robotic, *15*, 15–16
  3D printed, 21–23, *22*

S

Salutogenesis
  defined, 151–152
  engineering, 157, **157**
  and 4P medicine model, 152–154, **155**
  as guiding philosophy for healthspan optimization, 154–155
Scanning tunneling microscopy (STM)-based fabrication, 48–49, *49*
Seat of consciousness, 187–188
Second-generation F@H system, 31–32
Seeman, Nadrian, 103
Self-assembled drug delivery platforms, 53
Self-assembly, 51, 52–53, 178, **178**
Self-organization, in living matter, 176, 178, **178**
Semiconductor nanoparticles, 56–57
Sense of Coherence (SOC), 151–152
Single-strand polymers, 105
Small interfering RNA (siRNA) molecules, 52
Small single-stranded DNA (ssDNA), 119, 122
Smart contract molecular dynamics simulations, 72
Smart hydrogels (nanogels), 53
Smart programmable matter (4D printing), 66
  levels of, 67
Software platforms, molecular manufacturing, 65–66, *66*
  blockchain
    applications in manufacturing industries, 72–74
    definition and status, 69–72
    science, 72
  machine learning
    classical, 68–69
    quantum, 68–69
  molecular manufacturing economy, 74, **74**
  virtual reality prototyping
    digital twins and programmable smart matter, 66–67
    game engine platforms, 67–68
Solar cells, 3, 19
Space, molecular manufacturing and, 77
Space wrench, printed by 3D printers, 18
Spin chains, 57–58
Spin engineering, 57–58
Spintronics (spin electronics), 57
Stabilizer codes, 65
STARA (smart technology, artificial intelligence, robotics, and algorithms), 28
Star Trek Replicator, 25
Streptavidin, 111
Superorganisms, 190
Superposition, quantum property, 47
Surface-promoted replication and exponential amplification of DNA analogs (eSPREAD), 115
Sycamore 53-qubit quantum computer, 60
Symmetry-protected topological insulators, 63–64
Synthetic life form, 26
Systems biology, 150–151

T

Technology, 151, 190–191
Telemedicine, 151

Thalamus, 188
Thiolated DNA, 121
Third-generation F@H system, 32
Threshold, defined, 2
Time crystals, discrete, 59
Tissue engineering, 3D printing in, 23–24
Topological entanglement entropy, 63
Topological molecules, 54–55
Toric code, 63
TradeLens blockchain platform, 71
Transcranial Magnetic Stimulation (TMS), 192
Transiting Exoplanet Survey Satellite (TESS), 45
Transportation sector, 3D printers/printing in, 20, *21*
Trichoplax adhaerens, 174, 176
Two-dimensional electron gas, 62

U

UC Berkeley, 23
U.N. Intergovernmental Panel on Climate Change (IPCC), 2
Unified quantum/universal consciousness, 189, 191
Universal life, 183
Universal Mind, 194
University of California, 23
Urbee, 20

V

Velcro strips, DNA, 122
Venter, Craig, 26
Virtual reality headsets, 68
Virtual reality prototyping
  digital twins and programmable smart matter, 66–67
  game engine platforms, 67–68
Virtual self-assembly liquid, 56
von Neumann entropy, 63

W

Water, role of, 180–183

Z

Z-DNA, 101
Zowie (robot), 21
ZyVector, 48–49, *49*

9781138632004